Office Gynecology

This book is dedicated to our wives,
Laura and Pat

Preface

Decision making in office gynecology has become increasingly complex. Distilling clinical and laboratory information into an accurate diagnosis and formulating a successful treatment plan require experience. We have intentionally chosen actively practicing clinicians with demonstrated expertise as contributors. Each author was charged with clarifying a clinical management topic that is current and controversial.

The discipline of office gynecology demands continued education on the part of the practitioner—more so in recent years than ever. Pap smear reporting formats are in flux. The evaluation of urinary incontinence has become a science. Endocrine disorders are no longer gland problems. Women expect their providers of gynecologic care to be knowledgeable in the area of breast disorders. Preventive medicine for osteoporosis and cardiovascular disease are standards of care. The list is endless. Simultaneously, office technology has exploded. Ultrasonography, minimally invasive diagnostic procedures, ultra-accurate laboratory testing, and sophisticated endocrine manipulation are all rouline. Office gynecology will never be the same. Only the highest quality of care to our ambulatory patients is acceptable. It is our hope that this textbook will help provide the foundation for that state-of-the-art gynecologic care.

John V. Knaus
John H. Isaacs

Acknowledgments

Our sincere appreciation and thanks to Audrey R. Lichtenstein, whose tireless efforts kept the words processed and contributors aware of their responsibility.

Special thanks to Shirley A. Miller, R.N., and Judy A. Metken, R.N., who kept the office homefires burning during this project.

Our sincere appreciation to Stephen P. Isaacs and Kathleen M. Lafferty for their skillful manuscript editing. Our thanks to Kathy M. Arnold for her expert illustration and photographic advice.

We wish to thank Esther Gumpert and Laura Gillan at Springer-Verlag for their professional guidance and encouragement through all phases of this textbook's preparation.

Contents

Preface		vii
Acknowledgments		ix
Contributors		xiii
1	Gynecologic Health Care Screening *Fikret I. Atamdede and Jeanne W. Atamdede*	1
2	Management of the Abnormal Pap Smear *James R. Dolan and Michael D. Moen*	15
3	Chronic Vaginitis *Lauren F. Streicher and Fred A. Zar*	30
4	Abnormal Uterine Bleeding *Carolyn V. Kirschner*	49
5	Primary Amenorrhea and Delayed Onset of Puberty *Joan M. Leya*	56
6	Secondary Amenorrhea *Rodney J. Hoxsey*	62
7	Pelvic Inflammatory Disease *M. LeRoy Sprang*	76
8	Human Immunodeficiency Virus (HIV) and Hepatitis *Patricia Herrera and J. Paul O'Keefe*	97
9	Ectopic Pregnancy *James V. Brasch*	111
10	Office Ultrasonography of the Pelvis *Steven R. Goldstein*	121
11	Dysmenorrhea and Chronic Pelvic Pain *H. Jacob Saleh*	135
12	Advances in Oral Contraception *Michael J. Hickey and Jay H. Levin*	146

13	Management of Menopause, Including Hormonal Replacement Therapy	167
	Erica G. Sinsheimer, Pauline J. Shipley, and Jay J. Gold	
14	The Dominant Breast Mass	175
	John H. Isaacs	
15	Nipple Discharge	184
	Antonio J. Bravo	
16	Complications of Lactation	190
	John M. Hobart and Cheryl L. Dorenbos-Hobart	
17	Female Sexual Problems and the Gynecologist	211
	Domeena C. Renshaw	
18	Recurrent Urinary Tract Infections	221
	Mark J. Schacht	
19	Office Evaluation and Treatment of Urinary Incontinence	230
	Peter K. Sand	
20	Genital Herpes	257
	Timothy L. Sandmann	
21	Dystrophy and Human Papilloma Virus–Associated Disorders of the Vulva	260
	John V. Knaus	
22	Bartholin Duct Cysts and Abscesses: A Rational Approach to Treatment	266
	David W. Doty	
23	Lesbian Health Issues	276
	Ruth Schwartz	
24	Hirsutism and Virilism	287
	Ian S. Tummon	
25	Rape	295
	Teresita M. Hogan	
26	Surveillance of Gynecologic Malignancies	307
	Suzanne Bergen	
Index		315

Contributors

Fikret I. Atamdede, M.D.
Assistant Professor of Obstetrics and Gynecology
Division of Gynecologic Oncology
University of California at Los Angeles
Los Angeles, CA

Jeanne W. Atamdede, R.N.
Cedars-Sinai Medical Center
Los Angeles, CA

Suzanne Bergen, M.D.
Assistant Professor of Obstetrics and Gynecology
Division of Gynecologic Oncology
Loyola University of Chicago
Stritch School of Medicine
Maywood, IL

James V. Brasch, M.D.
Clinical Associate Professor of Obstetrics and Gynecology
University of Washington School of Medicine
Seattle, WA
Attending Physician, Holy Family Hospital
Spokane, WA

Antonio J. Bravo, M.D.
Attending Physician, Department of Obstetrics and Gynecology
Elmhurst Memorial Hospital
Elmhurst, IL

James R. Dolan, M.D.
Assistant Professor of Obstetrics and Gynecology
Division of Gynecologic Oncology
Loyola University of Chicago
Stritch School of Medicine
Maywood, IL

Cheryl L. Dorenbos-Hobart, R.N., B.S.N., F.A.C.C.E., CLE
Childbirth and Breastfeeding Education Coordinator
Good Shepherd Hospital
Barrington, IL

David W. Doty, D.O.
Chairman, Obstetrics and Gynecology Committee
Meadowview Regional Hospital
Maysville, KY

Jay J. Gold, M.D.
Clinical Professor of Medicine
Adjunct Professor of Obstetrics and Gynecology
University of Illinois at Chicago
College of Medicine
Chicago, IL
Chief, Section of Endocrinology
Department of Medicine
St. Francis Hospital, Evanston, IL

Steven R. Goldstein, M.D.
Assistant Professor of Obstetrics and Gynecology
New York University
School of Medicine
New York, NY

Patricia Herrera, M.D.
Fellow in Infectious Diseases
Loyola University of Chicago
Stritch School of Medicine
Maywood, IL

Michael J. Hickey, M.D.
Director, Hinsdale Center for Reproduction
Hinsdale Hospital
Hinsdale, IL
Clinical Instructor of Obstetrics and Gynecology
Loyola University of Chicago
Stritch School of Medicine
Maywood, IL

John M. Hobart, M.D.
Assistant Professor of Obstetrics and Gynecology
Northwestern University Medical School
Chicago, IL
Director, Obstetrics and Maternal-Fetal Medicine
Evanston Hospital
Evanston, IL

Teresita M. Hogan, M.D.
Clinical Instructor of Emergency Medicine
University of Illinois at Chicago School of Medicine
Chicago, IL
Education Director, Emergency Department
St. Francis Hospital
Evanston, IL

Rodney J. Hoxsey, M.D.
Assistant Professor of Obstetrics and Gynecology
Northwestern University Medical School
Chicago, IL
Senior Attending and Director
Division of Reproductive Endocrinology and Infertility
Evanston Hospital
Evanston, IL

John H. Isaacs, M.D.
Professor of Obstetrics and Gynecology
Loyola University of Chicago
Stritch School of Medicine
Maywood, IL
Gynecologist-in-Chief
St. Francis Hospital
Evanston, IL

Carolyn V. Kirschner, M.D.
Assistant Professor of Obstetrics and Gynecology
Division of Gynecology Oncology
Rush University
School of Medicine
Chicago, IL

John V. Knaus, D.O.
Clinical Associate Professor of Obstetrics and Gynecology
Loyola University of Chicago
Stritch School of Medicine
Maywood, IL
Chairman, Department of Obstetrics and Gynecology
Chief, Section of Gynecologic Oncology
St. Francis Hospital
Evanston, IL

Jay H. Levin, M.D.
Medical Co-Director
Hinsdale Center for Reproduction
Hinsdale Hospital
Hinsdale, IL

Joan M. Leya, M.D.
Chief, Section of Reproductive Endocrinology and Infertility
St. Francis Hospital
Evanston, IL

Michael D. Moen, M.D.
Chief Resident
Department of Obstetrics and Gynecology
Loyola University of Chicago
Stritch School of Medicine
Maywood, IL

J. Paul O'Keefe, M.D.
Professor of Medicine
Chief, Infectious Diseases
Loyola University of Chicago
Stritch School of Medicine
Maywood, IL

Domeena C. Renshaw, M.D.
Professor of Psychiatry
Loyola University of Chicago
Stritch School of Medicine
Maywood, IL

H. Jacob Saleh, M.D.
Assistant Professor of Clinical

Obstetrics and Gynecology
Northwestern University Medical School
Chicago, IL
Attending Physician, Evanston Hospital and St. Francis Hospital
Evanston, IL

Peter K. Sand, M.D.
Associate Professor of Obstetrics and Gynecology
Northwestern University Medical School
Chicago, IL Director of Urogynecology
Evanston Hospital
Evanston, IL

Timothy L. Sandmann, M.D.
Attending Physician, Department of Obstetrics and Gynecology
Wilson N. Jones Hospital
Sherman, TX

Mark J. Schacht, M.D.
Clinical Assistant Professor of Urology
University of Illinois at Chicago School of Medicine
Chicago, IL
Co-Director, Urogynecology
Humana Michael Reese Hospital
Chicago, IL
Attending Physican, Department of Urology
St. Francis Hospital
Evanston, IL

Ruth Schwartz, M.D.
Clinical Professor of Obstetrics and Gynecology
University of Rochester School of Medicine and Dentistry
Rochester, NY

Pauline J. Shipley, M.D.
Clinical Assistant Professor of Medicine
Loyola University of Chicago Stritch School Medicine
Maywood, IL
Attending Physician, Department of Medicinc

St. Francis Hospital
Evanston, IL

Erica G. Sinsheimer, M.D.
Clinical Assistant Professor of Medicine
Loyola University of Chicago Stritch School of Medicine
Maywood, IL
Attending Physician, Department of Medicine
St. Francis Hospital
Evanston, IL

M. LeRoy Sprang, M.D.
Assistant Professor of Clinical Obstetrics and Gynecology
Northwestern University Medical School
Chicago, IL
Senior Attending, Department of Obstetrics and Gynecology
Evanston and St. Francis Hospitals
Evanston, IL

Lauren F. Streicher, M.D.
Clinical Instructor in Obstetrics and Gynecology
Northwestern University Medical School
Chicago, IL

Ian S. Tummon, M.D.
Assistant Professor of Obstetrics and Gynecology
Division of Reproductive Medicine
University of Western Ontario School of Medicine
London, Ontario, Canada

Fred A. Zar, M.D.
Clinical Associate Professor of Medicine
Loyola University of Chicago Stritch School of Medicine
Maywood, IL
Chairman, Department of Medicine
Chief, Section of Infectious Diseases
St. Francis Hospital
Evanston, IL

1
Gynecologic Health Care Screening

Fikret I. Atamdede and Jeanne W. Atamdede

Principles of Screening

Unlike many other specialists, gynecologists are in a unique position to affect the health of their patients before a clinical disease takes hold. Approximately 60% of all office visits to general obstetricians/gynecologists are for the purpose of preventive medicine, and 80% of women see their obstetricians/gynecologists regularly for routine examination. One reason for this is that women are introduced to the medical field at a much earlier age than men due to either childbearing or the need for contraception. In addition, women are generally more aware of their bodies than most men.

History of Screening in Gynecology

In 1924 George Papanicolaou determined that cancer cells from the cervix could be identified in cytologic smears from the vagina.[1] In 1928 a Romanian pathologist, Aureli Babes, published a similar account[2] in which he described the diagnosis of cervical cancer using vaginal smear cytology. In a landmark article, published in 1941 and coauthored by Herbert Traut,[3] Papanicolaou noted that cancer cells from the cervix and endometrium could be identified in patients who had no clinical evidence of disease. A further refinement of this research occurred in 1947, when J.E. Ayre published[4] an article that stated that the efficiency of this test could be improved by obtaining the sample directly from the cervix. Thus, by the early 1950s the modern Pap smear was established, and an intensive effort was begun by the American Cancer Society in particular, and the medical community in general, to use this test for mass screening of women to prevent cervical cancer. Despite the success of this effort, however, doubt still remains as to the effectiveness of the Pap smear as a screening tool.

Preventive Medicine and Screening

Preventive medicine can be divided into three categories (Table 1.1):[5]

1. Primary prevention: Actions taken prior to the biological origin of disease.
2. Secondary prevention: Actions taken when a disease is detectable but has not caused actual sickness or disability.
3. Tertiary prevention: Services performed after illness has occurred to prevent further deterioration.

A primary goal of most health professionals is to avoid allowing a disease to progress to the point of having to provide tertiary prevention. Ideally, all disease processes could be handled by providing primary prevention with measures such as immunizations, examination of health hazards, and behavior modification. However, many health problems do not have effective primary prevention methods available. Extensive efforts must be made to provide effective secondary preventive measures. If a disease can be identified prior to causing actual illness, measures can be taken to pre-

TABLE 1.1. Level of care.[5]

Primary	Nutrition, exercise
	Hygiene, drug and alcohol use
	Seatbelt use, working environment
	Vaccinations, family planning
	Genetic counseling
Secondary	Estrogen replacement therapy
	Pap smears
	Breast examination, mammogram
	Endometrial biopsy for post- or perimenopausal bleeding
Tertiary	Treatment of pelvic inflammatory disease (PID)
	Surgery for disease
	Chemotherapy

vent further progression of the disease. Unfortunately, many diseases do not have a modifiable precursor stage. Thus, screening has to be defined within this context.

Principles of Screening

Screening is defined as the attempt to identify, in an asymptomatic individual, a disease process that, if left untreated, would progress to apparent disease. However, a screening test must be a "valid" test. Cochrane and Holland[6] outlined several considerations for instituting a screening program:

1. Ethical: When a patient presents with an illness, a physician does what he or she can, whether there is a lack of medical knowledge or not. However, if the medical profession initiates an active screening program, there should be conclusive evidence that screening can alter the natural history of the disease in a significant proportion of those screened.
2. Scientific: Whenever possible, the inherent hypothesis of every screening test, that is, that early detection followed by early therapy alters the natural history of the disease, should be tested in an objective randomized manner.
3. Financial: The question "Is it worth it?" must be addressed. If a screening test benefits a very small proportion of the population while costing society in general a significant amount of resources, the test's cost to society may not be justified.

They suggest several criteria for evaluating screening tests:

1. Simplicity: The test should be easy to administer and should be able to be performed by paramedical personnel.
2. Acceptability: The test must be acceptable in its ease and discomfort.
3. Accuracy: The test must reasonably detect the condition being investigated.
4. Cost: The benefits to the individual and society must outweigh the burden placed upon limited resources.
5. Precision: The test results should be consistently reproducible.
6. Sensitivity: The test should have a reasonably low level of false-negative results.
7. Specificity: The test should have a reasonably low level of false-positive results.

Another criterion that must be addressed is the predictive value of a test. The screening test must reasonably predict the disease when positive and rule out the disease when negative. This differs from specificity and sensitivity in that the predictive value is not intrinsic to the test, but rather to the population being tested. Ideally, a screening test should have a 100% sensitivity and specificity, should reflect the population being tested, and should be fully predictive. Because this set of circumstances rarely, if ever, occurs, the other factors cited come into play when deciding about instituting appropriate screening methods.

Any screening program must address the ability to provide the test to all eligible persons and, once performed, the ability to act upon the results. It must adequately monitor those screened, as well as identify population at risk and encourage their participation.

Gynecologic Cancer Screening

Vulvar Carcinoma

Vulvar carcinoma generally involves women 65 years or older. To date there is no formal screening method for detecting this cancer.

A careful examination of the vulva, with appropriate biopsies for suspicious lesions, should be a part of every pelvic examination. This is especially important in the patient with a vulvar dystrophy. Although lichen sclerosis is not a preinvasive lesion, its symptoms and appearance may mask a significant lesion. On the other hand, hyperplastic dystrophy, especially associated with atypia, may be associated with progression to carcinoma. However, vulvar carcinoma is generally associated with a long preinvasive stage, which allows a practitioner to perform simple therapeutic measures and provides some time to make a diagnosis. Therefore, careful yearly examination of the vulva is adequate screening for vulvar carcinoma. Consideration for biyearly examinations should be given to patients with atypical vulvar dystrophy.

Vaginal Carcinoma

Primary vaginal carcinoma is rare, representing less than 1% of all gynecologic cancers. It is usually associated with cervical carcinoma or with exposure to DES in utero. Because of its rare incidence, it does not have defined screening recommendations. However, a vaginal pool sample along with the standard Pap smear provides an excellent means to test for this cancer. A screening interval of once a year is considered adequate.

Cervical Carcinoma

The modern Pap smear, used in screening for cervical carcinoma, is the archtypical model for all screening tests in many ways. However, this test was never subjected to the close scrutiny that other screening tests now routinely undergo. No prospective, randomized, objective study as to the effectiveness, specificity, sensitivity, or predictive value of the test was ever done prior to its widespread acceptance and use. Fortunately, over time the Pap smear has proved to be an accurate, inexpensive, rapid, and reasonably reproducible test. Its specificity and sensitivity are reasonable, and the test is acceptable to most patients.

Sensitivity and Specificity

One advantage of the Pap screening is that it has proved to be very specific. It is commonly accepted that most competent cytology laboratories will have a false-positive rate of only 2% to 5%.[7] This high specificity, however, has a correspondingly low sensitivity. The false-negative rate can range from 2% to 50%, with a generally accepted average of about 15% to 20% for high-grade lesions and a somewhat higher rate for low-grade lesions.[8] It must be emphasized that the Pap smear is a screening tool and not a diagnostic test. If a lesion is seen, a biopsy should be done. False-negative rates of around 50% are reported

		Actual disease state	
		Disease present	Disease not present
Test result	Positive	True positive (TP)	False positive (FP)
	Negative	False negative (FN)	True negative (TN)

$$\text{Sensitivity} = \frac{TP}{TP + FN}$$

$$\text{Specificity} = \frac{TN}{FP + TN}$$

$$\text{Predictive Value} = \frac{TP}{TP + FP} \quad \text{or} \quad \frac{TN}{FN + TN}$$

FIGURE 1.1. Definition of sensitivity, specificity, and predictive value.

for Pap smears of frankly invasive cervical cancers.[8,9]

Specimen Collection

The results of the Pap smear are dependent on cytologic interpretation. To maximize the results, careful attention must be given to technique when obtaining the specimen. An adequate history is essential. The patient's last menstrual period, presence of an IUD, and recent infection or manipulation of cervix such as a biopsy or Pap within the last 6 weeks should be documented. Any previous surgery or treatment such as cryotherapy, laser therapy, conization, radiation therapy, or hormone therapy should also be noted. Because a large amount of blood or a heavy discharge may make interpretation of the sample difficult, one should avoid obtaining a Pap smear during a woman's period or when the patient has evidence of a vaginitis. Of course, if a lesion is seen, it should be evaluated, usually with a biopsy.

The Pap smear should be done prior to any manipulation of the cervix. Because of this, the speculum exam and sampling must be done prior to the bimanual examination. An adequate-sized speculum is inserted into the vagina to allow visualization of the cervix. Lubricating gels should be avoided, and only enough water to facilitate insertion should be used. Both lubricating jelly and excessive water cause artifacts that can complicate cytologic interpretation.

Excessive discharge can be gently wiped away with a cotton swab, although this is not recommended on a routine basis. The Pap smear is a sampling of exfoliated cells in the cervical mucosa and secretions; if these cells are wiped away, an adequate specimen may not be obtained.

Obtaining an adequate sample from the endocervix is very important. The ideal method is to aspirate the endocervix with a pipette and smear this specimen onto the slide.[10,11] An alternative method of obtaining a specimen involves using an endocytobrush, but this should not be used in pregnant patients.[12] If an endocytobrush is unavailable, then a moistened cotton swab can be used to obtain the endocervical sample.

The exocervix should be sampled with a cervical scraper. The original wooden device designed by Ayre is still the standard of care, although many other devices are available in various shapes and materials. The cervix should be gently scraped; excessive force may result in blood or tissue fragments that may make the cytologic interpretation difficult. A vaginal pool sample is no longer recommended except in certain circumstances. Cells in the fornix tend to be necrotic, denigrated, or mixed with other cells and bacteria that make interpretation difficult.

Once the specimen is obtained, it must be smeared on a carefully identified slide. The ideal thickness of the sample is a single cell layer. With practice, this is not difficult to achieve. Separate slides of the endocervical and exocervical samples may be used, or they may both be combined on a single slide. The method chosen is dependent on the laboratory used and on cost factors. The use of two slides reduces the possibility of air-drying artifact, but with practice and care this problem can be minimized when using a single slide.

Finally, the specimen must be carefully and promptly fixed. Air drying induces significant, irreversible artifacts in the specimen. As soon as the specimen is smeared, it should be fixed by using any commercially available fixative.

Screening Criteria

Until the Walton report in 1976, the issues of who should be screened and how often were rarely formally addressed.[13] The general recommendation at that time was that all sexually active women be screened yearly or biyearly. The Walton report attempted to address these recommendations on a more pragmatic level. Basing their analysis strongly on a cost-benefit ratio, the authors of this report suggested that all sexually active women be screened annually from age 18 to 35, then every five years until age 60. More frequent screening should not be discouraged for a woman over age 35 if she had any risk factors. However, subsequent reports showed that a 5-

1. Gynecologic Health Care Screening

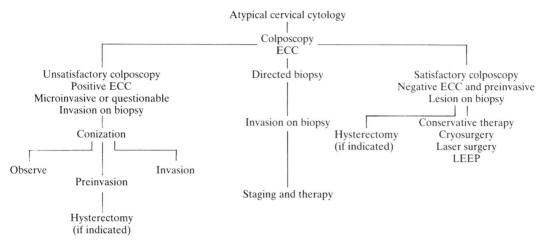

FIGURE 1.2. Management of atypical cervical cytology.*
*Modified from: Gusberg SB, Shingleton HM, Deppe G, eds., *Female Genital Cancer*, p. 209. © 1988 Churchill-Livingstone, New York. Reproduced with permission.

year interval between Pap smears does not provide adequate screening.[7,14-17] The inherent low sensitivity of the test is compensated by more frequent sampling, and 5 years between screenings is too long an interval.

In 1988 the American Cancer Society published a recommendation that all women at least 18 years old and/or sexually active should have an annual Pap smear. If a woman has had three successive negative, normal Pap smears and has no risk factors, then the interval could be increased to every 3 years. The problem with this recommendation is the difficulty in obtaining an accurate sexual history to assess risk factors.[18]

The American Board of Obstetrics and Gynecology recommendations suggest that all sexually active women, or women who are 18 years or older, have yearly Pap smears. The interval between Pap smears may be increased based on the physician's assessment of individual risk. At such time, a joint, informed decision can be made with the patient regarding an adequate interval.[5]

Interpretation of Results

Considering the very high specificity of Pap smears, a positive Pap must not be ignored; on the other hand, the moderately low sensitivity of the PAP smear implies that all lesions must be evaluated regardless of the PAP smear results. Figures 1.2 and 1.3 outline follow-up routines in nonpregnant and pregnant patients.[19]

The system used for reporting Pap smear results has recently changed. Until 1989 the reporting of Pap smear results varied widely from laboratory to laboratory. The most common system utilized the Papanicolaou classification (Table 1.2). Another method was based on descriptive classification (Table 1.3).[20] However, in 1988 the National Cancer Institute developed what is now known as the Bethesda system (Table 1.4).[21] The two stated goals of this system are to eliminate the Papanicolaou classes (I–V), since they did not necessarily have a close relationship to histologic diagnostic terminology, and to make a

TABLE 1.2. Papanicolaou's classification system.

Class I (negative)	Only normal cells are present.
Class II (negative)	No signs of malignancy; some atypical cells present.
Class III (doubtful)	Cells with atypical features present, suggestive of but not diagnostic of malignancy.
Class IV (positive)	Isolated atypical cells are present.
Class V (positive)	Numerous atypical cells or cell groups are present.

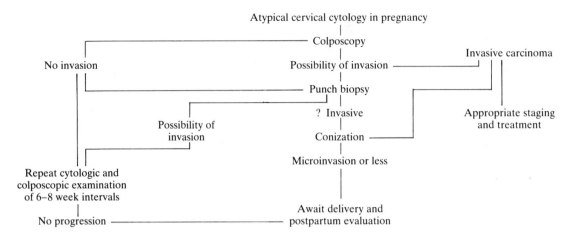

FIGURE 1.3. Management of atypical cervical cytology in pregnancy.*
*Gusberg SB, Shingleton HM, Deppe G, eds., *Female Genital Cancer*, p. 210. © 1988 Churchill-Livingstone, New York. Reproduced with permission.

TABLE 1.3. Richard's descriptive cytologic classification.

Inadequate for diagnosis
Essentially normal findings
Atypical cells present suggestive of (Specify)
Cytological finding consistent with:
    CIN Grade 1 (mild dysplasia)
    CIN Grade 2 (moderate dysplasia)
    CIN Grade 3 (severe dysplasia or carcinoma in situ)
    Invasive squamous cell carcinoma
    Endometrial carcinoma
    Other cancer (Specify)

TABLE 1.4. Bethesda classification system.

Statement on Specimen Adequacy
    Satisfactory
    Less than optimal
    Unsatisfactory
General Categorization
    Within normal limits
    Other: See descriptive diagnosis; further action required
Descriptive Diagnosis
    Infection
    Reactive and reparative changes
    Epithelial cell abnormalities
        Squamous cell
            Atypical squamous cell of undetermined significance
            Squamous intraepithelial lesion (SIL)
            Low-grade SIL, encompassing:
                Cellular changes associated with human papilloma virus (HPV)
                Mild dysplasia/CIN I
            High-grade SIL, encompassing:
                Moderate dysplasia/CIN II
                Severe dysplasia/CIN III
                Carcinoma in situ/CIN III
            Squamous cell carcinoma
    Glandular cell changes (abnormal atypical, or adenocarcinoma)
    Nonepithelial malignant neoplasm (Specify)

statement on the report as to the adequacy of the specimen. Whether these goals are being achieved, and whether the new system is an improvement, is being debated in the literature.[22,23] This system is now the recommended format for reporting Pap smear results. However, the general principles do not change: If an abnormal Pap smear is reported, the abnormality must be explained; and if a negative PAP smear is reported, any lesions still need to be fully evaluated.

Endometrial Cancer Screening

Endometrial carcinoma has now become the most common gynecologic cancer in the United States. Approximately 23,000 new cases are reported each year. Despite early symptomatology and a less aggressive progression than other gynecologic cancers, about 3,000 deaths per year are attributed to this disease.[24]

Recognized risk factors for endometrial

carcinoma are related to prolonged estrogen stimulation of the endometrium: obesity, low parity, menstrual irregularity, early menarche, late menopause, diabetes, and noncyclic use of exogenerous estrogens.[25]

Endometrial carcinoma does not have a clearly defined preinvasive stage. Although hyperplasia, especially with atypia, places a woman at higher risk for developing endometrial carcinoma, the relationship of these lesions to endometrial carcinoma is not as clear as is CIN to cervical carcinoma. Since the screening methods for endometrial carcinoma are relatively invasive and may be expensive, they may not be acceptable to the patient or society. With these considerations, the question of who should be screened becomes important.

Screening Methods

With refinement of the Pap smear technique, it is believed that, in addition to sampling squamous cells from the epithelial surfaces in the vagina, endometrial cells are also detected in the vaginal mucus. However, less than 40% of patients with known endometrial carcinoma will have a Pap smear positive for this disease.[26] In precursor lesions, the sensitivity is even less, only 10% to 20%.[7] This low sensitivity of the Pap smear, therefore, makes it an unreliable screening test for endometrial carcinoma. However, positive results should not be ignored. The presence of endometrial cells on a Pap smear in a postmenopausal woman requires further evaluation because 5% to 15% of these women will have significant endometrial pathology.[27,28]

Since abnormal endometrial cells are not routinely shed to the cervical or vaginal mucus, several devices are available to obtain direct endometrial cytologic samples. Some devices obtain cells by direct abrasion of the endometrium (Isaacs cell sampler, Mi-Mark Helix). Other devices collect samples by either suction or lavage (Jet-washer, Gravlee). All generate excellent specimens,[29,30] but, because they involve insertion through the cervix, they can cause significant patient discomfort. In addition, all abnormal findings require histologic confirmation, resulting in repeated instrumentation and further discomfort. For these reasons, cytologic sampling of the endometrium is not widely accepted for general screening.

Until recently, the only means of obtaining a histologic biopsy specimen from the endometrium was to perform a formal dilation and curettage (D&C). Since this can be an uncomfortable procedure, anesthesia and an operative suite are generally required. Several devices available today are small and flexible enough to be inserted through the cervix without anesthesia while still producing an adequate sample for histologic evaluation. This method of office endometrial biopsy is very accurate, with a false-negative rate of only about 5% to 10%. This compares very favorably with the 1% to 5% false-negative rate of the D&C.[31-33]

The office biopsy has several disadvantages, however, especially in the context of its use as a screening tool. It is a fairly time-consuming test and can cause significant patient discomfort. A formal histologic reading of the specimen can be expensive. Finally, although the incidence is small, there is risk to the patient from infection. If the above methods of mass screening for endometrial cancer are not feasible, then directed screening may be acceptable.

Screening Criteria

As stated earlier, there are recognized risk factors for endometrial carcinoma. Since mass screening is not practical, screening dependent on these risk factors may be more feasible. Identification of a woman at high risk for developing this disease must be defined. Table 1.5 suggests the criteria for identifying the high-risk woman. A woman who is over the age of 50; is obese; has a family history of cancer; and has a history of irregular bleeding, a history of exogenous unopposed estrogen use, or multiparity should probably be screened annually or biannually.

The Pap smear alone is not a reliable screening test; intrauterine cytologic sampling or the office endometrial biopsy method can be used. Proper evaluation should not be postponed if the patient presents with specific

TABLE 1.5. Risk factors for endometrial carcinoma.

1. Age ≥ 50
2. Family history of cancer, especially endometrial or breast
3. Upper socioeconomic class
4. Obesity
5. Low or no parity
6. Failure of ovulation
   a. Polycystic ovary
   b. Amenorrhea
   c. History of poly- and/or hypermennorhea
7. Chronic estrogen intake
8. Diabetes

complaints such as postmenopausal bleeding, heavy intra- or intermenstrual bleeding, or, for a postmenopausal woman, the presence of endometrial cells on a Pap smear. An abnormal cytologic screening requires histologic confirmation, while an abnormal screening biopsy is usually satisfactory for diagnosis and institution of therapy.

The role of hysteroscopy as a screening test is not well defined.[34,35] Considering the expense and the time required for a thorough examination of the endometrium, the use of the office hysteroscope is probably not justified in a screening role. In addition, transvaginal ultrasonography for assessment of endometrial thickness and fluid may be useful in defining patients who are candidates for endometrial sampling.[31,32,36] The role of sonography as a screening tool has not been defined.

Ovarian Carcinoma

If there is one disease that requires an effective screening method, it is ovarian carcinoma. Of approximately 20,000 women diagnosed with ovarian carcinoma each year, as many as 12,000 will die.[24] It is the sixth most common cancer in women and results in about 5% of all deaths due to cancer. The present overall survival from ovarian carcinoma is 38%. However, if survival is broken down by stage of disease, it is noted that patients with Stage I have an 80% to 100% survival rate, stages II and IIIa a 30% to 40% survival rate, stage IIIb a 20% survival rate, and stages IIIc and IV approximately 5% survival rate.[37] Unfortunately, 60% to 70% of ovarian cancers are not diagnosed until stage III or after. A significant reason for this is the lack of symptomatology. Many patients present with only vague abdominal discomfort, dyspepsia, flatulence, and bloating. These same symptoms are commonly found in benign conditions.

A need for a screening study that would alert the practitioner early in the course of the disease is clearly needed, since there is no identifiable preinvasive phase.

Screening Methods

The classic screening method for ovarian carcinoma has been the standard pelvic exam. However, this has not been particularly successful. To detect a single case of ovarian cancer, it is estimated that approximately 10,000 pelvic examinations need to be performed.[38]

In 1981 Bast and co-workers reported on a monoclonal antibody associated with ovarian carcinoma[39]; this became known as the CA-125 serum marker. It was initially thought that this would be the long sought-after screening tool for ovarian carcinoma, but several factors make the CA-125 a less than ideal screening tool. As noted in Table 1.6, many benign gynecologic conditions, as well as various gastrointestinal and pulmonary diseases, can significantly elevate the level of CA-125.[40] In addition, there are circumstances in which the CA-125 may be negative despite significant disease. For example, mucinous ovarian cancer is often negative for CA-125.[41] Thus the CA-125 carries an overall sensitivity of approximately 80%, with specificity of approximately 50%, when an elevation of >35 U/mL is used as the cutoff for abnormal. Virtually 100% specificity can be achieved with an increase of the cutoff to 200 U/mL, but sensitivity decreases significantly.[42]

The CA-125 is useful as a tumor marker in patients with an established diagnosis of ovarian carcinoma. An increase in the level of CA-125 can also predict clinical recurrence of tumor. CA-125 is also useful in the evaluation of a pelvic mass in the postmenopausal woman. Figure 1.4 illustrates a protocol for

1. Gynecologic Health Care Screening

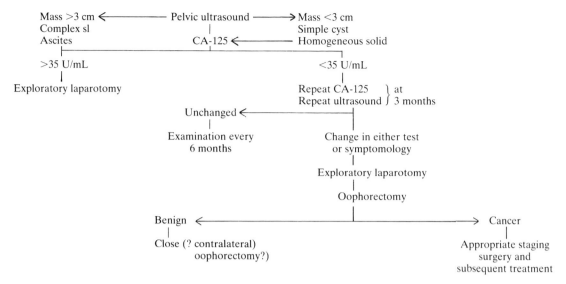

FIGURE 1.4. Evaluation of pelvic mass in a postmenopausal woman.*
*Modified from: Curry SL, *Early Diagnosis and Monitoring of Ovarian Cancer*. Advise and Consent, Inc., 1990, p. 4.

TABLE 1.6. Serum CA-125 levels.[40]

Group	>35 U/mL	>65 U/mL
Controls		
Male	0.7%	0.4%
Female	1.4%	0.0%
Nongynecological cancers		
Pancreatic	58.6%	44.8%
Lung	32.0%	24.0%
Breast	12.0%	8.0%
Colorectal	22.5%	16.9%
Misc. gastrointestinal	26.7%	20.0%
Misc. nongastrointestinal	25.0%	25.0%

the use of CA-125 when evaluating a postmenopausal pelvic mass.

The ultrasound has been shown to be an accurate means of assessing pelvic structures and has been suggested as a screening tool for ovarian carcinoma.[44-46] With the introduction of the vaginal probe, the accuracy of ultrasound has increased significantly.

In 1989 Cambell et al[44] published a report in which transabdominal ultrasound was used to screen for early cancer of the ovary. Approximately 5% of their screened patients had a positive ultrasound, but only five patients, or 0.1%, were actually found to have cancer.

Use of transabdominal ultrasound as a screening tool has disadvantages. It is time-consuming and requires a full bladder, which may be uncomfortable for the patient. In contrast, the transvaginal ultrasound requires an empty bladder, is usually quicker, and provides a much higher level of detail of the ovarian structure. The ovaries can be identified in nearly all premenopausal patients and in 60% to 80% of postmenopausal patients.[45] As a result, several studies have explored the use of the transvaginal ultrasound as a screening tool.[47-51]

The sensitivity of transvaginal ultrasound for ovarian abnormalities is nearly 100%. However, the specificity ranges between 50% and 75% for malignancies.[45,48,51] Because ovarian carcinoma is a low-prevalence disease, this moderate level of specificity makes the transvaginal ultrasound an impractical screening tool for the general population. High specificity is especially important in detecting ovarian caracinoma, because a highly invasive procedure must be performed to make a definitive diagnosis. Therefore, as with endometrial carcinoma, no ideal method of screening for ovarian carcinoma exists and screening criteria need to be established.

TABLE 1.7. Risk factors of ovarian carcinoma.

Family history:
   Ovarian carcinoma (high risk)
   Breast, colon, endometrial, or thyroid (moderate risk)
Nulliparity or low parity
No oral contraceptive use
Pelvic irradiation
High socioeconomic class
High-fat diet
Exposure to talc, asbestos

Screening Criteria

There are several risk factors for developing ovarian carcinoma (Table 1.7).[52] Of these, a family history and low parity seem to be most important. Because the incidence of ovarian carcinoma increases after age 40, screening should start at that age unless there is a family history of early-age cancer; in these cases, screening should be started at or about the age of the earliest reported cancer. Figure 1.5 suggests a multimodal screening program for these patients.[52]

However, only 40% of patients with ovarian carcinoma have definable risk factors. The majority of patients who will develop this cancer will not have been screened for this disease based on these criteria. Clinical judgment is invaluable, and vigilance in evaluation must be maintained.

Breast Carcinoma

Breast cancer is the second leading cause of cancer death for women in the United States, surpassed only by lung cancer. Approximately 120,000 new cases of breast cancer are diagnosed each year, resulting in about 40,000 deaths per year.[24] Although there have been many improvements in the care of these patients, early detection is of prime importance in treating this disease. Several studies have shown that the incidences of death from

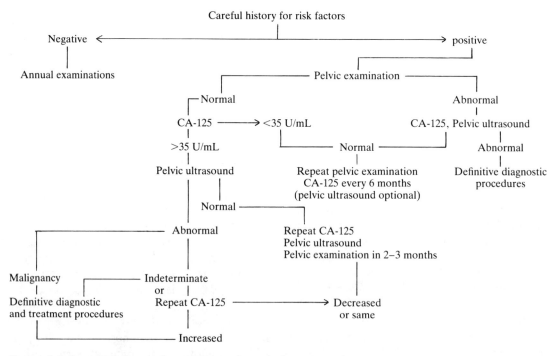

FIGURE 1.5. Screening for ovarian carcinoma.*

*Modified from: Orr JW, *Early Diagnosis and Monitoring of Ovarian Cancer*. Advise & Consent, 1990, pp. 17–20. Reproduced with permission.

breast cancer can be reduced in a screened population.[53-56] The major obstacle to effective screening for breast cancer is patient and physician noncompliance.

Screening Methods

The breast self-examination must be the foundation of any breast cancer screening program. It has several advantages. It is a no-cost test with high accessibility; virtually all women should be able to perform the test, except under highly unusual circumstances. It makes the patient aware of her normal breast structure so that subtle changes may be detected early. In this manner, the sensitivity of this test increases over time. In addition, the examination educates the patient in the importance of periodic breast self-examination and continued vigilance. The disadvantages of the test are few. It has a low sensitivity and specificity and may cause undue alarm when an abnormality is found.[57] Furthermore, there may be an overreliance on the test on the part of both the patient and the physician.

The clinical breast examination should be a part of every physical examination. However, there is a surprising reluctance on the part of health care providers to perform this important screening service. A carefully performed breast examination, by an experienced provider, can identify lesions as small as 1 cm.[58] Although no breast mass should be ignored, simple and relatively low-cost diagnostic procedures can be performed in the office that can determine the significance of a lesion.

The sensitivity of the clinical breast examination is somewhat difficult to define. In a large study, a false-negative rate of approximately 40% was noted.[59] However, if one looks at the false-negative rate of palpable lesions (lesions felt to be noncancerous on examination), a rate of approximately 6% to 10% is found.[60] This implies a high specificity when palpable lesions are noted.

The clinical examination is not very accurate for lesions of less than 1 cm or in large, pendulous breasts. Since the rate of survival is better with a smaller lesion,[61] it is highly desirable that these lesions be reliably detected.

In terms of cost availability and accuracy, the mammogram is unequaled in detecting these small lesions. The mammogram can identify very small lesions and can be helpful in the differential diagnosis. Sensitivities of 85% to 95% have been reported, and reports show that 20% to 50% of identified cancers are discovered only by the use of mammogram.[62] However, care should be taken to avoid an over-reliance on the mammogram. One report has shown that up to 16% of breast cancers would have been missed if the mammogram alone had been used.[63] Thus, in the presence of a palpable mass, a negative mammogram does not rule out cancer.

One problem with the mammogram is the lack of compliance in the use of the screening tool from both patients and physicians. The mammogram is a mildly uncomfortable test, probably comparable to a Pap smear. A major fear of patients is the risk of exposure to radiation. Early, nondedicated machines produced relatively significant doses of radiation. With improvements in the machines and in the sensitivity of the film, radiation doses have been minimized (0.1–0.4 rads).[58,64] Although there is a theoretical risk of increased breast carcinoma from even these low doses, the increase of survival resulting from screening far outweighs this risk.

Physician resistance results from a number of reasons, such as concerns about inadequate sensitivity, radiation risk, cost, and the perception that patients would not accept the procedure.[64] The sensitivity and radiation issues have already been addressed. Cost is a realistic concern. Some centers charge up to $100 per test. However, with appropriate planning, this expense can be decreased, and if one factors in the reduced cost of management for early cancers, cost diminishes in importance. Finally, a recent study has shown that an overwhelming majority of women would accept the test if it were recommended by their physician.[65]

Screening Guidelines

All women should be screened for breast cancer. Although definite risk factors exist for

TABLE 1.8. Breast cancer risk factors.

Factor	Increased risk
Primary relative with breast cancer	1.2–3.0
Menarche <12 years old	1.3
Menopause >55 years old	1.48–2.0
First child after age 35	2.0–3.0
Nulliparous	3.0
Contralateral breast cancer	5.0
Uterine cancer	2.0
Atypical lobular hyperplasia	4.0
Lobular carcinoma in situ	7.2

breast cancer (Table 1.8), many new breast cancers arise from patients without these factors. Thus the American Cancer Society has accepted the following screening protocol for breast cancer:

1. All women age 20 and above should be instructed in breast self-examination and encouraged to perform this regularly.
2. All women should have a physical examination of the breast every 3 years from age 20 to age 40, and annually thereafter.
3. All women should have a baseline mammographic examination between the ages of 35 and 40.
4. A mammogram should be performed every 1 to 2 years from ages 40 to 50.
5. Annual mammograms should be performed after age 50.

For patients at high risk for developing breast cancer, clinical judgment must be used in determining testing and follow-up. Earlier mammographic surveillance and more frequent physical examination of the breast may be warranted. Patients screened in this manner have shown a significant reduction in mortality from breast cancer.

References

1. Papanicolaou GN. New Cancer Diagnosis. *Proceedings Third Race Betterment Conference.* Battle Creek, Mich: Race Betterment Foundation; 1928:528–534.
2. Babes A. Diagnostique du cancer uterin par les frottis. *Presse Med.* 1928;36:451–454.
3. Papanicolaou GN, Traut HF. The diagnostic value of vaginal smears in carcinoma of the uterus. *Am J Obstet Gynecol.* 1941;42:193–206.
4. Ayre JE. Selective cytology smear for diagnosis of cancer. *Am J Obstet Gynecol.* 1947;53:609–617.
5. Preventive and screening services. In: *Precis III, An Update in Obstetrics and Gynecology.* The American College of Obstetricians and Gynecologists; 1986:51–55.
6. Cochrane AL, Holland WW. Validation of screening procedures. *Br Med Bull.* 1971;27:3–8.
7. Campion MJ, Reid R: Screening for gynecologic cancer. *Obstet Gynecol Clin of N Am.* 1990;17:695–727.
8. Van Der Graaf Y, Vooijs GP, Gaillard HLJ, et al. Screening errors in cervical cytology screening. *Acta Cytol.* 1987;31:434–438.
9. Fetherston WC. False-negative cytology in invasive cancer of the cervix. *Clin Obstet Gynecol.* 1983;26:929–937.
10. Richard RM, Vaillant HW. Influence of cell collection techniques upon cytological diagnosis. *Cancer.* 1965;18:1474.
11. Garite TJ, Feldman MJ. An evaluation of cytologic sampling techniques: a comparative study. *Acta Cytol.* 1978:22–83.
12. Taylor, PT, Andersen WA, Barker SR, et al. The screening Papanicolaou smear: contribution of the endocervical brush. *Obstet Gynecol.* 1987:70–734.
13. Walton RJ, Blanchet M, Boyes DA, et al. Cervical cancer screening programs (Department of National Health and Welfare Task Force Report). *Can Med Assoc J.* 1976;114:1003.
14. Celentano DD, Klassen AC, Weisman CS, Rosenshein NB. Duration of relative protection of screening for cervical cancer. *Prev Med.* 1989;18:411.
15. Boyes DA, Worth AJ, Anderson GH. Experience with cervical screening in British Columbia. *Gynecol Oncol.* 1981;12:143–155.
16. IARC Working Group on Evaluation of Cervical Cancer Screening Programs. Screening for squamous cervical cancer: Duration of low risk after negative results of cervical cytology and its implications for screening policies. *Br Med J.* 1986;293:659–664.
17. Clark EA, Anderson JW. Does screening by "Pap" smear help prevent cervical cancer? a case-control study. *Lancet.* 1979;2:1–4.
18. Paganini-Hill A, Ross RK. Reliability of recall of drug usage and other health-related information. *Am J Epidemiol.* 1982;16:114–122.

19. Principles of diagnosis. In: Gusberg SB, Shingleton HM, Deppe G, eds. *Female Genital Cancer.* New York: Churchill-Livingstone; 1988:197–221.
20. Richart, RM. Screening techniques for cervical neoplasia. *Clinical Obstet Gynecol.* 1979;22: 701–712.
21. National Cancer Institute Workshop. The 1988 Bethesda system for reporting cervical/vaginal cytologic diagnosis. *JAMA.* 1989;262:931.
22. Herbst AL. The Bethesda system for cervical/vaginal cytologic diagnosis. *Clin Obstet Gynecol.* 1992;35:22–27.
23. Kutman RJ, Malkasian GD, Soldis A, Solomon D. From Papanicolaou to Bethesda: the rationale for a new cytologic classification. *Obstet Gynecol.* 1991;77:779.
24. American Cancer Society. Cancer statistics 1991. *CA.* 1991;41:1.
25. The Epidemiology of Endometrial Cancer. In: Sciarra JJ, Lurain JR, eds. *Gynecology and Obstetrics.* Vol. 4, rev. ed. Philadelphia: JB Lippincott; 1991: chap. 13, 1–10.
26. Zucker PK, Kasdon EJ, Feldstein ML. The validity of Pap smear parameters as predictors of endometrial pathology in menopausal women. *Cancer.* 1986;58:2258–2263.
27. Ny ABP, Reagan J, Hawliczek CT, Wentz BW. Significance of endometrial cells in the detection of endometrial carcinoma and its precursors. *Acta Cytol.* 1974;18:356–361.
28. Yancey M, Magelssen D, Demaurez A, Lee R. Classification of endometrial cells on cervical cytology. *Obstet Gynecol.* 1990;76:1000.
29. An-Foraker SH, Kawada CY, McKinney D. Endometrial aspiration studies on Isaacs cell sampler with cytohistologic correlation. *Acta Cytol.* 1979;23:303–308.
30. Koss LG, Schreiber K, Oberlander SG, Moussouris HF, Lesser M. Detection of endometrial carcinoma and hyperplasia in asymptomatic women. *Obstet Gynecol.* 1984;64:1–11.
31. Hofmeister FJ. Endometrial biopsy: another look. *Am J Obstet Gynecol.* 1974;118:773.
32. Chambers JT, Chambers SK. Endometrial sampling: when? where? why? with what? *Clin Obstet Gynecol.* 1992;35:28–39.
33. Kaunitz AM, Masciello A, Ostrowski M, Rovira EZ. Comparison of endometrial biopsy with the endometrial pippelle and vabra aspirator. *J Reprod Med.* 1988;38:427.
34. Loffer FD. Hysteroscopy with selective endometrial sampling compared with D&C for abnormal uterine bleeding: the value of a negative hysteroscopic view. *Obstet Gynecol.* 1989;73:16.
35. Gimpelson R, Rappold H. A comparative study between panoramic hysterescopy with directed biopsies and dilatation and curettage. *Am J Obstet Gynecol.* 1988;158:489.
36. Carlson JA, Arger P, Thompson S, Carlson EJ. Clinical and pathologic correlation of endometrial cavity fluid detected by ultrasound in the postmenopausal patient. *Obstet Gynecol.* 1991;77:119.
37. Epithelial Ovarian Cancer. In: Berek JS, Hacker NF, eds. *Practical Gynecologic Oncology.* Baltimore: Williams and Wilkins; 1989: 327–364.
38. Advanced Epithelial Ovarian Cancer. In: DiSaia PJ, Creasman WT, eds. *Clinical Gynecologic Oncology.* 3rd ed. St. Louis: C.V. Mosby; 1989:325–416.
39. Bast RC, Feeney M, et al. Reactivity of a monoclonal antibody with human ovarian carcinoma. *J Clin Invest.* 1981;68:1331.
40. Bast RC, Klug TL, et al. A radioimmunoassay using a monoclonal antibody to monitor the course of epithelial ovarian cancer. *NEJM.* 1983;309:883.
41. Kabawat SE, Bast RC, Welsh WR, et al. Immunopathologic characterization of a monoclonal antibody that recognizes common surface antigens of human ovarian tumors of serous, endometrioid and clear cell types. *Am J Clin Pathol.* 1983;79:98.
42. Schwartz PE. Interpretation of CA-125 values. In: *Early Diagnosis and Monitoring of Ovarian Cancer.* Woodbridge, NJ: Advise and Consent, Inc; 1990:7–9.
43. Curry SL. Early diagnosis and management of a postmenopausal woman with an adnexal mass. In: *Early Diagnosis and Monitoring of Ovarian Cancer.* Woodbridge, NJ: Advise and Consent, Inc; 1990:3–5.
44. Cambell S, Bhan V, Royston P, Whitehead MI, Collins WP. Transabdominal ultrasound screening for early ovarian cancer. *Br Med J.* 1989;299:1363.
45. Fleischer AC: Transabdominal and transvaginal sonography of ovarian masses. *Clin Obstet Gynecol.* 1991;34:433.
46. Herrmann UJ, Locher GW, Goldhirsch A. Sonographic patterns of ovarian tumors: prediction of malignancy. *Obstet Gynecol.* 1987; 69:777.
47. Higgins RV, van Nagell JR, Donaldson ES, et al. Transvaginal sonography as a screening

method for ovarian cancer. *Gynecol Oncol.* 1989;34:402.
48. DePriest PD, van Nagell JR. Transvaginal ultrasound screening for ovarian cancer. *Clin Obstet Gynecol.* 1992;35/1:40.
49. van Nagell JR, dePriest PD, Puls LE, et al. Ovarian cancer screening in asymptomatic postmenopausal women by transvaginal sonography. *Cancer.* 1991;68:458.
50. Rodriguez MH, Platt LD, Medearis AL, et al. The use of transvaginal sonography for evaluation of postmenopausal ovarian size and morphology. *Am J Obstet Gynecol.* 1988;159:810.
51. Sassone AM, Timor-Tritsch IE, Artner A, et al. Transvaginal sonographic characterization of ovarian disease: evaluation of a new scoring system to predict ovarian malignancy. *Obstet Gynecol.* 1991;78:70.
52. Orr JW. Evaluation and monitoring of patients at risk of ovarian cancer. In: *Early Diagnosis and Monitoring of Ovarian Cancer.* Woodbridge, NJ: Advise and Consent; 1990:17–20.
53. Baker LH, Chin TDY, Wagner KV. Progress in screening for early breast cancer. *J Surg Oncol.* 1985;30:96.
54. Tabar L, Gad A, Holmbery LH, et al. Reduction in mortality from breast cancer after mass screening with mammography. *Lancet.* 1985:829.
55. Strax P: Results of mass screening for breast cancer in 50,000 examinations. *Cancer.* 1976; 37:30.
56. Verbeek ALM, Holland R, Sturmans F, et al. Reduction of breast cancer mortality through mass screening with modern mammography. *Lancet.* 1984:1272.
57. O'Malley MS, Fletcher SW. Screening for breast cancer with breast self-examination. *JAMA.* 1987;257:2197.
58. Winchester DP, Bernstein JR, Paige ML, Christ ML. *The Early Detection and Diagnosis of Breast Cancer.* American Cancer Society Monograph, 1988.
59. Gilbertson VA. The earlier detection of breast cancer. *Seminar Oncol.* 1974;1:87.
60. Winchester DP, Senor S, Immerman S, et al. A systematic approach to the evaluation and management of breast masses. *Cancer.* 1983; 51:2535.
61. Russo J, Frederick J, Ownby HE, et al. Predictors of recurrence and survival of patients with breast cancer. *Oncol J Club.* 1988;1:2.
62. Baker L: Breast cancer detection demonstration project: five year summary report. *CA.* 1982;32:194.
63. Moskowitz M. Screening for breast cancer: how effective are our tests? *CA.* 1983;33:76.
64. Cooper RA. Mammography. *Clin Obstet Gynecol.* 1989;32/4:768.
65. Fox SA, Klos DS, Tsou CV. Underuse of screening mammography by family physicians. *Radiology.* 1988;166:431.

2
Management of the Abnormal Pap Smear

James R. Dolan and Michael D. Moen

Anatomy and Pathology of the Cervix

The cervix is composed of fibromuscular and elastic tissue, and is lined by columnar and squamous epithelium. The vaginal portion of the cervix is generally lined by a nonkeratinizing squamous stratified epithelium, whereas the mucosa of the endocervical canal is composed of a single layer of mucin-secreting columnar epithelium.

The border between the stratified squamous epithelium of the ectocervix and the mucin-secreting columnar epithelium of the endocervix is called the *squamocolumnar junction*. Through a process termed *squamous metaplasia*, the columnar epithelium at the original squamocolumnar junction becomes replaced by a broad zone of metaplastic squamous epithelium. This process results in the formation of the *transformation zone*, which is a doughnut-shaped area on the portio of the cervix bounded peripherally by the original squamocolumnar junction and centrally by the new squamocolumnar junction.

The progression of this squamous transformation of the endocervical columnar mucosa is primarily dependent on the proliferation of undifferentiated subcolumnar *reserve cells* of the endocervical epithelium and the gradual transformation into a fully mature squamous epithelium. Local environmental factors, such as the acid pH of the vagina after puberty, trauma, chronic irritation, or cervical infection, can also play a role in the development and maturation of the transformation zone by stimulating repair and remodeling.[1]

Conversion of columnar to squamous epithelium is permanent and occurs at the greatest rate during fetal life, adolescence, and during the first pregnancy.[2]

The term *dysplasia* actually means an *abnormality of development* and is used to designate a proliferation of cytologically abnormal cells in the squamous epithelium of the cervix. This proliferation of cells represents the earliest precancerous changes within the epithelium of the cervix.[3] The vast majority of precancerous lesions of the cervix develop in the metaplastic epithelium at the squamocolumnar junction of the transformation zone, occurring twice as often on the anterior lip of the cervix as on the posterior lip.[4-7] These lesions often extend into the endocervical canal in the area of the new squamocolumnar junction but rarely onto the native squamous epithelium of the cervix outside the transformation zone.[4]

Dysplastic cells have histologic features similar to cervical cancer cells: nuclear enlargement, increased nuclear to cytoplasmic ratio, hyperchromaticity, irregular nuclear membranes, and multinucleation. Dysplasia is frequently subdivided into *mild*, *moderate*, and *severe* forms based on the degree and thickness of the epithelial changes.[1] When the atypical cells extend through the full thickness of the epithelium without invading the basement membrane, the term *carcinoma in situ* is used. The key distinguishing feature between dysplasia, carcinoma in situ, and invasive carcinoma is that the atypical cells do not invade the basement membrane in dysplasia and carcinoma in situ.[8]

The term *cervical intraepithelial neoplasia (CIN)* has been used to encompass all forms of cervical cancer precursor lesions, including dysplasia and carcinoma in situ. The CIN terminology underscores the fact that all cervical cancer precursors form a continuum.[9] In the original CIN terminology, lesions were graded 1 to 3 based on the extent to which epithelium was replaced by undifferentiated basiloid cells. In the modified CIN terminology, lesions are divided into two instead of three grades. The key histologic feature used to distinguish between low- and high-grade lesions is the presence of abnormal mitotic figures.[10]

The term *squamous intraepithelial lesion (SIL)* is the most recent modification in cervical precancer terminology. Patients with squamous intraepithelial lesions may have one of three courses: *regression*, *persistence*, or *progression*. Epidemiologic studies and long-term follow-up studies consistently report that the risk of progression to more significant stages of disease increases with increasing grades of lesions. The process of progression from dysplasia to carcinoma in situ and invasive cancer usually takes several years.[11-19]

The risk for progression of any individual squamous intraepithelial lesion is unknown, and lesions of all grades of severity have the potential to progress to invasive cancer if left untreated.[9] Although invasion can occur at any phase of CIN, it is more likely to occur in higher-grade lesions. DNA ploidy studies have revealed the presence of aneuploidy as a distinguishing feature in predicting which CIN lesions have a higher probability of progressing to invasive cancer. The percentage of CIN lesions that are aneuploid increases as the grade of the lesion increases.[20-23] The best histologic correlate associated with aneuploid lesions is the presence of abnormal mitotic figures, which are found in 85% of CIN lesions that are aneuploid.[23,24]

Risk factors for CIN and squamous cell carcinoma of the cervix include early age at first intercourse and multiple sexual partners, suggesting that exposure of the cervical transformation zone to carcinogenic factors during the time of the squamous metaplastic process results in an increased risk of developing cervical neoplasia.[2]

It is thought that cervical neoplasia is initiated by sexually transmitted carcinogens. Several studies have implicated a number of *human papilloma virus (HPV)* subtypes as major sexually transmitted cofactors in the etiology of cervical neoplasia.[25-27] Human papilloma viruses are double-stranded DNA viruses that infect epithelial cells of the skin and mucous membranes and produce local epithelial proliferation.

Three categories of HPV-related infections are described: clinically apparent disease, which includes condyloma acuminata, flat condyloma, and inverted condyloma; subclinical disease, which can be diagnosed by colposcopy; and latent disease, which is diagnosed by detection of HPV DNA.[8] Epithelial cells infected with HPV have a characteristic perinuclear cytoplasmic clearing and nuclear atypia. Cells that show these morphologic changes are pathognomonic for HPV infection and are termed *koilocytes*.[1]

Initial studies using southern blot hybridization techniques to detect the presence of HPV DNA revealed that approximately 10% of patients with normal Pap smears had evidence of HPV infection.[25,28-30] The prevalence of HPV DNA and abnormal Pap smears in association with cervical intraepithelial neoplasia has been well documented. HPV DNA has been detected in 80% to 90% of specimens of patients with cervical intraepithelial neoplasia and invasive squamous cell carcinoma.[25-27,30] HPV DNA typing studies have revealed low-risk and high-risk subtypes. Subtypes associated with lower risk for development of invasive cancer are types 6, 11, 42, 43, and 44. These subtypes are rarely found in invasive cancer. High-risk subtypes include 16 and 18, which are found in at least 80% to 90% of all high-grade lesions and cancers.[25,27,30]

Cytologic Screening

The principal goal of Papanicolaou smear screening is to detect occult carcinoma and precancerous abnormalities that may lead to invasive carcinoma. The Pap smear is a screening device that involves cytologic sampling of the transformation zone of the cervix to detect cellular abnormalities. Pap smear screening is used to select patients for further evaluation and to determine the presence or absence of significant cervical pathology.[31]

The use of cytology in detecting cervical cancer and precursors was first proposed by Papanicolaou in the 1940s.[32,33] Several studies have documented a dramatic reduction in the incidence of cervical cancer and death rate from cervical cancer after the institution of community-wide Pap smear screening.[34-38]

Terminology

The primary purpose of diagnostic terminology is to communicate to the referring physician information that can be used for patient management. The terminology originally proposed by Papanicolaou involved five classes of cytologic diagnoses ranging from *normal* to *consistent with malignancy*.[39] In 1973 the World Health Organization developed guidelines for reporting cytologic diagnoses that used the terms *dysplasia* and *carcinoma in situ*.[40] The National Cancer Institute Workshop held in Bethesda, Maryland, in 1988 developed the *Bethesda system* terminology for reporting cervical cytology results.[41] Recommendations from this conference included the following: The cytopathology report is a medical consultation; the Papanicolaou classification for reporting consultations is not acceptable in the modern practice of diagnostic cytopathology; and the Bethesda system should serve as a guideline for cytopathology reports of cervical/vaginal specimens. Terminology used in the Bethesda system closely correlates with histopathologic terminology.

The terms *low-grade* and *high-grade* squamous intraepithelial lesion (SIL) are used to designate cytologic changes to correlate with low- and high-grade cervical intraepithelial neoplasia.[41]

Further revisions in the Bethesda system were made at a second conference held in 1991. Major features of the revised Bethesda system are that it requires an estimate of the adequacy of the specimen for diagnostic evaluation, a general categorization of the specimen as being normal or abnormal, and a descriptive diagnosis that includes evidence of infection, inflammation, reactive changes, and epithelial cell abnormalities. Table 2.1 lists the components of the revised Bethesda system.[42]

Screening Interval

The optimal screening interval for Pap smears remains controversial, although a consensus now exists for recommending an annual Pap smear beginning at the time of initiation of sexual activity.[43] Several case control studies indicate that Pap smear screening offers substantial benefits in reducing the risk of cervical cancer and that smears at 2- to 3-year intervals offer about the same protection as smears at 1-year intervals.[44-48] Although these studies suggest that virtually all the protection offered by annual Pap smear screening can also be achieved with intervals of 2 to 3 years, annual programs may still be necessary to ensure compliance with a 2- to 3-year regimen due to poor attendance with annual Pap smear programs. Table 2.2 lists recent cervical cancer screening recommendations.[49-51]

The current Pap smear recommendation from the American Cancer Society is supported by the American College of Obstetricians and Gynecologists. It states that "all women who are or have been sexually active or have reached the age of 18 years [should] have an annual Pap test and pelvic examination. After a women has had three or more consecutive satisfactory normal annual examinations, the Pap test may be performed less frequently at the discretion of her physician."[52]

TABLE 2.1. Revised Bethesda system, 1991.[42]

Adequacy of specimen
  Satisfactory for evaluation
  Satisfactory for evaluation but limited by . . .
  Unsatisfactory for evaluation . . .
General Categorization (optional)
  Within normal limits
  Benign cellular changes: See descriptive diagnosis
  Epithelial cell abnormality: See descriptive diagnosis
Descriptive Diagnoses
  Benign Cellular Changes
    Infection
      Trichomonas vaginalis
      Fungus consistent with Candida
      Predominance of coccobacilli
      Actinomyces
      Cellular changes associated with herpes simplex virus
      Other
    Reactive Changes
      Reactive cellular changes associated with:
        Inflammation
        Atrophy with inflammation
        Radiation
        Intrauterine contraceptive device
        Other
  Epithelial Cell Abnormalities
    Squamous Cell
      Atypical squamous cells of undetermined significance
      Low-grade squamous intraepithelial lesion (encompassing cellular changes of human papilloma virus, mild dysplasia, CIN 1)
      High-grade squamous intraepithelial lesion (encompassing moderate and severe dysplasia, CIN 2, CIN 3, carcinoma in situ)
      Squamous cell carcinoma
    Glandular Cell
      Endometrial cells
      Atypical glandular cells of undetermined significance
      Endocervical adenocarcinoma
      Endometrial adenocarcinoma
      Extrauterine adenocarcinoma
      Adenocarcinoma, not otherwise specified
  Other Malignant Neoplasms: Specify
  Hormonal Evaluation (vaginal smears only)

TABLE 2.2. Pap smear screening recommendations.

The National Institutes of Health Consensus Conference (1980)[49]
  Pap smear screening should begin when a woman becomes sexually active or reaches age 18.
  Screening should be done every 1 to 3 years after two normal annual smears.
  Screening should be discontinued at age 60 if previous screening has been adequate and Pap smears have been consistently negative.
The Canadian Task Force (1982)[50]
  Pap smear screening should begin when a woman becomes sexually active or reaches age 18.
  Screening should be done annually until age 35 followed by rescreening every 5 years.
  Screening should be discontinued at age 60 if previous screening has been adequate and Pap smears have been consistently negative.
The American Cancer Society (1988)[51]
  Pap smear screening should begin when a woman becomes sexually active or reaches age 18.
  Screening should be performed every year until three or more annual Pap smears have been normal and then can be performed less frequently if recommended by a physician.
  There is no upper age limit on testing of mature women.

Technique

The original technique described by Papanicolaou involved a sampling of cells from the vaginal pool only. It became apparent that this method was inadequate and resulted in a high false-negative rate. In 1947 Ayre reported the use of a wooden spatula to scrape cells from the surface of the cervix. Because the majority of cervical cancer precursors involve the transformation zone, spatulas were designed to sample this area. This approach resulted in a lower false-negative rate than had been reported in smears that used cells from the vaginal pool only.[53,54]

Sampling of the endocervical canal is also important in reducing the false-negative rate, because the central border of the transformation zone extends into the endocervical canal. A significantly higher number of epithelial abnormalities is found in smears with endocervical cells compared with smears without endocervical cells. The absence of an endocervical component results in a higher false-negative rate. The use of the cytobrush for endocervical sampling has been shown to decrease the false-negative rate by increasing the number of smears with endocervical cells present.[55-65]

A Pap smear begins with examination of the

vagina and cervix with a speculum. A wooden *Ayre-type spatula* is rotated around the cervix twice with firm pressure. Cells obtained on the spatula are spread on a glass slide. The *cytobrush* is rotated 180 to 360 degrees in the external os and then rolled over the same slide. The cells obtained are then *immediately* fixed on the slide. With the advent of regulations for cytotechnologists by the U.S. federal government that limit the number of slides that can be screened each day, most cytopathology laboratories in the United States request that a single-slide technique, such as described above, be used for Pap smears.

Every specimen sent for cytopathologic evaluation should be accompanied by the patient's pertinent clinical history, including age, menstrual history, menopausal status, hormonal therapy, previous surgery or radiation exposure.

Despite the improvements offered by the wooden spatula and cytobrush, most studies continue to report a false-negative rate in detecting intraepithelial abnormalities on Pap smear of about 20%.[65-69] Several factors are involved in the failure to detect cervical cancer and cancer precursors by the present Pap smear system. These factors include poor technique in obtaining a smear, errors in laboratory screening, errors in interpretation of the smear, and errors in the clinical evaluation of the cytologic report.[31] Obtaining two smears at every examination results in a significantly lower false-negative rate but also a substantially increased cost.[69]

The optimal cancer detection system probably consists of a cervical smear and colposcopy. At this time, however, the cost of such a program precludes its institution. Annual Pap smear cytologic screening remains the most effective cervical cancer screening test for the present.[31,43]

Colposcopy

The primary method used to evaluate abnormal Pap smears is colposcopy with colposcopically directed cervical biopsies. Colposcopy was introduced in 1925 by Hinselmann but was not used extensively in the United States until the early 1970s.[70]

The main goals of the colposcopic examination are to identify the source of abnormal cells seen on the cytologic smear and to rule out invasive cancer. Identification of any abnormal areas facilitates the acquisition of histologic material (via directed biopsy) for definitive diagnosis and treatment. Since the majority of preinvasive lesions and cancers of the cervix originate in the transformation zone, this area is the focus of the colposcopic exam.[4-6]

The major indication for performing colposcopy is to evaluate abnormalities detected on cytologic examination of the cervix. Patients who have Pap smears showing evidence of human papilloma virus infection *without* dysplasia or intraepithelial lesions should also have a colposcopic examination performed. A large percentage of these patients will have a coexisting cervical intraepithelial neoplasia.[71] Patients who have Pap smears that show atypical squamous cells or atypical endocervical cells should also have a colposcopic examination performed. Studies have shown that 30% to 40% of these patients have a coexisting cervical intraepithelial neoplasia.[72-74]

Technique of Colposcopy

First is the inspection of the vulva and anus; squamous neoplasia of the lower genital tract is often multicentric. Speculum examination of the vagina and cervix is then performed, looking for any preexisting areas of white epithelium (leukoplakia). Next, cotton balls soaked in 3% to 5% acetic acid are applied to the cervix for 60 to 90 seconds. Application of acetic acid dissolves and removes mucus and causes CIN lesions to become whiter than the surrounding epithelium (acetowhite). The colposcopic examination is performed by viewing the cervix through the colposcope at a magnification of 10–40X. Areas of white epithelium, punctation, mosaicism, and atypical vessels are identified. A green filter is then used to inspect the vascular pattern of the transformation zone. Biopsies of previously identified abnormal areas are performed, and the location of lesions and biopsy sites are recorded. Endocervical curettage is then performed in nonpregnant patients. Tissue, mucous, and clotted blood are removed from the endocervical canal, placed on lens paper, and put in fixative solution. This is especially important in patients who have received prior treatment, because the squamocolumnar junction is usually in the endocervical canal.[75] Hemostasis is then achieved using silver nitrate sticks or Monsel's solution. As the speculum is withdrawn, the lower vaginal walls are then inspected. Finally, bimanual and rectal examinations are performed to rule out palpable cancer within the endocervix, vaginal walls, or adnexa.

Colposcopic Terminology

Terminology used to describe the colposcopic examination and the abnormal transformation zone include abnormalities involving the surface epithelium (*leukoplakia* and *acetowhite epithelium*) and abnormalities involving the underlying vasculature (*punctation*, *mosaicism*, and *atypical vessels*).[76] The tissue basis for these findings is described in Table 2.3.[70,77]

Other factors considered during colposcopic examination include surface contour, color tone, and clarity of demarcation.[76,78,79]

Management of Intraepithelial Lesions

Methods used to treat CIN result in the physical destruction or removal of the transformation zone and include *excisional techniques* such as loop excision, laser conization, and cold knife conization, and *ablative techniques* including laser vaporization, electrocoagulation diathermy, and cryosurgery (Table 2.4). Cervical intraepithelial neoplasia is considered a continuum of a single disease entity. Lesions destined to undergo malignant transformation cannot be separated by histology from those without this potential. Therefore, treatment is similar for all grades of lesions and involves removal of the abnormal transformation zone by one of these methods.[3,11]

Prerequisites for the treatment of cervical intraepithelial neoplasia include establishment that a cervical intraepithelial neoplasm exists, establishment that invasion is not present, and an assessment of the distribution and size of the lesion to determine the method best suited to eradicate the lesion.

Colposcopy is used to determine the topography of the lesion and to direct biopsies to obtain histologic diagnosis. Once the presence of cervical intraepithelial neoplasia is established and invasion has been ruled out, a

TABLE 2.3. Tissue basis of colposcopic terminology.

Leukoplakia: White plaques that can be identified without colposcopy or application of acetic acid; result from increased nuclear DNA and nuclear/cytoplasmic ratio in cells of immature mucosa, which cause opacity of the surface of the abnormal area due to inability of light to traverse the thickened epithelium.

Acetowhite Epithelium: Areas of abnormal epithelium that turn white after application of acetic acid; thought to result from osmolar change caused by acetic acid which results in dissipation of water from abnormal cells; cell membrane collapse around enlarged nucleus causes interruption in light transmission and the abnormal area appears white.

Punctation: Red dots seen in the epithelium; results from visualization of blood vessels on end that have perforated through the lamina propria and extend to the surface of the epithelium.

Mosaicism: Network of blood vessels that appears as red lines surrounding areas of abnormal epithelium; results from extension and proliferation of blood vessels beyond the lamina propria in the epithelial layer.

Atypical Blood Vessels: Blood vessels that run parallel to the epithelial surface and do not branch; result from formation of new capillaries in epithelial layer.

TABLE 2.4. Treatment methods for CIN.

Excisional Techniques
    Loop electrosurgical excision
    Laser conization
    Cold-knife conization
Ablative Techniques
    Laser vaporization
    Electrocoagulation diathermy
    Cryosurgery

TABLE 2.5. Indications for diagnostic conization.

1. Inadequate colposcopy (squamocolumnar junction not entirely visualized).
2. Positive endocervical curettage.
3. Colposcopic suspicion of occult invasive carcinoma.
4. Discrepancy between cytology and histology.
5. Suspicion of adenocarcinoma in situ.
6. Microinvasion on biopsy.

treatment method is chosen based on the size and distribution of the lesion.

If evaluation results in visualization of the entire squamocolumnar junction, an adequate assessment of the endocervical canal is achieved, and histologic findings are compatible with the cytologic diagnosis, then the histologic findings from an adequate number of target biopsies can be accepted as the definitive diagnosis. Conservative treatment may then be undertaken.[80,81]

Situations in which further diagnostic evaluation with cold-knife conization are indicated include inability to visualize the entire squamocolumnar junction, positive findings on endocervical curettage, colposcopic suspicion of occult invasion even if target biopsies show only CIN III, significant discrepancy (two grades) between cytology and histology or persistent abnormal cytology with normal findings on colposcopy, cytologic suspicion of adenocarcinoma in situ, or presence of microinvasion on biopsy.[82,83] The indications for diagnostic conization are summarized in Table 2.5.

Therapy should be planned according to lesion topography rather than histologic grade. The protection afforded by complete ablation of the transformation zone or by conization is comparable with that provided by hysterectomy. The efficacy of treatment depends on the volume of tissue destroyed or removed rather than on the modality used.[4,84-87] Whether ablative or excisional techniques are used, it is important that the lateral margins of the treatment area extend at least 5 mm beyond the margins of the lesion and the depth of treatment be at least 7 mm to ensure adequate treatment.[2,82,89,90]

Most reports of invasive cancer following conservative therapy for CIN suggest that treatment failures resulted from failure to diagnose invasive cancer during the *initial* evaluation. These studies revealed that two-thirds of invasive cancers, after conservative therapy, occurred within 12 months and more than 90% within 2 years. This implies that the lesion was present at the time of treatment for presumed CIN.[86,91-94]

Although ablative methods such as laser vaporization, electrocoagulation diathermy, and cryosurgery have been consistently reported to have a greater than 85% to 90% success rate,[84,88,90,95-104] there has recently been a renewed interest in excisional techniques (particularly loop electrosurgical excision and laser conization) for treating cervical intraepithelial neoplasia to provide further histologic confirmation of the diagnosis and allow assessment of the adequacy of treatment. Such an approach might contribute to a lower failure rate by providing further confirmation that invasive cancer is not present and that the entire abnormality has been removed.

Concern that patients might be overtreated by the approach outlined above has heightened interest in methods to determine which lesions have the greatest potential for progression and which lesions will regress if left untreated.

Current research using DNA ploidy studies and HPV viral typing to determine the "low-risk" or "high-risk" potential of any particular lesion has yielded useful results. Patients whose lesions are polyploid and contain HPV types 6 or 11 viral DNA have minimal risk of developing cervical cancer. Patients whose lesions are aneuploid and contain HPV types 16 or 18 viral DNA have a high risk of developing invasive disease. Although information concerning ploidy and HPV type might be useful in determining which patients may be candidates for observation and which patients would benefit from more aggressive therapy, the methods of ploidy analysis and viral typing are specialized and are not presently being applied to routine clinical material.[25]

Attempts to determine which lesions have the greatest potential for progression and re-

TABLE 2.6. Pap smear management.

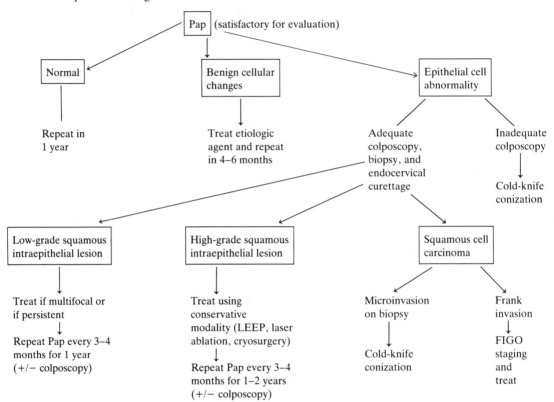

gression based on histology have been less successful. Despite the potential for progression of all grades of squamous intraepithelial lesions, the recent reclassification of lesions as low-grade and high-grade may be used to select those patients who can be followed with repeated cytologic examinations rather than removal of the transformation zone. This approach is supported by a recent study that reported a 64% regression rate for low-grade SIL.[105]

Table 2.6 outlines management decisions for the various levels of abnormalities identified on the Pap smear.

transformation zone (LLETZ), has recently gained popularity in the evaluation and management of patients with cervical pathology. Similar in the advantages offered by ablative techniques, loop excision can be performed on an outpatient basis with local anesthesia and minimal operating time. Hemostasis is usually achieved with the loop. If necessary, the coagulation mode can be used to stop any further bleeding. Healing is similar to that attained with laser ablation. The advantage of loop excision over ablative techniques is that it also provides a histologic specimen, which allows confirmation of the diagnosis and an assessment of the adequacy of treatment. Success rates greater than 90% are consistently reported.[87,106–110]

Treatment Methods

Loop Excision Procedure

The loop electrocoagulation excision procedure (LEEP), or large loop excision of the

Laser Vaporization

Use of the carbon dioxide laser for ablation of the abnormal transformation zone has become increasingly popular in the United States. As with other ablative techniques, efficacy depends on the volume of tissue destroyed rather than on the specific modality employed. Advantages of the laser include appropriateness for treating large and irregular lesions, precise tissue destruction, and a wound that heals with less fibrosis or scarring than cryosurgery. Laser surgery can be performed on an outpatient basis with local anesthesia and is well tolerated. A major disadvantage of the laser compared with cryosurgery is the cost involved. Success rates greater than 90% are consistently reported.[82,88,90,95-97,111]

Laser Conization

The carbon dioxide laser has also been used as a scalpel to perform excisional conization similar to that performed with a cold knife. Success rates greater than 90% have been reported.[112-114] Advantages of laser conization over cold-knife conization include better hemostasis, less scarring with better healing, and fewer complications.[114-116] Disadvantages of laser conization compared with laser vaporization are operating time and the need for general or regional anesthesia. Laser conization provides a specimen, and histologic diagnosis can be confirmed. The tissue thermal damage produced by the laser, however, may interfere with an accurate histologic evaluation of the specimen.[97,110,117-120]

Cold-Knife Conization

Cold-knife conization is primarily used for diagnosis rather than therapy. Indications for conization have been previously mentioned (Table 2.5). The major advantage of cold-knife conization is that it allows accurate histologic evaluation of the entire transformation zone. Disadvantages include the necessity for general or regional anesthesia; a longer operating time; and a complication rate for hemorrhage, sepsis, stenosis, infertility, and cervical incompetence ranging from 2% to 12%.[110,121,122]

Electrocoagulation

Electrocoagulation diathermy was first reported in 1966 and has been used extensively in Australia. It is not commonly used in the United States. Its effect is obtained by passing high-voltage electricity through the target tissue. This produces two different destructive effects: *fulguration* and *electrocoagulation*. Fulguration is produced by holding the electrodes just above the surface, allowing a spark of electricity to arc from the energy source to the tissue. The heat produced is so intense that tissue at the site of sparking undergoes rapid desiccation and charring. Electrocoagulation is produced by inserting an electrode below the epithelial surface and allowing current to dissect along planes of electroconductivity, causing heat denaturation and enzyme degradation of structural proteins. Success rates greater than 90% have been reported, and correlate with the surface area of the abnormal epithelium destroyed, and are unrelated to histologic grade of the lesion. Disadvantages of electrocoagulation include inability to accurately estimate the depth of destruction and tendency to produce significant scarring during the healing process.[82,84,98,99]

Cryosurgery

Cryosurgery was first used to treat cervical intraepithelial neoplasia in 1968. The mode of action involves hypothermia with resultant cryonecrosis. Cryonecrosis tends to be patchy and is accompanied by a wide zone of sublethal injury to adjacent tissues. Thawing results in the release of histamines and other vasoactive peptides, thereby producing local edema. The use of a freeze-thaw-refreeze technique increases the reliability of cell death. Also, to achieve a more effective heat transfer and a deeper zone of cryonecrosis, the probe is coated with a water-soluble lubricant.

Several series reporting results of cryosurgery to treat CIN reveal a primary success rate of 85% to 90%. Success is related to the extent of the lesion and volume of tissue destroyed. The advantages of cryosurgery are that it can be performed on an outpatient basis with no anesthesia, is fairly well tolerated, and has a low cost. Disadvantages include inability to adequately visualize the new squamocolumnar junction after treatment and patient discomfort due to post-treatment vaginal discharge.*

Post-Treatment Follow-up

Cervical healing after excisional or ablative removal of the transformation zone usually requires about 8 weeks. Thus, follow-up is usually initiated at 3 to 4 months post-therapy with cytologic examination. Repeat cytology is performed every 3 to 4 months for the first year after therapy and, if normal, yearly thereafter. Some authors have suggested the use of colposcopy during the first year of follow-up to enhance detection of recurrences and treatment failure.[124,125]

Most recurrences of CIN after initial therapy result from inadequate removal of the entire lesion or transformation zone during primary treatment.[4,85] After appropriate re-evaluation to ensure the absence of an invasive lesion or involvement of the endocervical canal, re-treatment with removal of the transformation zone can be performed using any of the excisional or ablative techniques.

Follow-up for patients who undergo treatment by excisional methods and have involvement of the margins of the specimen is still controversial. It appears that the healing process may play a role in successful treatment of these patients. At present, follow-up with cytology and colposcopy appears to be acceptable for these patients.[126,127]

References

1. Ferenczy A, Winkler B. Cervical intraepithelial neoplasia and condyloma. In: Kurman RJ, ed. *Blaustein's Pathology of the Female Genital Tract*, 3rd ed. New York: Springer-Verlag; 1987:177–217.
2. DiSaia PJ, Creasman WT. Preinvasive disease of the cervix, vagina, and vulva. In: *Clinical Gynecologic Oncology*, 3rd ed. St. Louis, Mo: CV Mosby Company; 1989:1–48.
3. Koss LG. Dysplasia. A real concept or a misnomer? *Obstet Gynecol*. 1978;51:374–379.
4. Abdul-Karim FW, Fu YS, Reagan JW, et al. Morphometric study of intraepithelial neoplasia of the uterine cervix. *Obstet Gynecol*. 1982;60:210–214.
5. Saito K, Saito A, Fu YS, et al. Topographic study of cervical condyloma and intraepithelial neoplasia. *Cancer*. 1987;59:2064–2070.
6. Richart RM. Colpomicroscopic studies of cervical intraepithelial neoplasia. *Cancer*. 1966;19:395–400.
7. Richart RM. Colpomicroscopic studies of the distribution of dysplasia and carcinoma in situ on the exposed portion of the human uterine cervix. *Cancer*. 1965;18:950–954.
8. Wright TC, Richart RM. Pathogenesis and diagnosis of preinvasive lesions of the lower genital tract. In: Hoskins WJ, Perez CA, Young RC, eds. *Principles and Practice of Gynecologic Oncology*. Philadelphia, Pa: JB Lippincott Company; 1992:509–536.
9. Richart RM. Natural history of cervical intraepithelial neoplasia. *Clin Obstet Gynecol*. 1968;10:748–784.
10. Richart RM. A modified terminology for cervical intraepithelial neoplasia. *Obstet Gynecol*. 1990;75:131–133.
11. Koss LG, Stewart FW, Foote FW, et al. Some histological aspects of behavior of epidermoid carcinoma in situ and related lesions of the uterine cervix. *Cancer*. 1963;16:1160–1211.
12. McIndoe WA, McLean MR, Jones RW, et al. The invasive potential of carcinoma in situ of the cervix. *Obstet Gynecol*. 1984;64:451–458.
13. Fox CH. Biologic behavior of dysplasia and carcinoma in situ. *Am J Obstet Gynecol*. 1967;99:960–974.
14. Hall JE, Walton L. Dysplasia of the cervix: a prospective study of 206 cases. *Am J Obstet Gynecol*. 1968;100:662–671.
15. Nasiell K, Roger V, Nasiell M, et al. Behavior of mild cervical dysplasia during long-term follow-up. *Obstet Gynecol*. 1986;67:665–669.
16. Nasiell K, Nasiell M, Vaclavinkova V, et al. Behavior of moderate cervical dysplasia during long-term follow-up. *Obstet Gynecol*. 1983;61:609–614.
17. Barron BA, Richart RM. A statistical model of the natural history of cervical carcinoma based on a prospective study of 557 cases. *JNCI*. 1968;41:1343–1353.

---

*References 82, 86, 100, 102–104, 111, 123.

18. Barron BA, Richart RM. A statistical model of the natural history of cervical carcinoma. II. Estimates of transition time from dysplasia to carcinoma in situ. *JNCI*. 1970;45:1025–1030.
19. Richart RM, Barron BA. A follow-up study of patients with cervical dysplasia. *Am J Obstet Gynecol*. 1969;105:386–393.
20. Reid R, Fu YS, Hershman RR, et al. Genital warts and cervical cancer. VI. The relationship between aneuploid and polyploid lesions. *Am J Obstet Gynecol*. 1984;150:189–199.
21. Fu YS, Reagan JW, Richart RM. Precursors of cervical cancer. *Cancer Surv* 1983;2:359.
22. Fu YS, Huang I, Beaudenon S, et al. Correlative study of HPV, DNA histopathology and morphometry in cervical condyloma and intraepithelial neoplasia. *Int J Gynecol Pathol*. 1988;7:297.
23. Bergeron C, Ferenczy A, Shah KV, et al. Multicentric human papilloma virus infections of the female genital tract: correlation of viral types with abnormal mitotic figures, colposcopic presentation, and location. *Obstet Gynecol*. 1987;69:736–742.
24. Winkler B, Crum PC, Fujii T, et al. Koilocytotic lesions of the cervix: the relationship of mitotic abnormalities to the presence of papillomavirus antigens and nuclear DNA content. *Cancer*. 1984;53:1081–1087.
25. Lorincz AT, Reid R, Jenson B. Human papillomavirus infection of the cervix: relative risk associations of common anogenital types. *Obstet Gynecol*. 1992;79:328–337.
26. Reid R, Stanhope R, Herschman BR. Genital warts and cervical cancer. I. Evidence of an association between subclinical papillomavirus infection and cervical malignancy. *Cancer*. 1982;50:377–387.
27. Reid R, Greenberg M, Jenson B. Sexually transmitted papillomaviral infections. I. The anatomic distribution and pathologic grade of neoplastic lesions associated with different viral types. *Am J Obstet Gynecol*. 1987;156:212–222.
28. Kiviat NB, Koutsky LA, Paavonen JA. Prevalence of genital papillomavirus infection among women attending a college student health clinic or a sexually transmitted disease clinic. *J Infect Dis*. 1989;159:293–302.
29. Lorincz AT, Temple GF, Patterson JA. Correlation of cellular atypia and human papillomavirus deoxyribonucleic acid sequences in exfoliated cells of the uterine cervix. *Obstet Gynecol*. 1986;68:508–512.
30. van de Brule A, Claas E, Maine M, et al. Application of anticontamination primers in the polymerase chain reaction for the detection of human papillomavirus genotypes in cervical scrapes and biopsies. *J Med Virol*. 1989;29:20–27.
31. Koss LG. The Papanicolaou test for cervical cancer detection. *JAMA*. 1989;261:737–743.
32. Papanicolaou GN. A survey of the actualities and potentialities of exfoliative cytology in cancer diagnosis. *Ann Med*. 1948;30:661.
33. Papanicolaou GN, Traut HF. Diagnostic value of vaginal smears in cancer of the uterus. *Am J Obstet Gynecol*. 1941;42:193–206.
34. Miller AB, Lindsey J, Hill GB. Mortality from cancer of the uterus in Canada and its relationship to screening for cancer of the cervix. *Int J Cancer*. 1976;17:602–612.
35. Christopherson WM, Mendez WM, Ahuja EM, et al. Cervix cancer control in Louisville, Kentucky. *Cancer*. 1970;26:29–38.
36. Canadian Task Force. Cervical cancer screening programs. I. Epidemiology and natural history of carcinoma of the cervix. *Can Med Assoc J*. 1976;114:1003–1033.
37. Day NE. Effect of cervical cancer screening in Scandinavia. *Obstet Gynecol*. 1984;63:714–718.
38. Christopherson WM. Trends in mortality from uterine cancer in relation to mass screening. *Acta Cytol*. 1977;21:5–9.
39. Papanicolaou G. *Atlas of Exfoliative Cytology*. Boston, Mass: Commonwealth Fund by University Press; 1954.
40. Riotton G, Christopherson WM. Cytology of the female genital tract. In: *International Histological Classification of Tumors #8*. Geneva: WHO; 1973.
41. National Cancer Institute Workshop. The 1988 Bethesda system for reporting cervical/vaginal cytologic diagnoses. *JAMA*. 1989;262:931–934.
42. National Cancer Institute Workshop. The revised Bethesda system for reporting cervical/vaginal cytologic diagnoses: Report of the 1991 Bethesda workshop. *J Repro Med*. 1992;37:383–386.
43. Nelson JH, Averette HE, Richart RM. Cervical intraepithelial neoplasia (dysplasia and carcinoma in situ) and early invasive cervical carcinoma. *CA*. 1989;39:157–178.
44. La Vechia C, Francesschi S, Decarli A, et al. "Pap" smear and the risk of cervical neoplasia: Quantitative estimates from a case-control study. *Lancet*. 1984;2:779–782.

45. MacGregor E, Moss SM, Parkin DM, et al. A case-control study of cervical cancer screening in northeast Scotland. *Br Med J.* 1985; 290:1543–1546.
46. Wangsuphachart V, Thomas DB, Koetsawang A, et al. Risk factors for invasive cervical cancer and reduction of risk by "Pap" smears in Thai women. *Int J Epidemiol.* 1987;16:363–366.
47. Olesen F. A case-control study of cervical cytology before diagnosis of cervical cancer in Denmark. *Int J Epidemiol.* 1988;17:501–508.
48. Shy K, Chu J, Mandelson M, et al. Papanicolaou smear screening interval and risk of cervical cancer. *Obstet Gynecol.* 1989;74:838–843.
49. Cervical cancer screening: Summary of an NIH consensus statement. *Br Med J.* 1980;281:1264–1266.
50. Canadian Task Force. Cervical cancer screening programs: summary of the 1982 Canadian task force report. *Can Med Assoc J.* 1982; 127:581–589.
51. Fink DJ. Change in American Cancer Society checkup guidelines for detection of cervical cancer. *CA.* 1988;38:127–128.
52. Pearse WH. Consensus report on frequency of Pap smear testing. *American College of Obstetricians and Gynecologists, Newsletter.* 1988;32:3.
53. Richart RM, Valiant HW. Influence of cell collection techniques upon cytologic diagnosis. *Cancer.* 1965;18:1474–1478.
54. Ayre JE. Cervical cytology in diagnosis of early cancer. *JAMA.* 1948;136:513–517.
55. Rubio CA, Kock Y, Stormby N, et al. Studies on the distribution of abnormal cells in cytological smears. VII. Cervical brush versus plastic and wooden spatulas. *Gynecol Oncol.* 1990;39:167–170.
56. Elias A, Linthorst G, Bekker B, et al. The significance of endocervical cells in the diagnosis of cervical epithelial changes. *Acta Cytol.* 1983;27:225–229.
57. Vooijs PG, Elias A, van der Graaf Y, et al. Relationship between the diagnosis of epithelial abnormalities and the composition of cervical smears. *Acta Cytol.* 1985;29:323–328.
58. Boon ME, de Graaff Guilloud JC, Rietveld WJ. Analysis of five sampling methods for the preparation of cervical smears. *Acta Cytol.* 1989;33:843–848.
59. Kristensen GB, Holund B, Grinsted P. Efficacy of the cytobrush versus the cotton swab in the collection of endocervical cells. *Acta Cytol.* 1989;33:849–851.
60. Lai-Goldman M, Nieberg RK, Mulcahy D, et al. The cytobrush for evaluating routine cervicovaginal endocervical smears. *J Repro Med.* 1990;35:959–963.
61. Hoffman MS, Hill DA, Gordy LW, et al. Comparing the yield of the standard Papanicolaou and endocervical brush smears. *J Repro Med.* 1991;36:267–269.
62. Pretorius RG, Sadeghi M, Fotheringham N, et al. A randomized trial of three methods of obtaining Papanicolaou smears. *Obstet Gynecol.* 1991;78:831–836.
63. Koonings PP, Dickinson K, d'Ablaing G, et al. A randomized clinical trial comparing the cytobrush and cotton swab for Papanicolaou smears. *Obstet Gynecol.* 1992;80:241–245.
64. Murata PJ, Johnson RA, McNicoll KE. Controlled evaluation of implementing the cytobrush technique to improve Papanicolaou smear quality. *Obstet Gynecol.* 1990;75:690–695.
65. Selvaggi SM. Spatula/cytobrush vs. spatula/cotton swab detection of cervical condylomatous lesions. *J Repro Med.* 1989;34:629.
66. van der Graaf Y, Vooijs GP, Gaillard HLJ, et al. Screening errors in cervical cytologic screening. *Acta Cytol.* 1987;31:434–438.
67. Boon ME, Alons-van Kordelaar JJM, Rietveld-Scheffers PEM. Consequences of the introduction of combined spatula and cytobrush sampling for cervical cytology: improvements in smear quality and detection rates. *Acta Cytol.* 1986;30:264–270.
68. Kawaguchi K, Nogi M, Ohya M, et al. The value of the cytobrush for obtaining cells from the uterine cervix. *Diagn Cytopathol.* 1987; 3:262.
69. Beilby JOW, Bourne R, Guillebaud J, et al. Paired cervical smears: A method of reducing the false-negative rate in population screening. *Obstet Gynecol.* 1982;60:46–48.
70. Burke L, Antonioli DA, Ducatman BS. Tissue basis of colposcopy. In: *Colposcopy Text and Atlas.* East Norwalk, Conn: Appelton and Lange; 1991:1–6.
71. Cartwright PS, Dao AH, Reed GW. Human papillomavirus in women: a three-year experience in a county hospital colposcopy clinic. *J Repro Med.* 1992;37:167–169.
72. Rader JS, Rosenzweig BA, Spirtas R, et al. Atypical squamous cells: a case-series study of the association between Papanicolaou smear

results and human papillomavirus DNA genotype. *J Repro Med.* 1991;36:291–297.
73. Lindheim SR, Smith-Nguyen G. Aggressive evaluation for atypical squamous cells in Papanicolaou smears. *J Repro Med.* 1990; 35:971–973.
74. Goff BA, Atanasoff P, Brown E, et al. Endocervical glandular atypia in Papanicolaou smears. *Obstet Gynecol.* 1992;79:101–104.
75. Urcuyo R, Rome RM, Nelson JH. Some observations on the value of endocervical curettage performed as an integral part of colposcopic examination of patients with abnormal cervical cytology. *Am J Obstet Gynecol.* 1977; 128:787–792.
76. Coppleson M. Colposcopic features of papillomaviral infection and premalignancy in the female lower genital tract. *Obstet Gynecol Clin North Am.* 1987;14:471–494.
77. Stafl A, Wilbanks GD. An international terminology of colposcopy: report of the nomenclature committee of the international federation of cervical pathology and colposcopy. *Obstet Gynecol.* 1991;77:313–314.
78. Reid R, Stanhope CR, Herschman BR, et al. Genital warts and cervical cancer. IV. A colposcopic index for differentiating subclinical papillomaviral infection from cervical intraepithelial neoplasia. *Am J Obstet Gynecol.* 1984;149:815–823.
79. Reid R, Scalzi P. Genital warts and cervical cancer. VII. An improved colposcopic index for differentiating benign papillomaviral infections from high-grade cervical intraepithelial neoplasia. *Am J Obstet Gynecol.* 1985;153:611–618.
80. Veridiano NP, Delke I, Tancer ML. Accuracy of colposcopically directed biopsy in patients with cervical neoplasia. *Obstet Gynecol.* 1981;58:185–187.
81. Stafl A, Mattingly RF. Colposcopic diagnosis of cervical neoplasia. *Obstet Gynecol.* 1973; 41:168–176.
82. Reid R: Preinvasive disease. In: Berek JS, Hacker NF, eds. *Practical Gynecologic Oncology*. Baltimore, Md: Williams and Wilkins; 1989:195–239.
83. McCord ML, Stovall TG, Summitt RL, et al. Discrepancy of cervical cytology and colposcopic biopsy: Is cervical conization necessary? *Obstet Gynecol.* 1991;77:715–719.
84. Chanen W, Rome RM. Electrocoagulation diathermy for cervical dysplasia and carcinoma in situ: A 15-year survey. *Obstet Gynecol.* 1983;61:673–679.
85. Benedet JL, Nickerson KG, White GW. Laser therapy for cervical intraepithelial neoplasia. *Obstet Gynecol.* 1981;58:188–191.
86. Richart RM, Townsend D, Crisp W. An analysis of long term follow-up results in patients with cervical intraepithelial neoplasia treated by cryosurgery. *Am J Obstet Gynecol.* 1980;137:823–826.
87. Wright TC, Gagnon S, Richart RM, et al. Treatment of cervical intraepithelial neoplasia using the loop electrosurgical excision procedure. *Obstet Gynecol.* 1992;79:173–178.
88. Townsend DE, Richart RM. Cryotherapy and carbon dioxide laser management of cervical intraepithelial neoplasia: a controlled comparison. *Obstet Gynecol.* 1983;61:75–78.
89. Anderson MC, Hartley RB. Cervical crypt involvement by intraepithelial neoplasia. *Obstet Gynecol.* 1980;55:546–550.
90. Wright VC, Davies E, Riopelle MA. Laser surgery for cervical intraepithelial neoplasia: principles and results. *Am J Obstet Gynecol.* 1983;145:181–184.
91. Townsend DE, Richart RM. Diagnostic errors in colposcopy. *Gynecol Oncol.* 1981; 12:s259–s264.
92. Pearson WE, Whittaker J, Ireland D, et al. Invasive carcinoma of the cervix after laser treatment. *Br J Obstet Gynecol.* 1989;96:486–488.
93. Townsend DE, Richart RM, Marks E, et al. Invasive cancer following outpatient evaluation and therapy for cervical disease. *Obstet Gynecol.* 1981;57:145.
94. Sevin B, Ford JH, Girtanner RD, et al. Invasive cancer of the cervix after cryosurgery. *Obstet Gynecol.* 1979;53:465–471.
95. Popkin DR. Treatment of cervical intraepithelial neoplasia with the carbon dioxide laser. *Am J Obstet Gynecol.* 1983;145:177–180.
96. Benedet JL, Miller DM, Nickerson KG. Results of conservative management of cervical intraepithelial neoplasia. *Obstet Gynecol.* 1992;79:105–110.
97. Baggish MS, Dorsey JH, Adelson M. A ten-year experience treating cervical intraepithelial neoplasia with the CO_2 laser. *Am J Obstet Gynecol.* 1989;161:60–68.
98. Deigan EA, Carmichael JA, Ohlke ID, et al. Treatment of cervical intraepithelial neoplasia with electrocautery: a report of 776 cases. *Am J Obstet Gynecol.* 1986;154:255–259.

99. Richart RM, Sciarra JJ. Treatment of cervical dysplasia by outpatient electrocauterization. *Am J Obstet Gynecol.* 1968;101:200–205.
100. Charles EH, Savage EW. Cryosurgical treatment of cervical intraepithelial neoplasia. *Obstet Gynecol Surv.* 1980;35:539–548.
101. Popkin DR, Scali V, Ahmed MN. Cryosurgery for the treatment of cervical intraepithelial neoplasia. *Am J Obstet Gynecol.* 1978;130:551–554.
102. Crisp WE, Smith MS, Asadourian LA, et al. Cryosurgical treatment of premalignant disease of the uterine cervix. *Am J Obstet Gynecol.* 1970;107:737–742.
103. Benedet JL, Miller DM, Nickerson KG, et al. The results of cryosurgical treatment of cervical intraepithelial neoplasia at one, five, and ten years. *Am J Obstet Gynecol.* 1987;157:268–273.
104. Bryson SCP, Lenehan P, Lickrish GM. The treatment of grade 3 cervical intraepithelial neoplasia with cryotherapy: an 11-year experience. *Am J Obstet Gynecol.* 1985;151:201–206.
105. Syrjanen K, Kataja V, Yliskoski M, et al. Natural history of cervical human papillomavirus lesions does not substantiate the biologic relevance of the Bethesda system. *Obstet Gynecol.* 1992;79:675–682.
106. Whiteley PF, Olah KS. Treatment of cervical intraepithelial neoplasia: experience with the low-voltage diathermy loop. *Am J Obstet Gynecol.* 1990;162:1272–1277.
107. Prendiville W, Cullimore J, Norman S. Large loop excision of the transformation zone (LLETZ). A new method of management for women with cervical intraepithelial neoplasia. *Br J Obstet Gynecol.* 1989;96:1054–1060.
108. Gunasekera PC, Phipps JH, Lewis BV. Large loop excision of the transformation zone (LLETZ) compared to carbon dioxide laser in the treatment of CIN: a superior mode of treatment. *Br J Obstet Gynecol.* 1990;97:995–998.
109. Luesley DM, Cullimore J, Redman CWE, et al. Loop diathermy excision of the cervical transformation zone in patients with abnormal cervical smears. *Br Med J.* 1990;300:1690–1693.
110. Turner RJ, Cohen RA, Voet RL, et al. Analysis of tissue margins of cone biopsy specimens obtained with "cold knife," CO_2 and Nd:YAG lasers and a radiofrequency surgical unit. *J Repro Med.* 1992;37:607–610.
111. Wetchler SJ. Treatment of cervical intraepithelial neoplasia with the CO_2 laser: laser versus cryotherapy. A review of effectiveness and cost. *Obstet Gynecol Surv.* 1984;39:469–473.
112. McIndoe GA, Robson MS, Tidy JA, et al. Laser excision rather than vaporization: the treatment of choice for cervical intraepithelial neoplasia. *Obstet Gynecol.* 1989;74:165–168.
113. Tabor A, Berget A. Cold-knife and laser conization for cervical intraepithelial neoplasia. *Obstet Gynecol.* 1990;76:633–635.
114. Bostofte E, Berget A, Larsen F, et al. Conization by carbon dioxide laser or cold-knife in the treatment of cervical intraepithelial neoplasia. *Acta Obstet Gynecol Scand.* 1986;65:199–202.
115. Kristensen GB, Jensen LK, Holund B. A randomized trial comparing two methods of cold-knife conization with laser conization. *Obstet Gynecol.* 1990;76:1009–1013.
116. Delmore J, Horbelt DV, Kallail KJ. Cervical conization: cold-knife and laser excision in residency training. *Obstet Gynecol.* 1992;79:1016–1019.
117. Howell R, Hammond R, Pryse-Davies J. The histologic reliability of laser cone biopsy of the cervix. *Obstet Gynecol.* 1991;77:905–911.
118. Fowler JM, Davos I, Leuchter RS, et al. Effect of CO_2 laser conization of the uterine cervix on pathologic interpretation of cervical intraepithelial neoplasia. *Obstet Gynecol.* 1992;79:693–698.
119. Larsson G, Alm P, Grundsell H. Laser conization versus cold-knife conization. *Surg Gynecol Obstet.* 1982;154:59–61.
120. Baggish MS. A comparison between laser excisional conization and laser vaporization for the treatment of cervical intraepithelial neoplasia. *Am J Obstet Gynecol.* 1986;155:39–44.
121. Sprang ML, Isaacs JH, Boraca CT. Management of carcinoma in situ of the cervix. *Am J Obstet Gynecol.* 1977;129:47–50.
122. Larsson G, Gullberg B, Grundsell H. A comparison of complications of laser and cold-knife conization. *Obstet Gynecol.* 1983;62:213–217.
123. Schantz A, Thormann L. Cryosurgery for dysplasia of the uterine ectocervix: a randomized study of the efficacy of the single- and double-freeze techniques. *Acta Obstet Gynecol Scand* 1984;63:417–420.
124. Paraskevaidis E, Jandial L, Mann EMF, et al.

Pattern of treatment failure following laser for cervical intraepithelial neoplasia: implications for follow-up protocol. *Obstet Gynecol.* 1991;78:80–83.
125. Hirai Y, Nishide K, Yamauchi K, et al. Cytomorphologic, cytometric and histomorphologic observations after laser therapy for cervical lesions. *J Repro Med.* 1992;37:267–272.
126. Andersen ES, Nielsen K, Larsen G. Laser conization: follow-up in patients with cervical intraepithelial neoplasia in the cone margin. *Gynecol Oncol.* 1990;39:328–331.
127. White CD, Cooper WL, Williams RR. Cervical intraepithelial neoplasia extending to the margins of resection in conization of the cervix. *J Repro Med.* 1991;36:635–638.

3
Chronic Vaginitis

Lauren F. Streicher and Fred A. Zar

Introduction

Vaginitis remains an exceedingly common clinical entity and may be a chronic cause of morbidity in a significant proportion of women. Approximately 10% of office visits to primary care physicians are for vaginal symptoms, accounting for over 10 million visits a year.[1] Vaginitis is the most common complaint seen in self-referred cases to gynecology clinics,[2] and 40% of women with vaginal symptoms will be diagnosed to have some form of vaginitis.[3]

The approach to the patient with chronic vaginitis requires an understanding of normal vaginal physiology and flora, infectious and noninfectious differential diagnoses applicable to their symptoms, and diagnostic approaches and therapies directed at minimizing if not eliminating recurrences.

Indigenous Flora

After transient colonization with maternal flora in the neonatal period followed by a predominantly anaerobic flora in the prepubescent years, the adult vaginal flora contains a predominance of 5 to 10 species, with lactobacilli being the most numerous.[1,4] Other organisms commonly found in quantities greater than 10^5 include species of *Eubacterium*, *Bacteroides*, *Peptococci*, *Corynebacterium*, coagulase-negative *Staphylococci*, and *Peptostreptococci*.[1] Up to 100 different species have been isolated from various women in population studies,[1] thus making the term *indigenous* flora instead of *normal* flora more appropriate. Sexual activity has been correlated with higher colonization rates for *Gardnerella vaginalis*, *Mycoplasma*, and *Ureaplasma*.[5] Oral contraceptives appear to have minimal effect on this flora.[6]

Physiologic Discharge

The presence of discharge is not necessarily pathologic, and to avoid unnecessary therapies, diagnostic procedures, and the emotional and social ramifications of the diagnosis of chronic vaginitis, it is crucial to differentiate this "physiologic discharge" from a disease state. Physiologic discharge consists of a combination of mucoid endocervical secretions; exfoliated vaginal epithelial cells; indigenous flora; and secretions from the sebaceous, sweat, Bartholin's, and Skene's glands, as well as transudates from the endometrium, fallopian tubes, and vaginal walls. These secretions may increase in volume with hyperestrogenic states such as pregnancy or exogenous hormone use. A physiologic increase in discharge may be observed with ovulation, due to increased water content of cervical mucus, and with coitus, due to increased vaginal wall exudate. In addition, anovulatory women have unopposed estrogen stimulation of endocervical glands resulting in greater than usual secretions. Physiologic secretions are odorless, clear to white in color, viscous and homogenous, and occasionally somewhat flocculent. There should be no pooling in the vagina, and leukocytes should be absent except for scant numbers during the secretory phase. The pH of physiologic secretions ranges from 3.5 to 5.0. This acid milieu is secondary to lactate production from the breakdown of glycogen deposited in the vaginal epithelial cells and through the metabolic activity of the lactobacilli normally present.

On exam, physiologic discharge should not be present at the introitus and should have no associated vulvar changes. Occasionally, women will perform overzealous cleansing of the vulva in an attempt to eradicate the discharge, which may result in some redness. As stated above, it is crucial to recognize this because patients often respond to instructions to lessen their efforts at topical cleansing decrease their bathing and douching intensity, and reassurance. Topical steroids after a water rinse may decrease vulvar irritation. Oral contraceptives occasionally cause a leukorrhea from excess cervical mucus with an "exacerbation" of physiologic discharge.

Diagnostic Approach to True Chronic Vaginitis

The approach to patients with suspected chronic vaginitis requires a thorough history and physical prior to diagnostic tests. The history should include a detailed description of the duration of symptoms and a sexual history with specific attention to recent changes in, or acquisition of, sexual partners. With recurrent disease, it is important to document responses to prior therapies as well as the occasional exacerbation of symptoms with treatment seen in the patient whose discharge is an allergic reaction to topical agents. A thorough description of the discharge, including color, consistency, and odor is also useful.

The physical exam should look for the cutaneous irritation of the vulva typically associated with vulvovaginal candidiasis, vestibular inflammation associated with vulvitis or vestibulitis, and signs of mucopurulent cervicitis. This exam is best performed with an unlubricated speculum in a patient who has not douched within 2 days, had intercourse within 2 days, or taken any vaginal contraceptive medication for 5 days.

Diagnostic Tests

The middle third of the vagina should be swabbed for assessment of pH. A pH of less than 5.0 is seen in physiologic discharge and vulvovaginal candidiasis. A more alkaline pH is classically seen in bacterial vaginosis and trichomonas vaginitis. If ejaculate is present in the vagina, inaccurate results may be found because this elevates the pH.

A drop of saline should be placed on a slide, and a swab of secretions should be dipped into it. Care should be taken not to disperse the drop because this may lead to water evaporation, creating a hypertonic solution that kills trichomonads. Under microscopy, lactobacilli usually appear as moderate numbers of clumped rods. Absence of these bacteria is suggestive of a pathologic state, prior antibiotic use, or frequent douching. A meticulous search should be made for clue cells (see bacterial vaginosis, below), motile trichomonads, and fungal elements.

Next, a potassium hydroxide (KOH) preparation should be made. This is best done by placing material from an applicator on a dry slide. The applicator should be held on an angle over the slide; 10% KOH should be dropped onto the applicator and from there onto the slide. The applicator should then be smelled for the typical fishy odor associated with bacterial vaginosis.[7] Light microscopy should be used to search for hyphal and yeast forms of Candida species. Given the extremely uncommon occurrence of vaginitis from aerobic bacterial infection, routine bacterial cultures are rarely warranted.

If signs of cervical inflammation are noted, appropriate specimens should be obtained for detection of herpes simplex, chlamydia, and gonorrhea.

Vulvovaginal Candidiasis

Epidemiology

As a cause of vaginitis in the United States, Candida species rank behind bacterial vaginosis and trichomonas vaginitis.[8] Yet approximately 75% of women will suffer one episode of vulvovaginal candidiasis in their lifetime,[9] and almost half of these will have at least one recurrence.[10] A subset of less than 5% of women will have chronic vulvovaginal candidiasis,[11] defined and discussed below.

From 85% to 90% of fungal vaginitis is due to *Candida albicans*. With rare exceptions, the remainder of cases are due to nonalbicans Candida species, most commonly *C tropicalis*[12] and *C glabrata*.[13] *C albicans* is indigenous vaginal flora in about 20% of asymptomatic premenopausal women[14] and is found in the mouth, rectum, and skin of 5 to 15% of those studied.[1] Although more than 200 strains of *C albicans* have been identified, there is no apparent difference in their ability to colonize the human vagina or cause true infection.[15,16] However, *C albicans* does adhere better to vaginal epithelium than other Candida species.[17]

An increase in the frequency of vaginal carriage of *C albicans* has been noted in pregnancy, diabetes, use of high-estrogen oral contraceptives or systemic glucocorticoids, women wearing tight or synthetic clothing, persons on systemic antibacterials, and women with intrauterine devices.[11]

Mechanism of Disease Production

The transition from yeast to hyphal or pseudohyphal form has been associated with symptomatic disease.[18] These latter forms have a heightened ability to adhere to and then penetrate the vaginal epithelial cell.[19] *C albicans* has been shown to produce bacteriocins that can suppress normal vaginal flora.[20] Vaginal epithelial cells from women with chronic vulvovaginal candidiasis do not show an increased ability to bind to *C albicans*.[21] In addition, an in vitro phenomenon that may have relevance to disease is that of switching colonies;[22,23] this occurs in cultures incubated at 24°C in which colonies shift from a white to opaque appearance. These opaque colonies demonstrate an increased resistance to flucytosine, an increased ability to adhere to vaginal epithelial cells, more efficient germination and mycelial formation, and an increased production of extracellular proteases.

Etiologic Factors

Systemic Antibacterials

Lactobacilli appear to be protective of colonization and subsequent infection from Candida species. Lactobacilli directly compete for nutrients with *C albicans*, stearically inhibit yeast adherence to vaginal epithelial cells, and elaborate bacteriocins that can inhibit yeast proliferation and growth.[11] Lactobacilli are susceptible to most oral antibiotics, and their numbers decrease dramatically with subsequent overgrowth of *C albicans* even after single doses of antimicrobials.[24,25,26]

Systemic Estrogens

Elevated systemic estrogens increase vaginal epithelial glycogen, which may act as a nutritional substrate for the overgrowth of Candida species.[27] An estrogen receptor in the cytosol of *C albicans* has been implicated in increased mycelial formation and adherence to epithelial cells in hyperestrogen states.[28] Pregnancy is associated with a carrier rate for *C albicans* of 10% in the first trimester that increases to over 50% in the third trimester.[29,30] From 60% to 90% of these carriers will develop active disease.[31] Not only is vulvovaginal candidiasis more common in pregnancy; it also appears to be more refractory to therapy. High-estrogen oral contraceptives are associated with more frequent vulvovaginal candidiasis, possibly due to elevated vaginal glucose concentrations.[32] However, low-estrogen preparations do not appear to be implicated in acute or chronic infection.[33]

Tight Clothing

Studies reveal that tight and occlusive clothing, as well as synthetic undergarments, can elevate the incidence of vulvovaginal candidiasis two- to threefold.[34–36]

Cell-Mediated Immunodeficiency

Vulvovaginal candidiasis may be the initial manifestation of human immunodeficiency virus (HIV) infection,[37] as well as a harbinger for successive Candida infection of the mouth and esophagus.[38] Exogenous immunosuppressants may also increase the incidence of vulvovaginal candidiasis. In addition to the above-mentioned estrogen effect, the mild cell-mediated immunodeficiency seen in the third trimester of pregnancy[39] probably contributes to vulvovaginal candidiasis.

Diabetes Mellitus

Vaginal colonization with Candida is more frequent in diabetic women than in nondiabetic women, yet symptomatic candidal infection appears with increased frequency only in uncontrolled diabetics.[11] It is exceedingly rare for premenopausal women to have diabetes mellitus as the sole predisposition to vulvovaginal candidiasis.[9]

Miscellaneous Etiologies

Authors have suggested a higher incidence of vulvovaginal candidiasis in women who are obese,[40] medically debilitated,[40] sexually active,[11] prone to candy binges,[9,41] or using intrauterine devices for contraception.[42]

Clinical Presentation

Unlike other vaginitides, vulvovaginal candidiasis is usually associated with significant perivaginal itching and/or irritation. Periurethral involvement may lead to dysuria, often characterized as external in nature. Although the quantity of discharge is typically not increased, up to 50% of women notice a change in their physiologic discharge, with clumps of material being present. A history of exacerbation the week prior to menses is not uncommon. Of note is the absence of malodor.

On exam, an abnormal discharge is noted in the vast majority of patients and the vaginal walls appear inflamed in approximately a quarter of patients.[43] Itching may lead to visible excoriations, and there may be classic satellite papules adjacent to the redness of the vulva. The discharge is often thick, is adherent to the vaginal walls, and may take on the consistency of curds of cottage cheese (Fig. 3.1). The examiner should also note the lack of offensive odor.

One of the most useful tests in evaluating the discharge is an assay of pH. A pH of 4.5 or less is extremely unusual in the other common vaginitides, bacterial vaginosis and trichomonas vaginitis. Caution is advised, however, because up to a third of patients with Candida infection may have a pH greater than 4.5.[43] Application of 10% KOH will not produce a fishy odor (negative whiff test) but

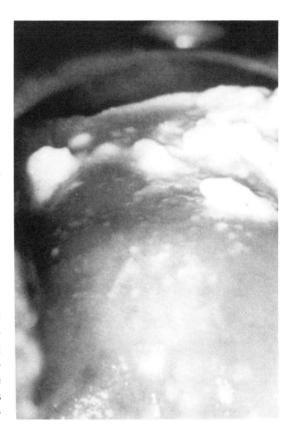

FIGURE 3.1. Clumps of cottage cheese–like discharge adherent to vaginal wall, commonly seen in vulvovaginal candidiasis. (Reproduced with permission from *Atlas of Sexually Transmitted Diseases*. New York: Gower Medical Publishing; 1989.)

TABLE 3.1. Topical antifungal therapy of acute vulvovaginal candidiasis.

Antifungal	Formulation	Dosage
Nystatin	100,000 U vaginal tab	1 tab × 14d
Clotrimazole	1% cream	5 g × 14d
	10% cream	5 g × 1
	100 mg vaginal tab	1 tab × 7d
	100 mg vaginal tab	2 tab × 3d
	500 mg vaginal tab	1 tab × 1
Miconazole	2% cream	5 g × 7d
	100 mg vaginal supp	1 supp × 7d
	200 mg vaginal supp	2 supp × 7d
	1200 mg vaginal supp*	1 supp × 1
Butoconazole	2% cream	5 g × 3d (6d if pregnant)
Terconazole	0.4% cream	5 g × 7d
	80 mg tab	1 tab × 3d
Tioconazole	2% cream*	5 g × 3d
	6.5% ointment	4.6 g × 1
Econazole*	150 mg vaginal tab	1 tab × 3d
Fenticonazole*	2% cream	5 g × 7d
Isoconazole*	300 mg vaginal tab	2 tabs × 1

*Not available in United States.

will demonstrate yeast and mycelial elements by microscopy in 60 to 90% of cases.[11] The saline preparation typically shows no white cells, but in up to 25% of cases, some are present.[43] Cultures for yeast are difficult to interpret given the prevalence of *C albicans* as indigenous flora. Culture should be relegated to cases in which one is highly suspicious of vulvovaginal candidiasis yet smears are not helpful or when one suspects recurrences are from fungi other than *C albicans*.

Therapy of Candida Vaginitis

Topical drugs available for the therapy of vulvovaginal candidiasis fall into two classes. The polyene class currently contains only one drug, nystatin. The azole class consists of a variety of medications whose recommended doses for the therapy of acute vulvovaginal candidiasis are listed in Table 3.1. Agents used less frequently include gentian violet, povidone-iodine, boric acid, and lactobacillus preparations. These therapies suffer from inadequately controlled studies of their efficacy or unacceptable patient intolerance.[44,45]

Etiology of Chronic Vulvovaginal Candidiasis

Sobel has defined chronic vulvovaginal candidiasis as at least four mycologically proven symptomatic episodes within a 12-month period.[46] Various mechanisms of these recurrences have been proposed.[46]

Reinfection from Gastrointestinal Reservoir

Even though 40% to 100% of women with vulvovaginal candidiasis will have demonstrable *C albicans* in the stool[47,48] and 80% of these strains will be identical to the vaginal strain,[49] it appears unlikely that this is a significant cause of recurrent disease. Attempts to eradicate the rectal focus with oral nystatin have failed to prevent recurrence.[50,51] One study shows efficacy of this approach, yet it may have included patients without recurrent disease in its cohort and it only followed patients up to 7 weeks of posttreatment.[52] This approach possibly fails due to the inability of oral nystatin to eradicate *C albicans* from the bowel.[53] In a study of ketoconazole maintenance therapy to prevent relapse, relapse occurred in spite of negative rectal cultures for yeast.[49]

Sexual Transmission

Male colonization with Candida is four times more common in sexual partners of women with vulvovaginal candidiasis.[54] Approximately 20% of male partners of women with chronic vulvovaginal candidiasis are also colonized.[50,54,55] The strains of sexual partners are usually the same.[49] No controlled studies have shown that routine treatment of male sexual partners prevents recurrent vaginitis. Of interest but of uncertain clinical significance, prostatic ejaculates have been found to be culture-positive for *C albicans*[56,57] as another potential source of reinfection.

Vaginal Relapse

Recurrent disease may be due to regrowth of fungi that, due to their intracellular location, have been incompletely eradicated from the vagina. With topical therapy, cultures at 30 days posttherapy reveal that the same Candida species is present in 20% to 25% of cases,[58] and well over two-thirds the organisms cultured are of the same strain as the original infection.[49] Quite probably the number of microorganisms causing disease is reduced sufficiently to alleviate symptoms, yet the organisms persist either in numbers too small to be cultured[58] or intracellularly in the cervix or vaginal mucosa.[19]

Antifungal Resistance

Drug resistance as a cause of chronic vulvovaginal candidiasis is rare, yet *C glabrata*, *C tropicalis*, and *Saccharomyces* species do demonstrate higher in vitro resistance to polyenes and azoles, which may be clinically relevant.[46] Susceptibility testing may be useful with these uncommon organisms. As yet, there is no evidence that long-term suppressive therapy for chronic vulvovaginal candidiasis selects out resistant strains.[49]

Cell-Mediated Immune Defects

Patients infected with HIV or treated with drugs that depress cell-mediated immunity are more prone to chronic vulvovaginal candidiasis than patients who are not. Although clinically difficult, exogenous immunosuppression should be reduced in refractory cases.

Of interest is that 40% to 70% of women with chronic vulvovaginal candidiasis appear to be anergic to Candida antigen by cutaneous testing.[46] A complex sequence of events that has been extensively evaluated by Witkin et al[59-61] may serve to explain this. The presence of an abnormal serum factor or local Candida-specific IgE antibodies in the vagina of women with chronic vulvovaginal candidiasis alters macrophage function and causes an increased production of prostaglandin E_2. This has a suppressive effect on interleukin-2 elaboration, which leads to a decrease in CD4 lymphocyte proliferation. These lymphocytes normally produce interferon-gamma, which has a suppressive effect on germ tube and mycelial formation of *C albicans*. Thus, the decreased number of CD4 cells may be permissive for the transformation of yeasts to the more invasive pseudohyphal and hyphal forms.

Anecdotal Etiologies

Other authors have suggested that women with chronic vulvovaginal candidiasis have zinc deficiencies;[62] allergic reactions to Candida antigen;[63,64] and abnormally high levels of glucose, arabinose, and ribose in the urine due to excessive ingestion of sweets.[41]

Therapy of Recurrent Vulvovaginal Candidiasis

Although uncommonly present, the identification and elimination of precipitating causes is the first step in the management of patients. Uncontrolled diabetes makes patients refractory to usual therapies, yet the yield of routine glucose tolerance tests, especially in asymptomatic premenopausal women, is extremely low and not routinely recommended.[9]

Exogenous cell-mediated immunosuppressives should be decreased or eliminated if possible. Women at risk for HIV infection from sexual contact, intravenous drug use, or transfusion from 1978 through 1985 should be screened for HIV antibody. Patients receiving high-estrogen oral contraceptives may benefit from alternate forms of contraception, yet there is little evidence to suggest that low-dose estrogen contraceptives are a predisposing factor to recurrent infection.[33] Occlusive clothing and synthetic underwear should be avoided. It is not uncommon for women with chronic vulvovaginal candidiasis to douche and bathe frequently in an attempt to remove the discharge physically or to alleviate pruritus. These practices should be discontinued because they have a deleterious effect on the protective indigenous vaginal flora.

Other maneuvers include patient counseling to decrease dietary increase of sweets, dairy products, and artificial sweeteners;[41] assays for blood zinc levels with supplementation if low;[62] and attempts at hyposensitization with laboratory strains of *C albicans*.[65] Without further studies, these approaches cannot be routinely recommended.

Antimicrobial therapy is the mainstay for patients without identifying precipitating events or those in whom disease persists in spite of their elimination. Infrequent recurrences (three or fewer per year) may be treated as single episodes (therapies outlined in Table 3.1). More frequent recurrences warrant long-term maintenance prophylaxis. For aesthetic reasons, long-term topical therapy has poor patient compliance. Recent studies have confirmed benefits from systemic therapy with the new azoles, ketoconazole and fluconazole. Awaiting confirmation is a single report of the efficacy of depot medroxyprogesterone.[66]

Ketoconazole is best given as 400 mg a day for an initial 14 days to resolve the acute infection, followed by low-dose suppression with 100 mg a day for 6 months.[67] Failure to give low-dose suppression results in a greater than 70% relapse rate, compared with a 5% relapse rate for those on daily suppression.[67] Fluconazole has shown promising results as therapy for acute vulvovaginal candidiasis.[68,69] When given as an initial dose of 150 mg followed by 150 mg every month, the number of recurrences of chronic vulvovaginal candidiasis decreases by 50% and the time to recurrence doubles.[46] More frequent doses, perhaps at weekly intervals, may reduce the relapse rate further.

It is important to remember the exceedingly rare hepatotoxicity associated with ketoconazole, seen in approximately 1 in 15,000 patients receiving the drug.[70] Symptoms and laboratory or clinical signs of hepatitis should prompt discontinuation of therapy.

Finally, and too often forgotten, patients with chronic vulvovaginal candidiasis require continuous support and counseling. The morbidity from this disease takes its toll on patients' self-esteem, sex life, and marital relations. Reassurance along with suppressive antimicrobial therapy should alleviate the medical and psychosocial problems in the vast majority of patients.

Trichomonas Vaginitis

Since the introduction of metronidazole in 1963, trichomonas vaginitis is a less common source of chronic vaginitis. However, it still comprises 20% to 30% of cases of vaginitis and affects approximately 3 million women per year.

The causative organism, Trichomonas vaginalis, is an anaerobic motile protozoan that grows optimally at a pH of 5.0 to 7.0. The organism is ovoid in shape, with a large nucleus and four flagella at its anterior pole. A fifth flagellum attaches posteriorly to a membrane that extends for two-thirds the length of the cell (Fig. 3.2). *T vaginalis* is 15 to 20 microns in length, slightly larger than a white blood cell, and three to four times the size of sperm.

Unlike vulvovaginal candidiasis, trichomonas vaginitis is primarily a sexually transmitted disease. From 30% to 40% of male sexual partners of infected women are carriers, and 85% of female partners of men with *T vaginalis* are also infected. Trichomonas is site specific in the human being, living exclusively in the urethra and lower genital tract. It rarely infects the endocervix.

Sexual intercourse, although the usual mode of transmission, is not always necessary

3. Chronic Vaginitis

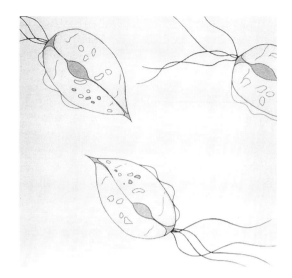

FIGURE 3.2. *Trichomonas vaginalis.* (Reproduced with permission from *Atlas of Sexually Transmitted Diseases.* New York: Gower Medical Publishing; 1989.)

eases, there can be an overlap of symptoms; it is often difficult to isolate those that are solely attributed to *T vaginalis*. As many as 25% of women who are infected are asymptomatic. Men are almost always asymptomatic carriers, and the infection is usually self-limited.

On physical examination, the vulva may be erythematous but usually appears normal. With introduction of the speculum, the typical yellow-green, frothy discharge is seen in 50% or fewer of patients.

The classical physical finding often described is the "strawberry cervix and vagina," which is secondary to inflammation of the cervix and vaginal walls (Fig. 3.3). There is subsequent swelling of papilla with microhemorrhage that then project through the discharge. This finding is present in only 2% of cases and, while specific for *T vaginalis*, is clearly not sensitive. If colposcopy is done, these punctate hemorrhages are seen in 45% of cases.[74]

since the trichomonad has been shown to survive in a moist environment outside the vagina for 1 to 3 hours. Infection from toilet seat splash has been documented.[71] Since trichomonas vaginitis is primarily a sexually transmitted disease, sexual activity and the form of contraception may alter the rate of *T vaginalis* infection. It has been shown that oral contraceptive users have a lower rate of trichomonas vaginitis than intrauterine device users or women who have had tubal ligations. In addition, users of the spermicidal agent Nonoxynol-9 have a 0.83 relative rate of infection compared to women who do not use spermicides.[72]

Clinical findings of trichomonas vaginitis have been well described. Although 50% to 70% of infected women complain of some abnormal discharge, only 30% describe the classic yellow-green, frothy, malodorous discharge.[73] Twenty-five percent of patients complain of dysuria, which is secondary to infection of the periurtheral glands. Other symptoms that are reported in varying degrees include vaginal discomfort, burning, dyspareunia, lower abdominal pain, and pruritus. Because trichomonas vaginitis often coexists with other vaginitides and sexually transmitted dis-

Diagnosis

The diagnosis of trichomonas vaginitis is usually suspected by clinical findings but should always be confirmed by microscopic examination.

It is helpful to first test the pH of the dis-

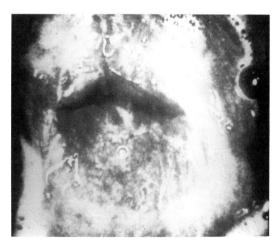

FIGURE 3.3. The "strawberry cervix" seen in trichomonas vaginitis. (Reproduced with permission from *Atlas of Sexually Transmitted Diseases.* New York: Gower Medical Publishing; 1989.)

charge using nitrazine paper. Although a pH less than 5.0 essentially rules out infection with T vaginalis, a pH greater than 5.0 does not differentiate between trichomonas vaginitis and bacterial vaginosis. It is important that the patient not douche, have intercourse, or use vaginal medication prior to the exam.

A saline wet mount of the discharge for the presence of motile trichomonads is only 60% sensitive when compared with culture.[75] Gently warming the slide increases motility and enhances the yield. Specificity is 100% and organisms are easier to see in symptomatic patients.[76]

White blood cells are almost always seen in the range of greater than 10 per high-power field. If the trichomonad is not motile, it can be differentiated from a PMN because of its ovoid shape and slightly larger size.

Cultures may be done but are indicated only if the diagnosis is suspected but the wet prep is persistently negative. Diamonds medium should be used and is available commercially, but it must be used fresh. The sensitivity of cultures is between 86% and 97%.[75] As strains of T vaginalis resistant to metronidazole develop, cultures may be more frequently indicated to test organisms' susceptibility.

T vaginalis may be seen on a Papanicolaou smear in approximately 50% of infected women, but false positives frequently exist (31%); thus, cytology should not be relied on for diagnosis.[75,77]

Identification of trichomonads by EIA, latex agglutination, or immunofluorescent monoclonal antibodies has a sensitivity of over 90%. In the future, when there is the ability to rapidly and inexpensively utilize these techniques in the office, these may become the standard of diagnosis.[75,78] Until the sensitivity, specificity, and cost of these new techniques are defined, the saline vaginal wet prep remains the first-line diagnostic tool.[79]

Routine Treatment

Fortunately, most strains of T vaginalis are susceptible to metronidazole and the other 5-nitroimidazoles. In the 1960s when metronidazole was first introduced, the standard treatment was 250 mg orally, three times daily for 1 week. The cure rate with this regimen is over 90%, and when male partners are treated simultaneously the effectiveness is even greater.

In 1980, 2-gram single-dose metronidazole therapy was introduced. Compliance is much higher, success rates are equivalent, and side effects are comparable to the 7-day regimen. In addition, the cost of treatment is significantly less.[80,81]

When used alone, metronidazole suppositories yield a cure rate of only 30% to 60% and cannot be recommended.[79]

It is essential to treat all male partners simultaneously to minimize treatment failures. To prevent transmission to sexual partners and to minimize future symptomatic episodes, treatment is recommended even in asymptomatic women.

Treatment Failures

Because absolute resistance to metronidazole has not been documented, treatment failures are usually secondary to patient noncompliance or failure to treat partners adequately. Completion of therapy is much greater with the single-dose regimen, and, because cure rates are equivalent, the 1-day regimen should be used initially unless also treating a concomitant bacterial vaginosis.

Noncompliance, and therefore treatment failure, increases with 7-day regimens because of metronidazole's unpleasant side effects, particularly gastrointestinal upset and a metallic taste. Persistent infection, however, is usually because of failure to treat the male partner simultaneously or because of failure to treat all partners.

Women who are still symptomatic after the first treatment will often be cured simply by repeating the standard treatment[82] or by extending 1-day therapy to 7-day therapy.[78]

An increasing cause of failure is due to relatively metronidazole-resistant strains of T vaginalis that require a higher concentration of metronidazole for cure. Absolute resistance has not been documented.[83] Highly resistant cases occur in approximately 1 of 2,000 patients, and repeat courses of standard therapy will not be successful.[79] These cases can usually be cured by increasing the metronidazole dose to 2.5 to 4 grams per day in divided doses, given for 7 to 13 days.[82]

Intravenous therapy of metronidazole-resistant *T vaginalis* has been shown to be successful in patients in which repeated oral therapy (standard and high doses of 4 g per day) has failed.[84] It is possible that treatment failure with oral regimens is not only due to metronidazole resistance but to poor oral absorption and inactivation of metronidazole by the vaginal flora. Therefore Betadine douches can also be done at the time of treatment to reduce vaginal flora.

Another regimen that has been shown to be effective in cases of persistent trichomonas vaginitis is the use of oral (500 mg TID) and topical (1 g vaginal suppository daily) metronidazole simultaneously for 14 days with daily rinsing of the vagina with 3% acetic acid.[85]

Adverse Reactions

Metronidazole is generally well tolerated; however, adverse reactions can occur. A metallic taste is commonly experienced, and up to 10% of patients complain of nausea. In addition, vertigo, headaches, diarrhea, peripheral neuropathy, seizures, and glossitis have been described. Metronidazole is also known to have a disulfiram-like effect, and patients should abstain from alcohol consumption during treatment.

There is a 7% incidence of leukopenia in patients after a 7-day course, but this has not been described from single-day regimens. Therefore a white blood count should be obtained prior to retreatment or high-dose therapy.

Allergies to metronidazole are rare, and effective alternative therapies are not known. Some effect is seen with topical clotrimazole, and in cases of allergy or during the first trimester of pregnancy when metronidazole is contraindicated this can be used. Cure rates are 48% to 66% when clotrimazole (100 mg) is used vaginally for 6 days.[86] The other five nitroimidazoles are currently not available in the United States.

Bacterial Vaginosis

Bacterial vaginosis is an entity primarily known for its confusing nomenclature rather than as the most prevalent infectious cause of vaginitis. The term *bacterial vaginosis* was proposed in 1984 for a syndrome that had previously been known by multiple identities. Prior to 1955, all vaginitis not caused by *T vaginalis*, yeast, or gonorrhea was known as "nonspecific vaginitis." In 1955 a suspected causative organism was identified, and the syndrome became known as haemophilus vaginitis.[87] A change in taxonomic position resulted in *Haemophilus vaginale* becoming *Corynebacterium vaginale* in 1963.[88] Further classification in 1980 named this organism *Gardnerella vaginale*.[89]

Soon after 1980, it became evident that rather than a single responsible etiologic agent, a polymicrobial, predominately anaerobic shift in vaginal bacteria was responsible for the characteristic malodorous discharge, and the term *anaerobic vaginitis* came into use. In 1984, by international convention the term *bacterial vaginosis* became the accepted nomenclature to describe this polymicrobial disturbance of the vaginal ecosystem.[90]

Pathophysiology

Bacterial vaginosis is a condition in which there is a shift from the lactobacillus-dominated aerobic milieu to a mixed flora comprised of anaerobic or faculatative aerobic bacteria and genital mycoplasmas. The result is at least a 10-fold increase in the number of anaerobes with marked reduction or absence of lactobacilli that results in a characteristic gray-white homogenous malodorous discharge.

A single organism that inhibits lactobacilli

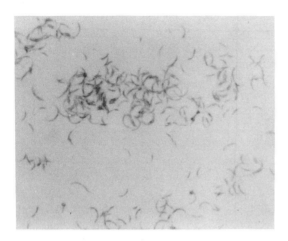

FIGURE 3.4. Gram stain of *Mobiluncus mulieris*. (Reproduced with permission from *Atlas of Sexually Transmitted Diseases*. New York: Gower Medical Publishing; 1989.)

and initiates anaerobic overgrowth has not been identified; however, several specific bacteria have been associated with bacterial vaginosis. *Gardnerella vaginalis* is a Gram-negative or Gram-variable rod seen in at least 92% of women with bacterial vaginosis.[91] However, it is commonly found in women who do not have bacterial vaginosis and may not be solely responsible for this condition.

Mobiluncus species (*M curtissi* and *M mulieris*), described in 1984,[92,93] are demonstrated in up to 96% of women with bacterial vaginosis and are curved anaerobic rods (Fig. 3.4). They are the only motile organisms seen in bacterial vaginosis and have six to eight flagella that are responsible for their corkscrew-like action on wet mount. Unlike *G vaginalis* all women with *Mobiluncus* species infection have clinical bacterial vaginosis, but not all women with bacterial vaginosis have demonstrable *Mobiluncus* species.[92]

Other organisms associated with bacterial vaginosis are *Mycoplasma hominis*, seen in 24% to 75% of women with bacterial vaginosis,[91] *Bacteriodes* species, and *Peptostreptococcus* species.

Unlike trichomonas vaginitis, there is a notable absence of white cells on wet prep and virtually no inflammation of the vaginal epithelium. It is the lack of inflammatory response that resulted in the term *bacterial vaginosis* rather than *bacterial vaginitis*, although *vaginal bacteriosis* is the grammatically correct expression.

The pathophysiology of bacterial vaginosis is not well understood because multiple organisms may be involved and no host factor has been identified that clearly increases susceptibility to bacterial vaginosis. In addition, it is not certain if a decrease in lactobacilli allows the overgrowth of anaerobic bacteria or if an increase of anaerobic bacteria causes a depletion of lactobacilli.

Epidemiology

Currently bacterial vaginosis is the most common infectious cause of vaginitis and comprises approximately 40% of all new cases of vaginitis.[91,94-96] There is a wide range of incidence, from 4% in asymptomatic college populations to 64% in sexually transmitted disease clinics; more typical patient populations show incidences in the 15% to 25% range.[94,97] Mixed infections are common and other vaginitides, cervical and pelvic infections, often coexist with bacterial vaginosis.

Multiple studies have attempted to elucidate factors that increase the risk of acquiring this illness. The only risk factors that have been well identified are the use of intrauterine devices, multiple sexual partners, prior history of trichomonas vaginitis, and nonbarrier contraceptive methods.[97]

The sexual transmission of bacterial vaginosis remains controversial. Clearly, bacterial vaginosis is associated with sexual activity, but bacterial vaginosis has been documented in the absence of sexual activity. Those who support bacterial vaginosis as a sexually transmitted disease cite that bacterial vaginosis increases with the number of sexual partners, is infrequent in monogamous women, and in some studies is nonexistent in virgins. In addition, the organisms associated with bacterial vaginosis have been documented in the urethras of male sexual partners of women with bacterial vaginosis and not in male controls. In contrast, some studies do report bacterial vaginosis in virgins. It has also been shown that bacterial vaginosis organisms are only present for a limited time after intercourse with an infected woman. Significantly, in many studies cure rates in women with bacterial vaginosis are not altered when the male partner is treated.[98,99]

Some investigators have proposed that semen deposition or other noninfectious factors associated with intercourse may alter the vaginal milieu in such a way that lactobacilli are depleted, resulting in an overgrowth of anaerobic bacteria.

Clinical Manifestations and Diagnosis

More than 50% of women with bacterial vaginosis are asymptomatic.[94,97] Typically the disease is manifested by a profuse, malodorous, thin, homogenous, gray-white discharge that is adherent to vaginal walls. Uncomplicated bacterial vaginosis is not associated with abdominal pain, pruritus, or dysuria. Vulvitis is not part of the syndrome because the organisms associated with bacterial vaginosis do not invade tissues or cause an inflammatory response. If inflammation is present, a mixed infection must be ruled out. The following four signs are the usual diagnostic criteria:

1. Clue cells: Clue cells are squamous vaginal epithelial cells whose borders are obscured by attached bacteria (Fig. 3.5). On wet mount, they should comprise at least 20% of epithelial cells. Clue cells are seen in 80% of patients and are specific for bacterial vaginosis. The majority of bacteria on clue cells are *G vaginalis*, but other bacteria can adhere as well.[100]
2. Discharge: The characteristic thin, homogenous discharge is seen in 69% of affected patients. The discharge adheres to but is easily wiped from vaginal walls, is gray-white, and is often profuse and malodorous (Fig. 3.6).
3. Increased vaginal pH: In essentially all patients with bacterial vaginosis, the vaginal pH is greater than 4.5. This is the most

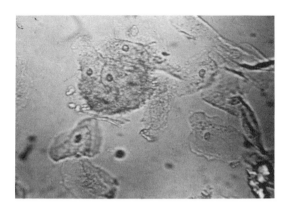

FIGURE 3.5. Large cell slightly to left of center is a typical clue cell. Stippled border is due to adherent bacteria, the majority of which are *Gardnerella vaginalis*. (Reproduced with permission from *Atlas of Sexually Transmitted Diseases*. New York: Gower Medical Publishing; 1989.)

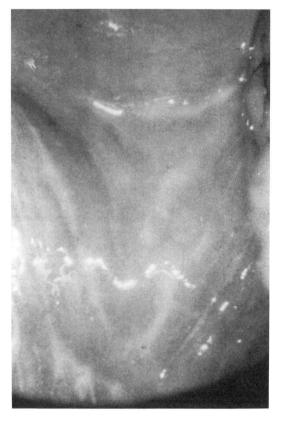

FIGURE 3.6. Typical discharge seen in bacterial vaginosis. Discharge is homogenous and vaginal wall is not inflamed. (Reproduced with permission from *Atlas of Sexually Transmitted Diseases*. New York: Gower Medical Publishing; 1989.)

sensitive finding but is poorly specific, because other conditions result in an elevation of vaginal pH.
4. Positive KOH test ("whiff test"): In at least 43% of women with bacterial vaginosis, a "fishy" odor is produced when vaginal fluid is mixed with 10% KOH. Amines produced by anaerobic bacteria volatize to produce malodorous putrescine, cadaverine, and trinethylamine. This odor is often produced by intercourse due to the alkalinity of semen causing volatization of the same amines.

The usually accepted criterion is the presence of any three of the above four signs.[94,97] Some authors feel that if clue cells are present and if the KOH test is positive, a third sign is not necessary for accurate diagnosis.[94]

Laboratory Diagnosis

Diagnosis is usually made by microscopic examination of a wet prep of a vaginal discharge. Efforts have been made, however, to standardize diagnosis using reproducible laboratory techniques.

Cultures are generally not useful. *G vaginalis* may be cultured successfully from almost all women with bacterial vaginosis, but it is also found in more than 50% of women without bacterial vaginosis. *Mobiluncus* species are a better marker for bacterial vaginosis but are more difficult to culture.

Gram stains are more useful and correlate well with clinical diagnosis.[91,96] Clue cells are identifiable on Gram stain. In addition, there is a decrease or absence of lactobacilli with predominance of Gram-variable coccobacilli representing with *G vaginalis* or *Bacteroides*. One advantage of the Gram stain is that a swab may be sent to a lab, which allows the less experienced clinician to use the microbiology lab's expertise. Also, criteria for the diagnosis of bacterial vaginosis by using the Gram stain can be reproduced reliably between different centers and microbiologists. A scoring system using reliable morphotypes from vaginal smears was recently proposed for this purpose.[101]

Many other diagnostic methods are being investigated but are useful for research purposes only. Gas chromotographic analysis of vaginal fluid is 90% sensitive and 97% specific for bacterial vaginosis. Thin-layer chromatography, enzymatic analysis, and DNA probes for *Mobiluncus* species are also being considered for clinical use.

Treatment

Metronidazole has been the most successfully utilized antibiotic for treatment of bacterial vaginosis. In addition to inhibiting the growth of anaerobic bacteria, metronidazole, since it is inactive against lactobacilli allows recolonization of that organism. The usual dose is 500 mg twice daily for 7 days, which yields an 82% cure rate 21 days after treatment. Shorter therapies have been studied.[99,102] A single 2-gram dose of metronidazole has an equivalent short-term efficacy rate, yet 3 weeks after treatment a significant number of patients have a recurrence. Other short-term therapies have been attempted, but the relapse rate has been unacceptable.

Recently a meta-analysis was performed on metronidazole treatment protocols.[103] There was no significant difference in cure rates, and the authors recommend a 2-gram single-dose therapy over the accepted 7-day regimen.

To avoid unpleasant gastrointestinal side effects, other routes of administering of metronidazole have been investigated. Vaginal metronidazole (500 mg BID for 7 days) has a 80% efficacy rate. Metronidazole-impregnated vaginal sponges and vaginal gels have been developed and are now available for clinical use.

Clindamycin (300 mg BID for 7 days) is an acceptable second-line treatment and is useful if the patient is unable to tolerate metronidazole or if therapy is necessary during the first trimester of pregnancy. Two percent clindamycin vaginal cream is now available for clinical use and has shown up to a 94% cure rate with daily use for 7 to 10 days.[91] Bacterial vaginosis should be treated during pregnancy because of the increased risk of obstetric complications. However, clindamycin should be

TABLE 3.2. Infectious sequellae of bacterial vaginosis.

Gynecologic
  Pelvic inflammatory disease
  Posthysterectomy vaginal cuff infections
  Urinary tract infections
  Infertility (secondary to mycoplasma)
Obstetric
  Preterm labor and delivery
  Premature reputure of membranes
  Chorioammonitis
  Postpartum endometritis
  Postcesarean endometritis
  Amnionitis
  Neonatal scalp abscess
  Increased pelvic inflammatory disease after first-trimester elective abortion

used during the first trimester or treatment delayed until the second trimester.

Some investigators are studying the use of vaginal lactate gels for treatment of bacterial vaginosis. Providing such a substrate for lactobacillus to grow may be particularly useful for maintenance therapy in the patient with multiple relapses.[91]

Treatment of the asymptomatic patient is somewhat controversial. At this time, the Centers for Disease Control does not recommend the treatment of asymptomatic women. However, serious gynecologic and obstetric sequellae have been documented in patients with bacterial vaginosis (Table 3.2).[91,94,96] Therefore many feel that all asymptomatic women, or at least those at particular risk such as obstetric patients and those women who are to undergo gynecologic surgery, should be treated.

Treatment Failures and Recurrence

The recurrence rate for appropriately treated bacterial vaginosis is as high as 80%. The reason for this recurrence is not well understood, but theories include (1) persistence of an unidentifiable host factor, (2) failure to reestablish a predominate lactobacillus flora, (3) reinfection by male partners, and (4) persistence of organisms that are inhibited but not killed by antibiotic therapy.

Recommendations for intractable disease include the treatment of male partners, a longer duration of therapy, and the use of lactate gel to reestablish normal flora. It is essential to rule out mixed infections; yeast is particularly a problem after prolonged antibiotic therapy.

Some recommend culturing the discharge and directing the choice of antibiotic against the dominant organism.[95] Except for resistant strains of *Mobiluncus*, resistance to metronidazole has not been documented.[104] In these rare resistant or refractory cases, treatment with clindamycin is recommended. Studies of clindamycin gel remain promising and may be the therapy of choice in the near future.

Miscellaneous and Less Common Causes of Vaginitis

Chronic vaginitis unresponsive to usual therapy is frustrating for both the patient and the physician, and it is imperative that both parties acknowledge that all vaginal discharge, vulvar itching, or burning are not necessarily symptoms of vaginitis. Self-diagnosis often results in inappropriate treatment such as frequent douching or use of over-the-counter medications, which then delays appropriate identification of the problem and in itself can result in chronic iatrogenic conditions.[105] Physicians are also guilty of prolonging or exacerbating symptoms by basing treatment on telephone consultations. It is important to remember that at least 10% of "vaginitis patients" have a normal or physiologic increase in discharge and need nothing more than reassurance. This is especially common in the patient with a new sexual partner who fears that she has been exposed to a sexually transmitted disease. Physiologic increase in secretions, discussed earlier in this chapter, is seen in many situations.

Hypoestrogenic states, such as postpartum vaginal atrophy or menopausal atropic vaginitis, can also result in abnormal discharge and vaginal irritation. These conditions respond well to topical or systemic estrogen therapy.

TABLE 3.3. Vulvar conditions unrelated to vaginitis.

Human papilloma virus (*Condylomata accuminata*)
Exophytic warts
Flat warts
Vestibulitis
Vestibular papillomatosis
Vulvodynia
Dermatoses
Psoriasis
Tinea
Lichen simplex
Vulvar dystrophies
Hyperplasia
Lichen sclerosis
Vulvar intraepithelial neoplasia

TABLE 3.4. Common chemical irritants.

Deodorant soaps	Iodine
Douches	Gentian violet
"Hygenic" sprays	Spermacides
Laundry detergents	Bubble bath
Petroleum jelly	Talcum powder
Perfumes	Perfumed, dyed toilet paper
Therapeutic creams and preparations	Deodorant sanitary pads or tampons
Boric acid	Swimming pool chemicals

Many vulvovaginal problems are truly pathologic and require treatment but do not fall under the mantle of vaginitis.[106] Conditions that cause vulvar itching or burning but are not vaginitis are outlined in Table 3.3.

After microscopic and physical examination of the patient with chronic vaginitis has ruled out vulvovaginal candidiasis, trichomonas vaginitis, bacterial vaginosis, physiologic discharge, cervicitis, and the other vulvar and vaginal conditions described, less common causes of vaginal irritation and discharge should be considered.

Vaginitis not caused by specific organisms is often due to allergic or inflammatory reactions. Inflammatory discharges causing vulvitis and/or vaginitis result either from a foreign object in the vagina or a chemical irritant applied to the vulvovaginal area. Failure to remove a tampon, diaphragm, cervical cap, pessary, or contraceptive sponge may cause a foul-smelling, copious, sometimes bloody discharge. Similar reactions are seen with retained condoms or other foreign objects (particularly toilet paper or, in the pediatric patient, small toys. Treatment is removal of the foreign object. In addition, Betadine douches and sitz baths are useful to alleviate the odor, irritation, and discharge.

Common irritants resulting in vulvovaginitis are shown in Table 3.4. Treatment is by identifying and removing the offending irritant. Sitz baths, Burrow's solution, and application of emollients will facilitate recovery.

True allergic reactions are rare. When most patients refer to allergic reactions to medications or soaps, they are really experiencing an inflammatory reaction of an irritant rather than a true cell-mediated immunity to an antigenic stimulus. However, allergic responses have been observed to semen, spermicides, local anesthesia, topical antibiotics, vaginal douches, fragrances, latex, and various antifungal creams in which IgE is elevated in vaginal fluids.[107]

Patients with irritant dermatitis generally complain of "burning," whereas patients with allergic dermatitis complain of itching with an urticarial reaction involving skin beyond the area of direct contact. Patch testing can be done to differentiate between the two entities.

Treatment consists of removal of the allergen and oral administration of H1 blockers. Application of 1% hydrocortisone cream and/or petroleum jelly may alleviate discomfort.

Allergic recurrent vaginitis has also been reported as a result of sexual transmission of IgE antibodies in women who complain of vaginitis after intercourse.[108] In these cases the seminal fluid demonstrates IgE antibodies reactive to vaginal secretions obtained from their partners. An allergic vaginitis has been induced by the transfer of IgE antibodies from men to women during coitus resulting in repeated sexually related episodes of vaginal pruritus and burning. Use of condoms has been curative.

A frequently overlooked source of vaginitis is cytolytic vaginosis, formerly known as Doderlein's cytolysis.[109] This condition is caused by an overgrowth of lactobacilli. It is frequently mistaken for vulvovaginal candidiasis because patients present with a white, cheesy discharge, pruritus, and a pH between 3.5 to 4.5. Many women who are treated for chronic vulvovaginal candidiasis are actually yeast-free but have an abundance of lactobacilli. Like vulvovaginal candidiasis, symptoms often increase in the luteal phase of the menstrual cycle. Diagnosis of this entity depends on a high index of suspicion, an absence of trichomonads, and the absence of Candida on wet smear with an abundance of lactobacilli seen. Treatment consists of increasing the pH of the vagina by douching with a solution of 30 to 60 grams of sodium bicarbonate to 1 liter of warm water two to three times per week and then once or twice a week as needed. An alternative to douching is the use of moistened tampons impregnated with baking soda, which should remain in the vagina for up to 6 hours.

References

1. Sparks JM. Vaginitis. *J Repro Med*. 1991; 36:745–752.
2. Josey WE. Vaginitis. Reducing the number of refractory cases. *Post Grad Med*. 1977;62: 171–174.
3. Fisher AM. Clinical aspects, vaginal discharge, vaginitis and pruritus vulvae. *Clin Obstet Gynecol*. 1981;8:241–246.
4. Levison ME, Corman LC, Carrington ER, et al. Quantitative microflora of the vagina. *Am J Obstet Gynecol*. 1977;127:80–85.
5. Shafer MA, Sweet RL, Ohm-Smith MJ, et al. Microbiology of the lower genital tract in postmenarchal adolescent girls: differences in sexual activity, contraception, and presence of non-specific vaginitis. *J Pediatr*. 1985;107:974–981.
6. Weinstein L. Vaginitis: an overview of a common condition. In: Sciarra JJ, ed. *Gynecology and Obstetrics*. Philadelphia, Pa: Harper & Row; 1987: Chapter 40, pp. 1–7.
7. Amsel P, Totten PA, Spiegal CA, et al. Nonspecific vaginitis: diagnostic criteria and microbial and epidemiologic associations. *Amer J Med*. 1983;74:14–22.
8. Centers for Disease Control. Non-reported sexually transmitted diseases. *MMWR*. 1979; 28:61–63.
9. Sobel JD. Epidemiology and pathogenesis of recurrent vulvovaginal candidiasis. *Am J Obstet Gynecol*. 1985;152:924–935.
10. Hurley R, De Louvois J. *Candida* vaginitis. *Postgrad Med J*. 1979;55:645–647.
11. Sobel JD. Vaginal infections in adult women. *Med Clin North Amer*. 1990;6:1573–1602.
12. Horowitz BJ, Edelstein SW, Lippman L. *Candida tropicalis* vulvovaginitis. *Obstet Gynecol*. 1985;66:229–232.
13. Rendo-Lopez V, Lynch M, Schmitt C, et al. *Torulopsis glabrata* vaginitis: clinical aspects and susceptibility to antifungal agents. *Obstet Gynecol*. 1990;76:651–655.
14. Drake TE, Maibach HI. *Candida* and candidiasis: cultural conditions, epidemiology, and pathogenesis. *Postgrad Med*. 1973;53:83–87.
15. Odds FC. Genital candidiasis. *Clin Exp Dermatol*. 1982;7:345–354.
16. Odds FC, Webster CE, Riley VC, et al. Epidemiology of vaginal *Candida* infection: significance of numbers of vaginal yeasts and their biotypes. *Eur J Obstet Gynecol Reprod Biol*. 1987;25:53–66.
17. King RD, Lee JC, Morris AL. Adherence of *Candida albicans* and other *Candida* species to mucosal epithelial cells. *Infect Immun*. 1980;27: 667–674.
18. Sobel JD. Pathophysiology of vulvovaginal candidiasis. *J Repro Med*. 1989;34(suppl): 572–580.
19. Garcia-Tamayo J, Castillo G, Martinez AJ. Human genital candidiasis: histochemistry, scanning and transmission electron microscopy. *Acta Cytologica*. 1982;26:7–14.
20. Monif GRG. Classification and pathogenesis of vulvovaginal candidiasis. *Obstet Gynecol*. 1985;152:935–939.
21. Trumbore DJ, Sobel JD. Recurrent vulvovaginal candidiasis: vaginal epithelial cell susceptibility of *Candida albicans* adherence. *Obstet Gynecol*. 1986;67:810–814.
22. Soll DR. High-frequency switching in *Candida albicans* and its relations to vaginal candidiasis. *Am J Obstet Gynecol*. 1988;158:997–1001.
23. Slutsky B, Buffo J, Soll DR. High-frequency switching colony morphology in *Candida albicans*. *Science*. 1985;230:666–669.
24. Caruso LJ. Vaginal moniliasis after tetracycline therapy. *Am J Obstet Gynecol*. 1964; 90:374.
25. Beveridge MM. Vaginal moniliasis after treatment of trichomonal infection with Flagyl. *Br J Vener Dis*. 1962;38:220.

26. Heary FJ. Recurrent *Candida* vulvovaginitis. Chemotherapy. 1982;28(suppl 1):48–50.
27. Milsom I, Forssman L. Repeated candidiasis: reinfection or recrudescence? A review. *Am J Obstet Gynecol*. 1985;152:956–959.
28. Powell BL, Frey CL, Drutz, DJ. Estrogen receptor in *Candida albicans*: a possible explanation for hormonal influences in vaginal candidiasis (Abstract 751). In: *Proceedings of the 23rd Interscience Conference on Antimicrobial Agents in Chemotherapy*, Las Vegas, Nev; 1983:222.
29. Morton RS, Rashid S. Candidal vaginitis: natural history, predisposing factors and prevention. *Proc R Soc Med*. 1977;70(suppl 4):3–12.
30. Hopsue-Havu VK, Gronroos M, Punnoren R. Vaginal yeasts in parturients and infestation of the newborns. *Acta Obstet Gynecol Scand*. 1980;59:73–77.
31. Bland PB. Experimental vaginal and curaneous moniliasis: clinical and laboratory studies of certain monilias associated with vaginal oral and cutaneous thrush. *Arch Dermatol Syhio* 1937;36:760–768.
32. Weid CL. Statistical evaluation of the effect of hormonal contraceptives on the cytologic smear pattern. *Obstet Gynecol*. 1966;27:327–331.
33. Davidson F, Oates, JK. The pill does not cause "thrush." *Br J Obstet Gynecol*. 1985;92:1265–1266.
34. Elegbe IA, Botu M. A preliminary study on dressing patterns and incidence of candidiasis. *Am J Public Health*. 1982;72:176–177.
35. Elegbe IA, Elegbe I. Quantitative relationships of *Candida albicans* infections and dressing patterns in Nigerian women. *Am J Public Health*. 1983;73:450–452.
36. Heidrich FE, Berg AO, Bergman JJ. Clothing factors and vaginitis. *J Fam Pract*. 1984;19:491–494.
37. Rhoads JL, Wright DC, Redfield RR, et al. Chronic vaginal candidiasis in women with human immunodeficiency virus infection. *J Amer Med Assoc*. 1987;257:3105–3107.
38. Imam N, Carpenter CCJ, Mayer KH, et al. Hierarchical pattern of mucosal *Candida* infections in HIV-seropositive women. *Amer J Med*. 1990;89:142–146.
39. Sridama V. Pacini F, Yang S-L, et al. Decreased levels of helper T cells: a possible cause of immunodeficiency in pregnancy. *N Engl J Med*. 1982;307:352–356.
40. McCue J. Evaluation and management of vaginitis: an update for primary care practitioners. *Arch Intern Med*. 1989;149:565–568.
41. Horowitz BJ, Edelstein SW, Lippman L. Sugar chromatography studies in recurrent *Candida* vulvovaginitis. *J Repro Med*. 1984;29:444–445.
42. Leegard M. The incidence of *Candida albicans* in the vagina of "healthy young women." *Acta Obstet Gynecol*. 1981;8:209–212.
43. Schaaf VM, Perez-Stable EJ, Borchardt K. The limited value of symptoms and signs in the diagnosis of vaginal infections. *Arch Intern Med*. 1990;150:1929–1933.
44. Clayton YM. Antifungal drugs in current use: a review. *Proc R Soc Med*. 1977;70(suppl 4):15–18.
45. Van Slyke KK, Miche VP, Rein MF. Treatment of vulvovaginal candidiasis with boric acid powder. *Am J Obstet Gynecol*. 1981;141:145–148.
46. Sobel JD. Pathogenesis and treatment of recurrent vulvovaginal candidiasis. *Clin Infect Dis*. 1992;14(suppl 1):S148–S153.
47. De Sousa HM, Van Uden N. The mode of infection in yeast vulvovaginitis. *Am J Obstet Gynecol*. 1960;80:1096–1099.
48. Miles MR, Olsen L, Rogers A. Recurrent vaginal candidiasis: importance of an intestinal reservoir. *J Am Med Assoc*. 1977;238:1836–1837.
49. O'Connor MI, Sobel JD. Epidemiology of recurrent vulvovaginal candidiasis. Identification and strain differentiation of *Candida albicans*. *J Infect Dis*. 1986;154:358–362.
50. Milne JD, Warnock DW. Effect of simultaneous oral and vaginal treatment on the rate of cure and relapse in vaginal candidosis. *Br J Vener Dis*. 1979;55:362–365.
51. Vellupillai S, Thin RN. Treatment of vulvovaginal yeast infection with nystatin. *Practitioner*. 1977;219:897–901.
52. Nystatin Multicenter Study Group. Therapy of candidal vaginitis: the effect of eliminating intestinal *Candida*. *Am J Obstet Gynecol*. 1986;155:651–655.
53. Davidson F, Mould RF. Recurrent genital candidosis in women and the effect of intermittent prophylactic treatment. *Br J Vener Dis*. 1978;54:176–183.
54. Rodin P, Kolator B. Carriage of yeasts on the penis. *BMJ*. 1976;1:123–124.
55. Davidson F. Yeasts and circumcision in the male. *Br J Vener Dis*. 1977;53:121–123.

56. Horowtiz BJ, Edelstein SW, Lippman L. Sexual transmission of *Candida*. *Obstet Gynecol*. 1987;69:883–886.
57. Ogunbanjo BO. Isolation of yeasts from male contacts of women with vaginal candidosis. *Genitourin Med*. 1988;64:135–136.
58. Odds FC. Candidosis of the genitalia. In: Odds FC, ed. *Candida and Candidosis*. 2nd ed. London: Bailliere Tindall; 1988:124–135.
59. Witkin SS, Hirsch J, Ledger WJ. A macrophage defect in women with recurrent Candida vaginitis and its reversal in vitro by prostaglandin inhibitors. *Am J Obstet Gynecol*. 1986;155:790–795.
60. Witkin SS, Jeremias J, Ledger WJ. A localized vaginal allergic response in women with recurrent vaginitis. *J Allergy Clin Immunol*. 1988;81:412–416.
61. Kalo-Klein A, Witkin SS. Prostaglandin E2 enhances and gamma interferon inhibits germ tube formation in *Candida albicans*. *Infect Immun*. 1990;58:260–262.
62. Edman J, Sobel JD, Taylor ML. Zinc status in women with recurrent vulvovaginal candidiasis. *Am J Obstet Gynecol*. 1986;155:1082–1085.
63. Kudelka NM. Allergy in chronic molilial vaginitis. *Ann Allergy*. 1971;29:266–267.
64. Palacios HJ. Hypersensitivity as a course of dermatologic and vaginal moniliasis resistant to topical therapy. *Am Allergy*. 1976;37:110–115.
65. Rosedale N, Browne K. Hyposensitization in the management of recurring vaginal candidiasis. *Ann Allergy*. 1979;43:250–253.
66. Dennerstein GJ. Depo-provera in the treatment of recurrent vulvovaginal candidiasis. *J Repro Med*. 1986;31:801–803.
67. Sobel JD. Recurrent vulvovaginal candidiasis: a prospective study of the efficacy of maintenance ketoconazole therapy. *N Engl J Med*. 1986;315:1455–1458.
68. Brammer KW. Treatment of vaginal candidosis with a single oral dose of fluconazole. *Eur J Clin Microbiol Infect Dis*. 1988;7:364–367.
69. International Multicenter Trial. A comparison of single-dose oral fluconazole with 3-day intravaginal clotrimazole in the treatment of vaginal candidiasis. *Br J Obstet Gynecol*. 1989;96:226–232.
70. Lewis JH, Zimmerman HJ, Benson GD, et al. Hepatic injury associated with ketoconazole therapy: analysis of 33 cases. *Gastroenterology*. 1984;86:503–513.
71. Burgess JA. Trichomonas vaginalis infection from splashing in water closets. *Br J Vener Dis*. 1963;39:248.
72. Barbone F, Austin H, Louv WC. A follow-up study of methods of contraception, sexual activity, and rates of trichomoniasis, candidiasis, and bacterial vaginosis. *Am J Obstet Gynecol*. 1990;163:510–514.
73. McLellan R, Spencer RM, Brockman M, et al. The clinical diagnosis of trichomoniasis. *Obstet Gynecol Gyn*. 1982;60:30–34.
74. Wolner-Hanssen P, Krieger JN, Stevens CE. Clinical manifestations of vaginal trichomoniasis. *JAMA*. 1989;261:571.
75. Krieger JN, Tam MR, Stevens CE, et al. Diagnosis of trichomoniasis: comparison of conventional wet-mount examination with cytologic studies, cultures and monoclonal antibody staining of direct specimens. *JAMA*. 1988;259:1223–1227.
76. Lossick JG. The diagnosis of vaginal trichomoniasis. *JAMA*. 1988;259:1230.
77. Thomason JL, Gelbart SM, Sobuin JF, et al. Comparison of four methods to detect *Trichomonas vaginalis*. *J Clin Microbiol*. 1988;26:1869.
78. Thomason JL, Gelbart SM. Trichomonas vaginalis. *Obstet Gynecol*. 1989;74:536–541.
79. Lossick JG, Kent HL. Trichomoniasis: trends in diagnosis and management. *Am J Obstet Gynecol*. 1991;165:1217–1222.
80. Hager WD, Brown ST, Kraus SJ, et al. Metronidazole for vaginal trichomoniasis: seven-day vs. single-dose regimens. *JAMA*. 1980;244:1219–1220.
81. Lossick JG. Single-dose metronidazole treatment for vaginal trichomoniasis. *Obstet Gynecol*. 1980;56:508–510.
82. Lossick JG, Muller M, Gorrell TE. In vitro drug susceptibility and doses of metronidazole required for cure in cases of refractory vaginal trichomoniasis. *J Infect Dis*. 1986;153:948–955.
83. Muller M, Meingassner JG, Miller WA, et al. Three metronidazole-resistant strains of *Trichomonas vaginalis* from the United States. *Am J Obstet Gynecol*. 1980;138:808–812.
84. Dombrowski MP, Sokol RJ, Brown WJ, et al. Intravenous therapy of metronidazole-resistant *Trichomonas vaginalis*. *Obstet Gynecol*. 1987;69:524–525.
85. Grossman JH, Galask RP. Persistent vaginitis caused by metronidazole-resistant trichomo-

nas. *Obstet Gynecol*. 1990;76:521–522.
86. Schnell JD. The incidence of vaginal *Candida* and *Trichomonas* infections and treatment of Trichomonas vaginitis with clotrimazole. *Postgrad Med J*. 1974;50:79–80.
87. Gardner HL, Dukes CD. *Haemophilus vaginalis* vaginitis. *Am J Obstet Gynecol*. 1955;69:962–976.
88. Zimmerman K, Turner GC. The taxonomic position of *"Haemophilus vaginalis."* *J Pathol Bacteriol*. 1963;85:213–219.
89. Greenwood JR, Pickett MJ. Transfer of *Haemophilus vaginalis* to a new genus, *Gardnerella*: *G vaginalis*. *Int J System Bacteriol*. 1980;30:170–178.
90. Westron L, Evaldson G, Holmes KK, et al. In: Mardh PA, Taylor-Robinson D, eds. *Bacterial Vaginosis*. Uppsala, Sweden: Almquist and Wiksell; 1984;259–260.
91. Spiegel CA. Bacterial vaginosis. *Clin Micro Rev*. 1991;4:485–502.
92. Thomason JL, Schreckenberger PC, Spellacy WN, et al. Clinical and microbiological characterization of patients with non-specific vaginosis associated with motile, curved anaerobic rods. *J Infect Dis*. 1984;149:801–809.
93. Spiegel CA, Roberts M. *Mobiluncus* gen nov., *Mobiluncus curtisii* sp nov., *Mobiluncus curtisii* subsp. *holmesii* subsp. nov., and *Mobiluncus* species, curved rods from the human vagina. *J System Bacteriol*. 1984;34:177–184.
94. Thomason JL, Gelbart Sm, Scaglione NJ. Bacterial vaginosis: current review with indications for asymptomatic therapy. *Am J Obstet Gynecol*. 1991;165:1210–1217.
95. Faro S. Bacterial vaginitis. *Clin Obstet Gynecol*. 1991;34:582–586.
96. Eschenback DA, Hillier S, Crithlow C, et al. Diagnosis and clinical manifestations of bacterial vaginosis. *Am J Obstet Gynecol*. 1988;158:819–828.
97. Amsel R, Totten PA, Spiegel CA, et al. Nonspecific vaginitis: diagnostic criteria and microbial and epidemiologic associations. *Amer J Med*. 1983;74:14–22.
98. Holst E. Reservoir of four organisms associated with bacterial vaginosis suggests lack of sexual transmission. *J Clin Micro*. 1990;28:2035–2039.
99. Eschenback DA, Critchiow CW, Watkins H, et al. A dose duration study of metronidazole for the treatment of nonspecific vaginitis. *Scand J Infect Dis*. 1983;40(suppl 1):73–80.
100. Cook RL, Reid G, Pond DG, et al. Clue cells in bacterial vaginosis: immunoflourescent identification of the adherent Gram-negative bacteria as *Gardnerella vaginitis*. *J Infect Dis*. 1989;160:490–496.
101. Nugent RP, Krohn MA, Hillier SL. Reliability of diagnosing bacterial vaginosis is improved by a standardized method of Gram stain interpretation. *J Clin Micro*. 1991;29:297–300.
102. Swedberg J, Steinder JF, Deiss F, et al. Comparison of single-dose vs. one week course of metronidazole for symptomatic bacterial vaginosis. *JAMA*. 1985;254:1046–1049.
103. Lugo-Miro UI, Green M, Mazur L. Comparison of different metronidazole therapeutic regimens for bacterial vaginosis. *JAMA*. 1992;268:92–95.
104. Spiegel CA. Susceptibility of *Mobiluncus* species to 23 antimicrobial agents and 15 other compounds. *Antimicrob Agents Chemotherapy*. 1987;31:249–252.
105. Hammill HA. Unusual causes of vaginitis (excluding trichomonas, bacterial vaginosis and *Candida albicans*). *Obstet Gynecol Clin N Amer*. 1989;16:337–345.
106. McKay M. Vulvitis and vulvovaginitis: cutaneous considerations. *Am J Obstet Gynecol*. 1991;165:1176–1182.
107. Witkin SS, Jeremias J, Ledger WJ. A localized vaginal allergic response in women with recurrent vaginitis. *J Allergy Clin Immunol*. 1988;81:412–416.
108. Witkin SS, Jeremias J, Ledger WJ. Recurrent vaginitis as a result of sexual transmission of IgE antibodies. *Am J Obstet Gynecol*. 1988;159:32–36.
109. Cibley LJ, Cibley LJ. Cytolytic vaginosis. *Am J Obstet Gynecol*. 1991;165:1245–1249.

4
Abnormal Uterine Bleeding

Carolyn V. Kirschner

Abnormal uterine bleeding occurs in women of any age. Pregnancy, infection, various birth control methods, endocrine abnormalities, neoplasm, blood dyscrasias, and systemic diseases are all part of the differential diagnosis of this abnormality. In the premenopausal woman, pregnancy-related disorders, infection, and birth control methods are the most common factors. In the perimenopausal years, abnormal bleeding often occurs due to perimenopausal ovarian dysfunction. In the postmenopausal woman, neoplasm must be excluded.

Menstrual disorders constitute a large percentage of a gynecologic practice. A study by Coulter et al[1] found that menstrual disorders constitute 21% of gynecology referrals. Of patients who are referred with menstrual disorders, almost 50% will undergo a hysterectomy in the next 5 years, 50% will have dilation and curettage, 12% will receive oral therapy, and 5% will have no active treatment for these symptoms. Therefore abnormal uterine bleeding is an important aspect of a gynecologic practice and one that should be understood by the clinician as much as possible.

Definitions for some of the terms associated with abnormal uterine bleeding are as follows:[2]

1. Menorrhagia or hypermenorrhea: excessive uterine bleeding both in the amount and duration of flow, occurring at regular intervals.
2. Hypomenorrhea: decreased menstrual flow at regular intervals.
3. Metrorrhagia: uterine bleeding at irregular intervals.
4. Menometrorrhagia: frequent irregular excessive uterine bleeding.
5. Polymenorrhea: frequent regular episodes of uterine bleeding occurring at intervals of less than 21 days.
6. Oligomenorrhea: infrequent irregular bleeding occurring at intervals of more than 45 days.
7. Perimenopausal bleeding: irregular bleeding occurring within the context of other climacteric symptoms.
8. Postmenopausal bleeding: any amount of vaginal bleeding that occurs at least 6 months after the last normal menstrual period.

Normal and Abnormal Menstrual Cycling

Normal menstrual cycling depends on the functional integrity of the hypothalamus, anterior pituitary gland, ovary, and uterus. Gonadotropin-releasing hormone is secreted in a pulsatile fashion from the hypothalamus and travels through the portal veins to the anterior lobe of the pituitary gland. The pituitary gland responds by producing and releasing two gonadatropins: follicle-stimulating hormone and luteinizing hormone. Negative and positive feedback regulate the release of these hormones according to levels of estrogen and progesterone produced by the ovary.

The initial menstrual cycles of young adolescent girls are often anovulatory, but ovulation is normally achieved by the age of 16. Toward the perimenopausal period, cycles become again anovulatory, with only 10% of women experiencing normal regular menstrual periods at age 50.[3]

Prostaglandins have recently been shown to be involved with the regulation of normal flow. Prostaglandin F_2 alpha causes vasoconstriction and myometrial contraction, whereas prostaglandin E_2 causes vasodilation and myometrial relaxation. Both of these prostaglandins have been recovered in endometrial tissue. Blood volumes in the menstrual cycle are also influenced by platelet aggregation and fibrin formation. In the perimenopausal patient, dysfunctional bleeding is felt to be due to a decreased sensitivity of the ovary to follicle-stimulating hormone and luteinizing hormone stimulation. As the oocyte responds to stimulation with a formation of the follicle but produces insufficient estrogen to provide the feedback necessary for ovulation, the patient then experiences estrogen breakthrough bleeding. Estrogen withdrawal bleeding is a second condition that may occur in the perimenopausal period. In this circumstance, estrogen levels build up slowly, leading to irregular endometrial proliferation with irregular menstrual shedding.

Anatomic disorders that result in abnormal bleeding and lesions of the cervix, vagina, and endometrium can all present with abnormal uterine bleeding. Postcoital bleeding and spotting tend to be associated with vaginal and cervical lesions. Leiomyomata of the uterine cavity may cause endometrial bleeding, particularly if they are located in the submucosal layer of the uterus. The endometrial tissue may become ischemic or necrotic and slough, which leads to abnormal bleeding or spotting.

Ovarian cysts or neoplasms can cause menstrual irregularity because of hormone overproduction or underproduction if the ovarian stroma is replaced by tumor mass.

Evaluation of Abnormal Uterine Bleeding

Initially a careful history and physical examination should be performed. Emphasis should be placed on obtaining evidence for or against psychological dysfunction or emotional stress, nutritional disorders, previous infection, type of birth control use, abnormal bleeding or bruising, evidence of other systemic disease. The physical exam should rule out abnormalities of the reproductive tract, including visible neoplasm. A Papanicolaou smear should be performed at the initial visit, and the patient should be given instructions for accurate recording of bleeding, such as the use of a bleeding calendar.

Physicians will often ask patients to quantitate the amount of bleeding experienced with menses in terms of the number of pads used. A study by Fraser et al,[4] however, found that there was no correlation between the number of pads used and the actual volume of menstrual flow. In light of the inaccuracy of this method of quantitation, it would seem prudent to base evaluation and treatment on a patient's subjective report of significant change in menstrual flow or a bothersome component to the menstrual flow.

Evaluation of the Premenopausal Female

In the premenopausal female, the most common cause of abnormal uterine bleeding is pregnancy related, and therefore a pregnancy test should be a part of the initial work-up of abnormal uterine bleeding in this age group. If the pregnancy test is positive, then the differential diagnosis includes threatened abortion, incomplete abortion, missed abortion, ectopic pregnancy, or normal pregnancy. Work-up at this point depends on the physical examination and symptoms. Ultrasound is often helpful to establish the presence or absence of a gestational sac and/or fetal pole. If an intrauterine pregnancy is confirmed on ultrasound, then conservative management of the bleeding is warranted. Bleeding will occur in approximately 25% of all pregnancies. Half of these will go on to result in a normal gestation; the remainder will abort. If the ultrasound does not confirm an intrauterine pregnancy, then consideration must be given to either missed abortion, incomplete abortion, or ectopic pregnancy. Quantitative HCG may be helpful in this circumstance; a doubling of the serum HCG occurs every 48 hours in a normal pregnancy.

Infection is another cause of abnormal uterine bleeding. If the history indicates possible exposure to sexually transmitted disease, cultures for gonorrhea and chlamydia should be performed. Vulvovaginitis is rarely a cause of abnormal bleeding.

Birth control methods are known to cause abnormalities of the menstrual cycle; in particular, the use of intrauterine devices cause menorraghia and intermenstrual bleeding. If the patient has significant abnormal bleeding due to the presence of an intrauterine device, then consideration should be given to an alternative form of birth control.

Oral contraceptives are a common treatment for a young woman with abnormal menstrual cycles, and yet a major deterrent to the use of birth control pills may be breakthrough bleeding. This is especially true of the low-dose formulations that are currently available. The incidence of breakthrough bleeding increases as the estrogen dose of the pill decreases.[5] Breakthrough bleeding often occurs in the first 3 months of pill use and will usually resolve spontaneously without treatment. Patients should be encouraged to continue pill use for 3 months to allow for spontaneous resolution of breakthrough bleeding. Speroff[5] recommends explaining to the patient that breakthrough bleeding in the first few months of pill use represents tissue breakdown as the endometrium adjusts from its usual thicker state to the relatively thin state present in a patient on oral contraceptives.

Breakthrough bleeding that occurs after several months of pill use may be due to progestin-induced decidualization. Because of the thin endometrium as a consequence of pill use, the endometrium is more fragile and prone to breakdown in asynchronous bleeding. The addition of extra estrogen at troublesome times in the pill cycle will usually correct this problem. Similarly, bleeding that occurs just before the end of the pill cycle is probably due to a relative estrogen deficiency; control of the bleeding may be accomplished with a course of additional estrogen. Speroff suggests 2.5 mg of conjugated estrogen or 20 micrograms of ethinyl estradiol administered daily for 7 days when the bleeding is present. The patient continues with her usual schedule of oral contraceptives. If the problem has not resolved after one course of estrogen, a second 7-day course of estrogen may be effective. After this point, however, a search must be undertaken for the presence of underlying organic pathology.

Blood dyscrasias are a rare cause of abnormal uterine bleeding in the premenopausal woman. The exception is the pediatric population, in which these disorders are somewhat more frequent. If the history suggests abnormal bruising or bleeding other than through the genital tract, a hematologic evaluation, including bleeding time and platelet count, should be undertaken.

Various systemic diseases may present with abnormal uterine bleeding. Usually there is physical evidence to confirm the presence of such disorders. Consideration should be given to thyroid disease, systemic lupus erythematosus, and nutritional abnormalities. A recent study comparing menstrual regularity in patients on various diets suggested that the probability of being menstrually regular was negatively associated with increased dietary fiber and magnesium intake.[6]

Other disorders include Cushing's disease and disorders of the anterior pituitary gland. Major liver disease may result in abnormal uterine bleeding because of a decrease in the ability to metabolize estrogen. Abnormal clotting factors as a result of liver disease and thrombocytopenia secondary to splenomegaly with sequestracian of platelets may also contribute to abnormal uterine bleeding.

Certain medications have also been associated with menstrual irregularities. These include phentolamine, tricyclic antidepressants, corticosteroids, major tranquilizers, propranolol, digoxin, and cyproheptadine hydrochloride.

Cigarette smoking may also affect menstrual patterns. Brown et al[7] evaluated 2115 women aged 18 through 49 via a questionnaire. There was a significant association between smoking habits and prolonged menstrual periods, heavy periods, frequent periods, irregular periods, intramenstrual bleeding, and dysmenorrhea. Interestingly, current smokers reported the worst experience, and ex-smokers occupied an intermediate position.

Exercise is known to have an effect on menstrual regularity. Beitins et al[8] have shown that the initiation of strenuous exercise frequently leads to corpus luteum dysfunction associated with insufficient progesterone secretion. Exercise may also decrease the luteal phase length. Therefore, due to a lack of positive feedback to estrogen and a decrease in luteal hormone secretion during the luteal phase, exercise may alter the neuroendocrine system. In addition, Loucks's[9] data confirm that the initiation of a high volume of aerobic training disrupts the menstrual cycle of many women. He feels that these women may be more susceptible to reproductive disruptions than other women and that caloric deficiency may be responsible for the observed disruption.

Tubal sterilization has long been suggested as a cause of menstrual irregularity. A recent study by Rulin et al[10], however, refutes this notion. In their study changes in menstrual parameters after tubal sterilization were analyzed; no change in menstrual cycle, duration of menstrual flow, or intramenstrual bleeding was found. Compared with the control group, however, the prevalence of dysmenorrhea did increase in sterilized women.

Management of Abnormal Uterine Bleeding

If all the above etiologies have been ruled out, then dysfunctional uterine bleeding due to irregular ovulation or the possibility of neoplasm remains. At this point an endometrial biopsy is usually indicated, particularly in patients aged 40 and older.

Office endometrial biopsy compares quite favorably in terms of complication rate, cost, and diagnostic accuracy with traditional dilation and curettage. Grimes[11] found both fewer complications with endometrial sampling than with dilation and curettage and adequate specimens in 96% of patients.

If the biopsy shows a proliferative endometrium, indicating a probable lack of ovulation, then the disorder is treated based on the fertility desires of the patient. If the patient is a nonsmoker who desires contraception, consideration can be given to the administration of oral contraceptives. If the patient desires pregnancy, then induction of ovulation may be instituted.

Biopsy-proven endometrial hyperplasia is treated, again based on the patient's age and fertility desires. Endometrial hyperplasia is thought to result from persistent unopposed estrogen stimulation. This would occur with anovulatory menstrual cycles or endogenously produced estrogen, such as would occur with polycystic ovarian disease, hormone-producing ovarian tumors (granulosa cell tumors, ovarian tumors, thecomas), or exogenous estrogen administration. In general, hyperplasia without atypia is not felt to be a precursor of adenocarcinoma. Severe cytological atypia, however, may be associated with either concurrent adenocarcinoma of the endometrium or a predisposition to development of adenocarcinoma in the future. Recommendations for treatment of endometrial hyperplasia, as outlined by DiSaia and Creasman,[12] are as follows:

1. The teenage girl should be treated with oral contraceptives for 6 months with resampling of the endometrial cavity at the end of this time.
2. A woman of childbearing age should be treated with oral contraceptives for 3 months, followed by resampling of the endometrium and then induction of ovulation. If the patient is not interested in childbear-

ing, continued use of estrogen-progestin artificial cycles is recommended.
3. A perimenopausal woman should be treated by hysterectomy or by progestins alone. Hysterectomy is recommended for patients with complex hyperplasia with atypia, and a trial of progestins is utilized for patients with lesser lesions. DiSaia and Creasman recommend Provera (20 mg daily for 10 days or monthly for 6 months). The endometrial cavity should be sampled every 3 months.
4. A postmenopausal woman should undergo hysterectomy unless strongly contraindicated.

It should be noted that Megace, 40 mg twice daily, recently has been utilized as an alternative to Provera. The advantage of Megace is a decrease in side effects such as depression and water retention, although at high doses increased appetite and subsequent weight gain have been reported.

Evaluation of the Perimenopausal and Postmenopausal Female

The incidence of abnormal vaginal bleeding increases as women approach the end of their reproductive years.[13] In the perimenopausal years, the most common cause of abnormal vaginal bleeding is anovulation. The most common cause of abnormal bleeding in the postmenopausal woman is genital atrophy. In both instances, however, malignancy must be excluded. After careful inspection and palpation of the vulva, vagina, cervix, and uterus and performance of a Pap smear, an endometrial biopsy should be performed. Often the cervix is stenotic in this age group, and it may not be possible to perform this as an office procedure comfortably. If the catheter does not initially pass easily, a single-tooth tenaculum may be placed on the anterior lip of the cervix to provide gentle countertraction that may then allow introduction of the catheter. Use of nonsteroidal antiinflammatory agents prior to the procedure may also be of benefit. If patient discomfort or stenosis precludes adequate sampling, however, formal dilation and curettage is indicated. Similarly, if endometrial biopsy is performed and is nondiagnostic or if bleeding continues despite what is considered adequate treatment, dilation and curettage is mandatory.

A diagnosis of genital atrophy in a postmenopausal patient is one of exclusion. If, after careful examination and endometrial sampling, findings do suggest genital atrophy, a trial of hormone replacement is indicated. Both systemic estrogen and local estrogen in the form of cream to the vulvovaginal area may alleviate atrophic changes.

Some patients with menorrhagia and no obvious etiology have benefited from nonsteroidal antiinflammatory drugs during the heaviest days of their menstrual cycles. The mechanism of action of these agents is thought to be the antiprostaglandin effect. With this regimen bleeding in women with menorrhagia may be reduced by as much as 50%.[14]

Bleeding from a benign neoplasm, such as a cervical or endometrial polyp, usually requires pathologic diagnosis. Endometrial polyps may be missed on office biopsy and may require formal dilation and curettage. Bleeding from an endometrial polyp will often cease after removal of the lesion. If the bleeding persists despite adequate dilation and curettage, consideration should be given to progestational agents. Treatment of hyperplastic endometrial lesions is as outlined above.

It is imperative that if bleeding persists despite adequate treatment or if diagnosis is obscure, repeat dilation and curettage and/or hysterectomy must be considered. Several studies have shown that the yield on dilation and curettage may be increased with the use of concurrent hysteroscopic examination. For this reason many practitioners routinely perform hysteroscopy in conjunction with dilation and curettage.

A diagnosis of endometrial hyperplasia with atypia or of endometrial carcinoma in the perimenopausal and postmenopausal age groups should be treated with hysterectomy, if at all possible. If surgery is absolutely contraindicated because of the patient's overall health, then treatment with progesterones may be instituted. Recently, premalignant lesions of the uterus have been treated with endometrial ablation under hysteroscopic guidance.[15] This procedure requires practice and skill, but it may be of benefit in selected cases. However, most of these procedures are performed under general anesthesia and in many ways are little different from hysterectomy.

Hysteroscopic evaluation of the endometrial cavity can be done as an office procedure or as an ambulatory surgery procedure. Direct visualization of the endometrium and endocervix, rather than the "blind" procedure involved with dilation and curettage, should theoretically limit the likelihood of incomplete diagnosis and treatment. Direct vision also allows for treatment not only when the uterus is normal in size and shape but also when the uterus is enlarged and distorted. Hysteroscopic ablation of the endometrium can then be performed either with electrocautery or with laser. Transcervical resection of the endometrium is another form of endometrial ablation. This is similar in concept to transurethral resection of the prostate in which the entire thickness of the endometrium is systematically excised with the electrocautery loop of a resectoscope. The resected endometrium can then be submitted for histologic assessment. Operative complications are uncommon but include uterine perforation and fluid overload. Magos et al[16] report a 90% improvement of menstrual symptoms using this technique. Rarely, when all the above methods fail, consideration must be given to hysterectomy. An occasional neoplasm may be diagnosed by this method.

The importance of endometrial sampling or curettage for diagnosis and therapy cannot be overemphasized. As the age of the patient increases so should the frequency with which this modality is utilized. Curettage of itself may be sufficient to correct an episode of uterine bleeding. Scommegna and Dmowski found that of patients who underwent curettage, 30% to 40% had recurrent episodes of bleeding.[17] If bleeding does recur, either submucous leiomyoma, polyps or neoplasm should be suspected. If the etiology of the bleeding remains obscure, then the possibility of hysterectomy must be considered.

Management of Excessive Bleeding

Occasionally, therapy must be directed toward the immediate control of acute uterine hemorrhage. If neoplasm seems unlikely, hormonal manipulation usually can be utilized to control excessive uterine bleeding rapidly. In the adolescent or reproductive-age female, the bleeding probably involves asynchrony of the endometrium with irregular shedding. Therapy is therefore directed toward restoring stability of the endometrium. The use of an oral contraceptive administered as one tablet four times each day for 5 to 7 days will usually result in cessation of bleeding within the first 12 to 24 hours. If this therapy is successful, then cyclic contraceptive pills should be prescribed for the next two or three cycles. If the flow is not diminished rapidly, dilation and curettage must be considered.[10]

An alternative management of acute uterine hemorrhage when estrogen deficiency is suspected is the use of intravenous conjugated estrogens, 25 mg up to every 4 hours. This should be followed by institution of oral contraceptives to allow for uniform shedding of the endometrium.

Conclusion

Abnormal uterine bleeding is a common gynecologic complaint. Every physician who treats women must be familiar with the diagnostic work-up of this disorder.

References

1. Coulter A, Bradlow J, Agass M, Martin-Bates C, Tulloch A. Outcomes of referrals to gynecology outpatient clinics for menstrual problems: an audit of general practice records. *Br J Obstet Gynaecol.* 1991;98:789–796.
2. Kempers RD. Dysfunctional uterine bleeding.

In: *Gynecology and Obstetrics.* New York: JB Lippincott Company, 1991.
3. Meldrum DR. Perimenopausal menstrual problems. *Clin Obstet Gynecol.* 1983;26:762–768.
4. Fraser IS, McCarron G, Markham R. A preliminary study of factors influencing perception of menstrual blood loss volume. *Am J Obstet Gynecol.* 1984;149:788–793.
5. Speroff L, Glass RH, Kase NG. *Clinical Gynecologic Endocrinology and Infertility.* Baltimore, Md: Williams & Wilkins, 1989.
6. Pedersen AB, Bartholomew MJ, Dolence LA, Aljadir LP, Netteburg KL, Lloyd T. Menstrual differences due to vegetarian and nonvegetarian diets. *Am J Clin Nutr.* 1991;53:878–885.
7. Brown S, Vessey M, Stratton I. The influence of method of contraception and cigarette smoking on menstrual patterns. *Br J Obstet Gynaecol.* 1988;95:905–910.
8. Beitins IZ, McArthur JW, Turnbull BA, Skrinar GS, Bullen BA. Exercise induces two types of human luteal dysfunction: confirmation by urinary free progesterone. *J Clin Endocrinol Metab.* 1991;72:1350–1358.
9. Loucks AB. Effects of exercise training on the menstrual cycle: existence and mechanisms. *Med Sci Sports Exerc.* 1990;22:275–280.
10. Rulin MC, Davidson AR, Philliben SG, Graves WL, Cushman LF. Changes in menstrual symptoms among sterilized and comparison women: a prospective study. *Obstet Gynecol.* 1989;74:149–154.
11. Grimes, DA. Diagnostic dilation and curettage: a reappraisal. *Am J Obstet Gynecol.* 1982;142:1–6.
12. DiSaia PJ, Creasman WT. *Clinical Gynecologic Oncology.* St. Louis, Mo: CV Mosby Company, 1989.
13. Nesse RE. Abnormal vaginal bleeding in perimenopausal women. *Am Fam Physician.* 1989;40:185–192.
14. Hall P, Maclachlan N, Thorn N, et al. Control of menorrhagia by the cyclo-oxygenase inhibitors naproxen sodium and mefenamic acid. *Br J Obster Gynaecol.* 1987;94:554.
15. DeCherney AH, Diamond MD, Lavy G, Polan ML. Endometrial ablation for intractable uterine bleeding: hysteroscopic resection. *Obstet Gynecol.* 1987;70:668.
16. Magos AL, Baumann R, Lockwood GM, Turnbull AC. Experience with the first 250 endometrial resections for menorrhagia. *Lancet.* 1991;337:1074–1078.
17. Scommegna A, Dmowski WP. Dysfunctional uterine bleeding. *Clin Obstet Gynecol.* 1973;16:221–254.

5
Primary Amenorrhea and Delayed Onset of Puberty

Joan M. Leya

Primary Amenorrhea

Primary amenorrhea is defined as lack of menarche by the age of 14 regardless of the stage of pubertal development. It is a sign of an underlying disorder that should be revealed. In general, etiologic factors responsible for secondary amenorrhea may also be associated with primary amenorrhea. Causes unique to the patient who has not experienced menarche will be discussed in detail in this chapter.

The history may provide the first clues toward a diagnosis. A family or personal history of genetic anomalies or growth problems may be elicited. Stress, weight loss, poor nutrition, and strenuous exercise may contribute to primary amenorrhea. Chronic illnesses such as cystic fibrosis[1] may be responsible. Any complaints of headaches, galactorrhea, or visual changes as well as exposure to radiation or chemotherapeutic (alkylating) agents should be noted.

The physical examination should begin with height and weight measurements. A body-mass index (kg/m^2) in which weight is corrected for height should be calculated. A range below the normal of 20 to 25 may be associated with amenorrhea.[2] Frisch[3] maintains that a specific percentage of total body fat is necessary for initiation as well as maintenance of menstruation. Obtaining a critical weight (47.8 ±0.5 kg for American girls) has also been theorized as being necessary for menarche to occur.[4] A short stature associated with a webbed neck and shield chest is clinically suggestive of Turner's syndrome.

Pubertal staging of breast and pubic hair development should be performed (Tables 5.1 and 5.2), and sense of smell should be evaluated.

Evidence of thyromegaly, galactorrhea, acromegaly, Cushingoid features, or hirsutism should be sought. Finally, an examination of the internal and external genitalia and inguinal area should be performed. Defects of the mullerian system (fallopian tubes, uterus, and upper vagina) are responsible for 17% to 50% of all cases of primary amenorrhea.[2,6,7] Absence of a vagina and uterus (confirmed by rectal examination or by ultrasound) denotes mullerian agenesis (Mayer-Rokitansky-Kuster-Hauser syndrome), the second most common cause of primary amenorrhea. In incomplete forms, rudimentary uterine horns may be present or discontinuity in the mullerian tract (that is, vaginal septae) may exist. If a mullerian defect is detected, intravenous pyelogram may be used to identify those patients with concurrent renal anomalies (30%).

Laboratory evaluation should commence with measurements of serum gonadotropins (follicle-stimulating hormone [FSH] and luteinizing hormone [LH], as well as thyroid-stimulating hormone (TSH) and prolactin. Pregnancy should be ruled out.

In the presence of hirsutism, ovarian and adrenal androgen levels (androstenedione, testosterone, and dehydroepiandrosterone sulfate

TABLE 5.1. Stages of breast development.[5]

Stage 1: Preadolescent; elevation of papilla only.
Stage 2: Breast bud stage; elevation of breast and papilla as a small mound, enlargement of areolar diameter.
Stage 3: Further enlargement of breast and areola, with no separation of their contours.
Stage 4: Projection of areola and papilla to form a secondary mound above the level of the breast.
Stage 5: Mature stage; projection of papilla only, resulting from recession of the areola to the general contour of the breast.

TABLE 5.2. Stages of female pubic hair development.[5]

Stage 1: Preadolescent; no pubic hair.
Stage 2: Sparse growth of long, slightly pigmented, downy hair, straight or only slightly curled, appearing chiefly along the labia.
Stage 3: Hair is considerably darker, coarser, and curlier. The hair spreads sparsely over the junction of the pubes.
Stage 4: Hair is now adult in type, but the area covered by it is still considerably smaller than in most adults. There is no spread to the medial surface of the thighs.
Stage 5: Hair is adult in quantity and type, distributed as an inverse triangle of the classic feminine pattern. The spread is to the medial surface of the thighs but not up the linea alba or elsewhere above the base of the inverted triangle.

[DHEAS] as well as 17-hydroxyprogesterone can suggest the etiology of primary amenorrhea. Ovarian malignancy has been associated with primary amenorrhea and testosterone levels greater than 200 ng/dL.[8] DHEAS levels greater than 700 mg/dL in association with 17-hydroxyprogesterone levels greater than 300 ng/dL can signal congenital adrenal hyperplasia due to 21-hydroxylase deficiency (a rare cause of primary amenorrhea). Further workup (such as dexamethasone suppression testing) may be necessary to rule out an adrenal tumor.

With lesser elevations of DHEAS, testosterone, and androstenedione, varying degrees of polycystic ovarian disease may be present and can, in rare cases, result in primary amenorrhea. Ultrasound examination may reveal multiple bilateral ovarian cysts less than 7 mm in diameter. An LH/FSH ratio greater than 2 may suggest the diagnosis as well.

FSH and LH levels in the low (<5 mIU/mL) or normal range are consistent with hypothalamic or pituitary disorders leading to primary amenorrhea. The gonadotropin-releasing hormone (GnRH) pulsatility required to initiate ovulatory cycles and subsequent menstruation may be disrupted by alterations in weight and nutrition as well as by increased levels of stress and exercise.[9]

The results of direct testing of the hypothalamus in females are not consistent; thus, diagnoses of GnRH abnormalities are made by exclusion.

Direct examination of the pituitary area via CT scanning with contrast should be performed if low or normal gonadotropins are detected and if TSH is normal. Hyperprolactinemia caused by a prolactin-secreting pituitary adenoma may be detected and is associated with primary amenorrhea.[10] Hyperprolactinemia in the absence of galactorrhea and a pituitary lesion should not preclude treatment, however.[11] The empty sella syndrome and other space-occupying lesions (that is, growth-hormone or ACTH producing) of the pituitary are, in rare cases, associated with primary amenorrhea and may also be detected. Chronic internal hydrocephalus causing primary amenorrhea has also been reported.[12] If hypercortisolism or growth-hormone hyposecretion or hypersecretion are suspected, appropriate hormonal testing should be performed initially.[13]

A rare cause of hypogonadotropism in females is associated with anosmia and is analogous to Kallmann's syndrome in the male. Maldevelopment of the olfactory lobes, abnormalities of the hypothalamus, and midline facial defects are associated with this congenital disorder.

If gonadotropin levels are elevated (LH and FSH >40 mIU/mL), chromosomal analysis should be performed. Normal (XX) chromosomes with elevated gonadotropins are most commonly associated with premature ovarian failure. Immune disorders should be sought in these patients, although the yield is likely to be higher in those with secondary amenorrhea due to premature menopause. Thyroid disturbances (Hashimoto's thyroiditis, subacute thyroiditis, and Graves' disease) as well as hypoparathyroidism, Addison's disease, insulin-dependent diabetes mellitus, and collagen vascular disease have been reported.[14,15] If normal ovarian follicles are detected on ultrasound, several rare causes of ovarian failure with normal XX karyotype should be considered. The resistant ovary syndrome (resulting from a defect at the gonadotropin receptor level on the ovary) has been described. A deficiency of the enzyme 17-hydroxylase at the ovarian and adrenal levels may exist, but high progesterone levels, low cortisol levels, and absent secondary sexual development as well as hypertension and hypokalemia (due to elevated 11-deoxycorticosterone levels) are present.

Karyotypic abnormalities have been detected in one-third of patients with primary amenorrhea.[6] An XY chromosome complement in a phenotypic female lacking mullerian structures and sexual hair denotes testicular feminization. Such patients have cryptorchid or inguinal testes and normal male levels of testosterone but an absence of any testosterone effect. A defect in testosterone binding at the receptor level is present. The production of mullerian-inhibiting factor by the testes is responsible for mullerian regression.

An XY-chromosome complement with normal female testosterone levels and a palpable mullerian system constitutes XY gonadal dysgenesis or Swyer's syndrome. Fibrous streaks representing the regressed/undeveloped testes should be resected to prevent malignant degeneration (that is, gonadoblastoma or dysgerminoma). In patients with testicular feminization, gonadectomy can be delayed until puberty is complete.

In the absence of a Y chromosome, gonadal dysgenesis of the Turner's type (X0) is most common. Mullerian structures are present, and the patient classically presents with hypergonadotropism, short stature, webbed neck, and shield chest. Mosaicism with lines containing a Y chromosome can occur; yearly pelvic examinations and alpha-fetoprotein and B-hCG levels should therefore be performed to help detect malignant degeneration in the rare patient in whom karyotyping fails to identify the Y chromosome.

Treatment is directed at the cause of amenorrhea. For hypothalamic amenorrhea, weight gain, stress reduction, exercise limitation, and nutritional counseling may be necessary. Hypothyroidism should be treated. Pituitary prolactinomas are responsive to bromocriptine therapy. Patients with a lack of ovarian function secondary to gonadal dysgenesis should be placed on estrogen/progestin replacement therapy, on a regimen similar to that of postmenopausal women. A Y-chromosome complement requires gonadectomy.

Congenital adrenal hyperplasia secondary to an adrenal enyzme defect responds to dexamethasone therapy. Polycystic ovarian disease associated with amenorrhea responds to cyclic oral contraceptive therapy in the patient requesting contraception. Clomiphene citrate for ovulatory purposes is used when ovulation and pregnancy are desired.

If a distal obstruction to normally shedding endometrium exists (that is, vaginal septum or imperforate hymen), a continuity in the genital tract must be created to prevent painful hematocolpos, retrograde menstruation, and endometriosis. A hysterectomy may be necessary if continuity cannot be achieved.

In case of an absent or hypoplastic vagina, progressive dilation of the vaginal dimple to create a functional vagina is preferable to surgical creation of a neovagina.

Pregnancy has been achieved with patients with hypergonadotropic[16] as well as hypogonadotropic amenorrhea.[17]

Delayed Onset of Puberty

Adolescents may become concerned if they do not reach pubertal milestones as quickly as their peers. A knowledge of the normal timetable of pubertal progression is essential because wide variations can occur (Table 5.3).[18]

TABLE 5.3. Current ages of pubertal stages in North American Females (years).[18]

Pubertal stage	Mean age (range: +/− 2SD)
Breast buds (B2)	10.9 (8.9 to 12.90)
Pubic hair (PH2)*	11.2 (9.0 to 13.4)
Menarche	12.7 (10.8 to 14.6)

* B and PH refer to Tanner breast and pubic hair stages.

In general, if the first sign of pubertal progression (breast budding) has not occurred by age 13 (>2 SD above the mean age of 10.9 years), a delay in puberty is recognized. In the majority of patients with delayed puberty no pathologic etiology is identified. This constitutional delay of puberty (CDP) is associated with the ultimate achievement of normal adult height and secondary sexual characteristics. CDP is a diagnosis of exclusion, however, and is often made retrospectively.

The work-up of delayed puberty starts with the patient's history; chronic illnesses such as cystic fibrosis and asthma can be associated with delayed onset of puberty,[19,20] as can recent stress.[21] A history of head trauma, surgical or radiation-induced castration, or anosmia should also be sought. Because CDP can be familial, the pubertal milestones of parents are important. Determination of bone age (using a radiograph of the left hand and wrist and comparing it with standards of maturation in a normal population) and staging of pubertal development (see again Tables 5.1 and 5.2),[2] as well as plasma concentrations of TSH, free T4, prolactin, luteinizing hormone (LH), follicle-stimulating hormone (FSH), dehydroepiandrosterone sulfate (DHEAS), and estradiol, are helpful.

In addition, karyotyping and CT examination of the pituitary and sella turcica should be performed. Recording height and weight on a standard growth curve as well as height velocity (cm/yr) over a 6-month period should be done. As little as a 5% deficit in weight for height can result in gonadotropin deficiency. Screening for occult disease, anemia, and nutritional factors that may affect growth consists of a complete blood count, urinalysis, and serum chemistry survey. Plasma somatomedin C levels should exceed 1.0 U/mL in early puberty and 1.5 U/mL in midpuberty.[22] A deficiency of growth hormone (GH), defined as a GH response less than 7 ng/mL following two standard GH stimulation tests,[23] may be associated with a delay in growth and puberty; this deficiency can be diagnosed if the patient is primed with estrogens prior to testing.[24]

The treatment of girls with delayed puberty secondary to hypergonadotropic hypogonadism (elevated FSH and LH, low estradiol) begins with the removal of the gonads if a Y chromosome is present, due to the risk of neoplastic degeneration. Hormonal replacement is designed to optimize growth as well as to lead to breast development and to induce menstruation. Small doses of ethinyl estradiol (0.05 to 0.10 mg/day) should be used to prevent excessive bone maturation and premature epiphyseal closure. The latter can ultimately lead to shortened adult stature. Estrogen replacement from days 1 to 25 of the month should be opposed by a progesterone derivative such as medroxyprogesterone acetate 10 mg/day during days 16 to 25 to decrease the risk of uterine and breast cancer.[25] Once puberty has been induced, treatment should be continued throughout the patient's lifetime to maintain the cardiovascular and bone benefits of hormone replacement therapy and to provide the sometimes added psychological benefit of monthly menstrual periods (particularly in adolescents). A standard regimen of 0.625 mg of conjugated estrogen on day 1 to 25 with 10 mg of medroxyprogesterone acetate on days 16 to 25 may be used.

Hypogonadotropic hypogonadism (low FSH and LH, low estradiol) can be associated with pituitary or hypothalamic tumors. Other pituitary hormone disturbances may be present (for example, growth hormone deficiency, hyperprolactinemia, or hypothyroidism). Rarely, gonadotropin deficiency associated with anosmia or resulting from head trauma may be encountered. Hypogonadotropic hypogonadism due to idiopathic gonadotropin deficiency may be difficult to distinguish from a constitutional delay of puberty, particularly in girls in whom GnRH-agonist testing is inconclusive.[26]

Although episodic increases in LH secretion

at night occur in normal girls prior to the onset of puberty, this is not a practical way to distinguish hypogonadism from CDP. Reassurance and patience are all that are required in the large majority of patients with CDP; however, self-image may be improved by medical management if the patient requests it. If short stature is a primary concern, an anabolic steroid such as depot testosterone (30 mg monthly) for 3 to 6 months may induce growth in the patient with CDP in whom bone age is characteristically delayed. Development of female secondary sexual characteristics as well as growth induction can be affected with depot estradiol cyprionate 0.5 mg to 1.5 mg monthly intramuscularly (IM). Conjugated estrogen (0.3 mg) or ethinyl estradiol (0.05 mg daily) orally for 3 to 6 months may also be used. Six months after completion of therapy, bone age, pubertal staging, and linear growth assessments should be made. By utilizing the Bayley-Pinneau table predicted height can be calculated using the patient's present height and bone age. If spontaneous puberty has not ensued, a second course of therapy may be useful. Pulsatile subcutaneous gonadotrophin-releasing hormone is another treatment option and one that may mimic pubertal progression in a more physiologic fashion; however, it is more costly and cumbersome than the aforementioned regimens.[28] Short stature in association with growth-hormone deficiency[29] can also be treated effectively. The presence of pubic or axillary hair in the absence of other clinical signs of puberty may simply be due to adrenarche and does not rule out pubertal delay. However, once breast buds have appeared in a girl with delayed puberty, puberty goes to completion in more than 95% of cases.[26]

References

1. Stead RJ, Hodson ME, Botten JC, et al. Amenorrhea in cystic fibrosis. *Clin Endocrinol.* 1987;26:187–195.
2. Franks S. Primary and secondary amenorrhea. *Br Med J.* 1987;294:815–819.
3. Frisch RE. Body fat, menarche, and reproductive ability. *Sem Reprod Endocrinol.* 1985; 3:45–54.
4. Sizonenko PC. Normal sexual maturation. *Pediatrician.* 1987;14:191–201.
5. Marshall WA, Tanner JM. Variations in pattern of pubertal changes in girls. *Arch Dis Child.* 1969;44:291–303.
6. Chryssikopoulos A, Grigorion O. The etiology in 77 primary amenorrhea patients. *Int J Fertil.* 1987;32:245–249.
7. Rao K, Pillai NV. Primary amenorrhea: analysis of 40 cases. *J Indian Med Assoc.* 1991; 89:42–43.
8. Larsen WG, Felmar EA, Wallace ME, et al. Sertoli-Leydig cell tumor of the ovary: a rare cause of amenorrhea. *Obstet Gynecol.* 1992; 79:831–832.
9. White CM, Hergenroeder AC. Amenorrhea, osteopenia and the female athlete. *Pediatr Clin North Am* 1990;37:1125–1141.
10. Howlett TA, Wass JAH, Grossman A, et al. Prolactinomas presenting as primary amenorrhea and delayed or arrested puberty: response to medical therapy. *Clin Endocrinol.* 1989; 30:131–140.
11. Hughes EG, Garner PR. Primary amenorrhea associated with hyperprolactinemia: four cases with normal sellar architecture and absence of galactorrhea. *Fertil Steril.* 1987;47:1031–1032.
12. Moeslein S, Dericks-Tan JSE, Lorenz R, et al. Double-stimulation with LH-RH in primary amenorrhea caused by chronic internal hydrocephylus: a case study. *Gynecol Endocrinol.* 1987;1:202–207.
13. Montini M, Pagani G, Gianola D, et al. Acromegaly and primary amenorrhea: ovulation and pregnancy induced by SMS 201–995 and bromocriptine. *J Endocrinol Invest.* 1990;13:193.
14. Rebar RW, Connolly HV. Clinical features of young women with hypergonadotropic amenorrhea. *Fertil Steril.* 1990;53:804–810.
15. Alper MM, Garner PR. Premature ovarian failure: its relationship to autoimmune disease. *Obstet Gynecol.* 1985;66:27–30.
16. Check JH, Nowroozi K, Chase JS, et al. Ovulation induction and pregnancies in 100 consecutive women with hypergonadotropic amenorrhea. *Fertil Stertil.* 1990;53:811–816.
17. Braat DDM, Schoemaker R, Schoemaker J. Life table analysis of fecundity in intravenously gonadotropin releasing hormone treated patients with normogonadotropic and hypogonadotropic amenorrhea. *Fertil Steril.* 1991;55: 266–271.

18. Tanner JM, Davies PWS. Reply (pubertal data for growth velocity charts). *J Pediatr.* 1986; 109:564–565.
19. Balfour-Lynn L. Effect of asthma on growth and puberty. *Pediatrician.* 1987;14:237–241.
20. Landon C, Rosenfeld R. Short stature and pubertal delay in cystic fibrosis. *Pediatrician.* 1987;14:253–260.
21. Eisenstein TD, Gerson MJ. Psychosocial growth retardation in adolescence. *J Adol Health Care.* 1988;9:436–440.
22. Cara JF, Rosenfield RL, Furlanetto RW. A longitudinal study of the relationship of plasma somatomedin-C concentration to the pubertal growth spurt. *Am J Dis Child.* 1987;41:562–564.
23. Frasier SD. A review of growth hormone stimulation tests in children. *Pediatrics.* 1974; 53:929–937.
24. Shalet, SM. Treatment of constitutional delay in growth and puberty. *Clin Endocrinol.* 1989; 31:81–86.
25. Sizonenko, PC. Delayed sexual maturation. *Pediatrician.* 1987;14:202–211.
26. Rosenfield RL. Diagnosis and management of delayed puberty. *J Clin Endocrinol Metab.* 1990;10:559–562.
27. Bayley N, Pinneau SR. Tables for predicting adult height from skeletal age: revised for use with the Greulich-Pyle hand standards. *J Pediatrics.* 1952;40:423.
28. Clayton PE, Shalet SM, Price DA. Endocrine manipulation of constitutional delay in growth and puberty. *J Endocrin.* 1988;116:321–323.
29. Wilson DM, Rosenfeld RG. Treatment of short stature and delayed adolescence. *Pediatr Clin North Am.* 1987;34:865–879.

6
Secondary Amenorrhea

Rodney J. Hoxsey

Introduction

Few things cause greater concern to a woman of reproductive age than the cessation of her regular monthly menstrual period. Between menarche and menopause the only physiologic causes of amenorrhea are pregnancy and lactation. Once pregnancy has been ruled out, the absence of menses for three or more of the previous cycle intervals should be considered pathologic. While the majority of patients with secondary amenorrhea have relatively simple problems that can be easily diagnosed and managed, the physician must be concerned with potential diseases and disorders that may have more serious consequences for the patient. Amenorrhea is a symptom and not a diagnosis in itself. Nonphysiologic amenorrhea is estimated to occur in 5% of women of reproductive age. In two-thirds of women menses occur at intervals of 28 ±3 days, with a normal range of 18 to 40 days.[1] Oligomenorrhea of greater than 40 days should be treated with the same significance as secondary amenorrhea; it indicates failure of the hypothalamic-pituitary-ovarian axis to induce cyclic changes in the endometrium that normally result in regular menses. Amenorrhea may also result from failure of the end organ (endometrium) to respond or from obstruction of the outflow tract.

To evaluate the patient thoroughly it is useful to categorize the causes of secondary amenorrhea. It is also important that the clinician understand the basic physiology of normal menstruation so that appropriate tests can be performed and interpreted to arrive at the diagnosis in the most expeditious manner.

Physiology of Menstruation

The only physiologic role of the endometrium is to support implantation of the embryo. Therefore the onset of menstrual bleeding is the sign of reproductive failure in that cycle. The hypothalamus, pituitary, and ovaries determine the hormonal changes that lead to regular development and shedding of the endometrium. Normal menstruation is dependent on the precise interaction of the hypothalamus, the pituitary, and the ovaries via neuropeptide and steroid hormones.

Hypothalamus

Gonadotropin-releasing hormone (GnRH), a decapeptide with a half-life of only a few minutes, is released from the arcuate nucleus of the median basal hypothalamus in a monotonous pulsatile pattern approximately every hour. The GnRH travels via the hypophyseal portal venous system to the anterior pituitary; there it stimulates the gonadotropes to synthesize and secrete follicle-stimulating hormone (FSH) and luteinizing hormone (LH). Maintenance of the normal ovulatory cycle is dependent on the frequency and amplitude of this GnRH pulse generator. This is modulated by the positive and negative feedback of the ovarian steroids and neurotransmitters that may act as intermediaries between the ovarian steroids and GnRH neurons.[2] Dopamine, β-endorphin, and serotonin inhibit GnRH release, whereas norepinephrine stimulates it. Depending on the endocrine environment, catecholestrogens may stimulate or inhibit GnRH.[3] Opiates from cortical brain centers and other metabolic products may also affect hypothalamic function positively or negatively.[4]

Many of the clinical manifestations of secondary amenorrhea can be explained by changes in the pulsatile release of GnRH by the hypothalamus. Changes in the frequency and amplitude of these pulses may cause the changes in the LH to FSH ratio seen in polycystic ovarian disease. Loss of rhythm of release can lead to hypothalamic amenorrhea. Continuous or frequent release can "down regulate" or deplete the pituitary of gonadotropins and cause "hypoestrogenic" amenorrhea. Certain drugs, weight loss, and exercise have all been shown to affect GnRH release. Hypothalamic trauma or tumors can also eliminate GnRH release and cause hypogonadotropic, hypoestrogenic amenorrhea.[5]

Pituitary

The anterior pituitary is the site of synthesis and pulsatile secretion of the glycoprotein gonadotropins, follicle-stimulating hormone (FSH) and luteinizing hormone (LH), in response to the GnRH pulses. Even without changes in GnRH pulse amplitude or frequency, the release of the gonadotropins varies throughout the menstrual cycle, peaking at the preovulatory surge. The release of FSH and LH is therefore modulated by both positive and negative feedback by estrogen, progesterone, and perhaps other products from the ovary.[6]

Basal levels of FSH are highest early in the follicular phase of the menstrual cycle. This brings about recruitment and development of several primordial follicles in the ovary. FSH's action on the granulosa cells of the follicles induces its own receptors, activates the aromatase enzyme necessary for the conversion of ovarian androgens to estrogen, and induces the formation of LH receptors in preparation for the LH surge that triggers ovulation. Basal levels of LH act on the theca interna to produce androgens that diffuse across the basement membrane of the developing follicle, where the aromatase activity of the granulosa cells convert them to estrogens. The estrogen, along with FSH, further stimulates granulosa cell proliferation. The relatively low levels of estrogen early in the follicular phase enhances pituitary synthesis and storage of FSH and LH but inhibits its release. As the estradiol level rises in the dominant follicle, FSH synthesis decreases, but because the pituitary becomes sensitized, greater amounts of LH are released with each GnRH pulse. The precise trigger for the preovulatory LH surge is not known; in addition to a critical level of estrogen, low levels of progesterone released by the luteinized follicle may be a stimulus. A preovulatory surge of FSH increases the number of LH receptors on the granulosa cells to induce synthesis and secretion of progesterone. The continued release of LH in the luteal phase of the menstrual cycle facilitates production of estradiol and progesterone by the corpus luteum. Increasing levels of circulating estradiol and progesterone cause β-endorphins to inhibit GnRH pulse frequency. As LH secretion decreases, the corpus luteum ceases to function, estrogen and progesterone levels fall, and the endometrium is shed.

Destruction of the pituitary gland by trauma, tumor, or infarction results in hypogonadotropic, hypoestrogenic amenorrhea. The abnormal ratio of LH to FSH seen in polycystic ovarian disease may be explained by the estrogen feedback mechanism on the pituitary. The relatively high circulating levels of estrogen from the peripheral conversion of androgens may inhibit the FSH production required to bring about full follicular maturation and LH receptor development. The higher basal levels of estrogen cause increased release of LH that induces further androgen production, which in turn inhibits follicular development. The result is hypergonadotropic, hyperestrogenic amenorrhea and the histologic picture of multiple follicles in various stages of arrested development.

Ovaries

The ovaries are formed by the migration of primordial germ cells from the primitive gut region of the developing fetus. Each ovum is surrounded by a layer granulosa cells that are surrounded by stroma or theca cells. There are approximately 7 million primordial follicles before birth; most undergo atresia until they are totally depleted in the fifth decade of a normal woman's life. Throughout a woman's reproductive life only about 400 follicles will become dominant follicles and release mature oocytes. The endocrine function of the dominant follicle defines the normal menstrual cycle.

Primordial follicle growth to a primary follicle is independent of gonadotropin stimulation. It is not known what regulates the number of follicles that undergo growth in each cycle, but it appears to be proportional to the number of follicles present. For example, removal of an ovary does not result in an early menopause. Also not known is how follicles are chosen. It appears that more resistant follicles are present at the end of the reproductive life, thus explaining the higher basal levels of gonadotropins and decreased fertility in older women. The primary follicles acquire FSH receptors, and as the FSH level rises in the first 7 days of the cycle the primary follicles begin to grow. Normal follicular maturation to the dominant follicle is dependent on a delicate balance between FSH and LH levels, the development of FSH and LH receptors, and the production of androgens and estrogens.[7]

The development of the dominant preovulatory follicle is characterized by the change from an androgen to an estrogen intrafollicular environment, which is dependent on high FSH and low LH. The LH binds to the theca cells that produce the androgen androstenedione from cholesterol. The androstenedione is then carried to the FSH-stimulated granulosa cells, which convert it to estradiol by aromatization. As the dominant follicle outgrows its cohorts, it sequesters more FSH, which increases aromatase production and enhances the conversion of androstenedione to estradiol. The higher estrogen further increases the FSH receptors. The disproportionately high surrounding androgen levels and the low intrafollicular estrogen concentration of the smaller follicles further inhibit their development. The rising estrogen from the dominant follicle exerts its negative feedback on the pituitary, which inhibits FSH production and deprives the smaller follicles of the FSH needed for their development. The dominant follicle requires less FSH because of its increased sensitivity and therefore continues growing. At the time of peak estrogen production by the dominant follicle, the pituitary has become sensitized for the LH surge. Ovulation occurs about 36 hours after the onset of the LH surge. The FSH surge that accompanies the LH surge increases the LH receptors on the granulosa cells, enabling them to produce progesterone. Continued LH production is necessary for normal corpus luteum function. Because of the negative feedback of estrogen and progesterone on LH production, corpus luteum function begins to wane 8 to 10 days after ovulation. This leads to orderly shedding of the endometrium about 14 days after ovulation. If pregnancy occurs, the corpus luteum is rescued by the placental human chorionic gonadotropin (hCG) that appears in the circulation about 6 to 7 days after ovulation, when implantation occurs. The hCG stimulates corpus luteum steroid production for approximately 7 to 8 weeks; then this function is taken over by the placenta.

Factors that may alter the responsiveness of the ovary to normal hypothalamic and pituitary signals include premature depletion of primordial follicles, lack of appropriate receptor development or response, and an inappropriate response, which might occur in the theca cell dominant polycystic ovary.

Uterus

The bleeding of menstruation is from the endometrium, which is totally estrogen dependent. If the endometrium "sees" no estrogen, it will remain atrophic and no bleeding will occur. Estrogen stimulates endometrial proliferation, which, if unopposed, will continue to proliferate to the point of hyperplasia and even adenocarcinoma. With continued growth the endometrium requires increasing amounts of estrogen to maintain its stability. Decreases in estrogen produce estrogen breakthrough or estrogen-withdrawal bleeding. This is the mechanism of bleeding in anovulatory women; this is also why an oligomenorrheic woman should be evaluated in the same manner as a woman with secondary amenorrhea due to chronic anovulation.

Progesterone antagonizes the further proliferation of the endometrium and brings about the differentiation into the secretory phase required for embryo implantation. Whenever progesterone is withdrawn from the estrogen-primed endometrium, bleeding will occur. Progesterone binds to endometrial progesterone receptors that are present only after they have been induced by estrogens.[8] Therefore an atrophic endometrium will not bleed after given a progesterone challenge. A patient who bleeds after taking progesterone proves that she is making estrogen. This is the basis for the progesterone or progestin challenge test, one of the first steps frequently taken by the physician confronted with the patient with amenorrhea.

Causes of Secondary Amenorrhea

Secondary amenorrhea implies that prior normal cyclic menstruation associated with ovulation has occurred. This is indeed true with the majority of patients. However, to assume that this is the case does little to define the problem for the individual patient and may lead to an erroneous initial diagnosis. It may also cause further delay in the appropriate evaluation and treatment of a chronic problem that had its onset at puberty. Dividing the causes of amenorrhea into categories representing the anatomic, ovarian, and endocrinolgic functions of the reproductive system is useful[9] (Table 6.1).

TABLE 6.1. Causes of secondary amennorrhea.

I. Anatomic Causes
   A. Pregnancy
   B. Intrauterine adhesions
      1. Traumatic
      2. Infectious

II. Ovarian Causes
   A. Menopause
   B. Iatrogenic
      1. Ovarian surgery
      2. Irradiation
      3. Chemotherapy
   C. Viral infections
   D. Genetic
   E. Autoimmune
      1. Polyglandular
      2. Ovarian autoantibodies
   F. Enzymatic defects (adult onset)
   G. Idiopathic

III. Endocrine Causes
   A. Neuroendocrine disorders
      1. Hypothalamic chronic anovulation
         a. Anorexia nervosa
         b. Simple weight loss
         c. Exercise
         d. Psychogenic stress
         e. Systemic illnesses
      2. Hypopituitarism
         a. Tumors
         b. Trauma or Ischemia
      3. Hyperprolactinemia
   B. Inappropriate feedback
      1. Androgen excess
         a. Adrenal
         b. Ovarian (PCOD)
      2. Androgen- or estrogen-producing neoplasms
      3. Liver & renal disease
      4. Obesity
   C. Other endocrine disorder
      1. Thyroid dysfunction
      2. Adrenal hyperfunction

Anatomic Causes

Anatomic causes of secondary amenorrhea, those that prevent menstrual bleeding, are largely due to destruction of the endometrium by trauma or infection with replacement by intrauterine adhesions (synechiae). Causes of primary amenorrhea such as congenital absence or obstruction of the uterus or vagina do not have to be considered when evaluating secondary amenorrhea.

Ovarian Causes

Ovarian failure as a cause of secondary amenorrhea includes disorders in which the ovaries are prematurely depleted of follicles and those in which the follicles do not respond to the normal endogenous gonadotropin stimulation. Primary amenorrhea may occur if the ovaries have been depleted of follicles prior to menarche.

Endocrine Causes

Endocrine dysfunction resulting in chronic anovulation causes secondary amenorrhea in women who have normal ovaries and who can be made to ovulate with appropriate therapy. Endocrine dysfunction is the most common cause of amenorrhea and includes patients with hypothalamic or pituitary dysfunction, adrenal and thyroid disorders, and inappropriate steroid feedback such as polycystic ovarian syndrome. These disorders may have their origin prior to menarche and might also manifest as primary amenorrhea.

Evaluation of the Patient

The most important aspect of the clinical evaluation of the patient who presents with secondary amenorrhea is the history and physical examination. In the majority of cases a working diagnosis can be established and confirming laboratory tests can be ordered during the patient's initial visit. A frank discussion with the patient regarding her possible problem and offering an immediate outline of the diagnostic steps to be taken should be presented. A follow-up consultation should be arranged to review the final diagnosis and to discuss treatment, which in many cases may be long term.

History

Because the patient is presenting with the complaint of cessation of her menstrual periods, it is very important to take a detailed menstrual history, including the age of menarche, previous cycle lengths, nature of prior menstrual flow, abnormalities preceding the amenorrhea, and dysmenorrhea. The history should also include whether or not the patient formerly experienced moliminal symptoms characteristic of ovulatory cycles such as abdominal bloating, breast tenderness, weight gain, and mood changes. Patients with amenorrhea due to intrauterine synechiae may continue to have premenstrual symptoms and dysmenorrhea on a cyclic basis, or they may develop cyclic pain secondary to the development of a hematometra from menstruation above scar tissue in the lower uterine segment. Usually patients with prior anovulatory bleeding do not experience molimina, and their cycles are irregular and unpredictable. Premature ovarian failure usually does not cause an abrupt cessation of menses. Preceding cycles may have become progressively more irregular, similar to those experienced by women entering menopause. In addition, patients frequently experience the typical hot flashes, flushing and insomnia common to menopause.

The history should include previous pelvic surgery, radiation, or chemotherapy. The nature of chronic illnesses or concurrent diseases should be explored, especially autoimmune diseases and other endocrine disorders. Any past or current treatment of psychiatric problems should be elicited. A pregnancy history with special attention to postpartum or postabortal complications such as curettage for retained placenta, hemorrhage, or infection should be obtained. This is especially important when menses fail to resume after a pregnancy termination. Hemorrhage and shock could result in pituitary infarction or Sheehan's syndrome. Asherman syndrome or intrauterine synechiae is most commonly associated with a pregnancy-related curettage.[10]

Past contraceptive use is rarely a cause of secondary amenorrhea. Although infections associated with intrauterine devices (IUDs) have been associated with Asherman syndrome, they are uncommon and should rarely cause amenorrhea. At one time, oral contraceptives were thought to cause amenorrhea in a small percentage of users after those users stopped taking them. This is now known not to be a cause but may have masked an underlying problem. Almost all women who developed post-pill amenorrhea had menstrual irregularities prior to the use of oral contraceptives[11], and in fact may have been placed on the pills to "regulate their periods." Long-acting intramuscular progestin therapy such as DepoProvera can cause prolonged amenorrhea of many months' duration, but menses will eventually resume without treatment.

A family history of amenorrhea, genetic anomalies, autoimmune diseases, or premature menopause is important. Many cases of premature ovarian failure appear to be X-linked chromosomal disorders.[12] Others may be due to the development of autoantibodies that may affect multiple organ systems.[13]

The review of systems is in many ways an inquiry into the patient's physiologic response to an alteration in her endogenous hormonal milieu. She performs her own endocrine bioassay by noting changes in hormone-sensitive organ systems. Increased androgens may cause acne, hirsutism, temporal balding, deepening of the voice, increased muscle mass, and decreased breast size. Besides the vasomotor changes caused by decreased estrogen, she may also have developed dysuria and dysmenorrhea or may have experienced stress fractures due to osteoporosis. Hyperprolactinemia caused by an enlarging pituitary prolactinoma may cause headaches or visual changes in addition to galactorrhea. Thyroid and adrenal disorders also have characteristic symptoms.

The history is not complete without an inquiry into the patient's lifestyle. Especially important are diet and exercise habits, including a history of binge eating, purging, or bulimia. Compulsive or competitive exercise with low body-fat composition has been associated with hypothalamic amenorrhea. Psychological and emotional stresses and changes in the environment may also contribute to chronic hypothalamic anovulation and amenorrhea.[14] Obesity, usually from childhood or puberty, is frequently seen in women with polycystic ovarian disease (PCOD). But it is important to keep in mind that PCOD is highly variable in its clinical presentation and that not all PCOD patients are hirsute or obese.

Physical Examination

A complete physical examination should be performed with special attention directed to hormone-sensitive end organs. Skin texture and hair distribution should be noted. Loss of skin turgor and thinning of secondary sexual hair occurs with ovarian failure. Scaly skin and brittle nails or the presence of striae may be a clue to thyroid or adrenal disease. Male secondary sexual terminal hair pattern may indicate an excess androgen source from PCOD or a virilizing ovarian tumor. The vulva and vagina, including the clitoris, are very sensitive to both estrogens and androgens. Atrophy is present in hypoestrogenic states whether of ovarian or hypothalamic-pituitary causes. The vagina is well stimulated with PCOD, but despite the excess androgen the clitoris is rarely enlarged.

The pelvic examination is completed by noting the presence or absence of estrogen-dependent cervical mucus, uterine enlargement, or ovarian masses. The uterus may be enlarged if a hematometra is developing, although this is rare with Asherman syndrome. The ovaries might be bilaterally enlarged with PCOD or they may be normal in size. As a possible estrogen- or androgen-producing neoplasm, unilateral enlargement must be taken seriously.

The body dimensions and habitus are also important to note. The extreme cachexia of severe hypothalamic amenorrhea states such as anorexia nervosa is obvious. A masculine habitus can be seen with PCOD, virilizing tumors, or Cushing's disease, but there is less muscle mass and thinner skin with striae with Cushing's disease than with PCOD or virilizing tumors. The stigmata associated with Turner's syndrome are not as obvious in the Turner mosaic patient, who is likely to present with secondary amenorrhea.

The breast development by Tanner staging and the presence of any nipple secretions should be assessed. The presence of areolar hair alone should not be considered a sign of masculinization. Loss of breast mass can occur with hypoestrogenic and virilizing states. Patients with PCOD have full or large breasts.

The remainder of the physical examination is directed toward illnesses elicited in the history. The thyroid should always be examined for enlargement or nodules.

Progesterone Challenge Test

Following the initial evaluation, a progesterone challenge may be indicated to assess the level of endogenous estrogen and the competence of the endometrium. A course of a progestational agent is administered using either intramuscular progesterone in oil (200 mg) or oral medroxyprogesterone acetate (10 mg of Provera) for 5 to 10 days. In the patient with a responsive endometrium and an intact outflow tract, the occurrence of any bleeding indicates the presence of estrogen and the likely diagnosis of chronic anovulation. Unfortunately, patients with Asherman syndrome or with very low levels of estrogen may have scanty withdrawal bleeding. If bleeding occurs, it usually begins 2 to 7 days after the conclusion of the medication. The progestational agent may occasionally trigger ovulation in an anovulatory patient, and the withdrawal bleeding may occur 2 weeks later. The patient with chronic anovulation and excessive endometrial stimulation may have alarmingly heavy withdrawal bleeding.

If the patient has a negative progesterone challenge test, then estrogen and progesterone may be administered; this should induce bleeding if the endometrium is normal. The usual dose is conjugated estrogen (2.5 mg of Premarin) orally for 25 days, with 10 mg of Provera given the last 10 days. This will usually distinguish the patient with hypoestrogenism from ovarian failure or hypothalamic amenorrhea from Asherman syndrome. The result of the progesterone challenge test is not always definitive and should not be used as the sole diagnostic test in evaluating the patient with secondary amenorrhea.

Laboratory Tests

Laboratory tests should be performed to confirm the clinical impression. Extensive endocrine assays are not indicated in the patient with cyclic dysmenorrhea and moliminal symptoms and a history of a pregnancy-related curettage. She would be better served by immediate evaluation of her endometrial cavity. A pelvic ultrasound might be useful if a hematometra is suspected. The definitive test for intrauterine synechiae is the hysterosalpingogram.[15] This is best performed using a cannula with an adjustable cervical cone or a suction cap so that the entire cervical canal and lower uterine segment can be visualized. A hysterosalpingogram should be done prior to definitive hysteroscopy. Having the patient record her basal body temperature may confirm that she is ovulatory and can allow scheduling of her definitive hysteroscopic procedure during the proliferative phase of the cycle.

Although it is generally possible to form strong clinical impressions about the etiology of the amenorrhea, without laboratory testing it is often difficult to distinguish chronic anovulation from ovarian failure. Hypothalamic chronic anovulation is a heterogeneous group of disorders with similar manifestations. Stress, diet, exercise, body-fat composition, environment, and lifestyle affect different individuals variably. The diagnosis of hypothalamic chronic anovulation is often by exclusion after laboratory testing.

Once pregnancy and anatomic causes of amenorrhea have been ruled out, laboratory tests should be ordered to distinguish between ovarian failure and endocrine causes of amenorrhea. Initial tests of basal levels of LH, FSH, prolactin, and thyroid-stimulating hormone (TSH) will allow the categorization of virtually all amenorrheic women.

An elevated TSH level may indicate primary hypothyroidism, which may be associated with amenorrhea even when the tetraiodothyronine (T4) and triiodothyronine (T3) levels are in the normal range.[16] Basal prolactin levels may also be elevated in hypothyroidism because of the increased secretion of thyrotropin-releasing hormone, which is stimulatory to prolactin release. Because of hyperplasia of the pituitary thyrotrophs, radiographic enlargement of the pituitary gland may also be seen.

If the initial prolactin level is elevated (greater than 25 ng/mL), the patient's medication history should be reviewed because certain psychotropic drugs may cause hyperprolactinemia. The prolactin level should be repeated in the morning with the patient in a relaxed, fasting state, because it may be falsely elevated by stress, exercise, anxiety, lack of sleep, or even a breast examination. If the prolactin level is again elevated, the patient should be evaluated for a pituitary prolactinoma. While it is true that the higher the prolactin level (greater than 100 ng/mL), the greater the chance of a pituitary tumor, tumors have been found with relatively low immunoreactive prolactin levels. A baseline radiographic test seems warranted. The most reliable imaging test available today is nuclear magnetic resonance test. An elevated prolactin level may be seen in about one-third of women with secondary amenorrhea; however, only one-third of patients with hyperprolactinemia will have galactorrhea because the hypoestrogenism associated with the amenorrhea prevents the breast from responding to the prolactin. When secondary amenorrhea and galactorrhea are present, about one-half of patients will have pituitary enlargement. About one-third of women with galactorrhea have normal menses and normal prolactin levels.[17]

Increased FSH levels (greater than 40 mIU/mL) indicate ovarian failure. All patients under the age of 30 with the diagnosis of ovarian failure should have karyotype determination because of the high incidence of malignancy in the gonad of an individual with Y-chromosome mosaicism. This is true even in the patient who seems normally developed. The only exceptions might be in the patient with a known cause of ovarian failure such as radiation or chemotherapy.[18] Amenorrhea with elevated gonadotropins after age 30 is termed premature menopause. Genetic evaluation is unnecessary because tumors in these patients rarely occur.

Repeat FSH, LH, and estradiol testing may be useful if there is any doubt. The LH is frequently in the normal range, especially in the early transition to the menopause. An occasional estradiol level above the menopausal range (greater than 50 pg/mL) is seen in the woman with some residual follicles. A high FSH level may occasionally stimulate the follicle to produce estrogen, which in turn enhances its own development. Pregnancies have been reported in these women who appear to have intermittent menopause. The resistant or insensitive ovary syndrome is a form of ovarian failure with elevated gonadotropins in which apparently normal follicles are present in the ovary. The mechanism is thought to be an absence or lack of response of the gonadotropin receptors.[19] A deep ovarian biopsy by laparotomy is the only way to determine the presence of follicles. Because pregnancies have occurred in patients with no follicles on biopsy and because pregnancies even in the presence of follicles are rare, it is not recommended that the patient be subjected to surgery.

Because of the association of premature ovarian failure with other autoimmune endocrine disorders, a general laboratory profile of other endocrine functions should be performed. Antithyroid antibodies should be determined to rule out thyroiditis. To check for hypoparathyroidism, calcium and phosphorus levels should be determined. To exclude hypoadrenalism and diabetes mellitus, morning fasting cortisol and glucose levels should be determined. Additional screening tests for autoimmune disorders include sedimentation rate, antinuclear antibody, rheumatoid factor, serum proteins, and complete blood count. Specific antiovarian antibodies might also be obtained through a reference laboratory.[20]

When the FSH, TSH, and prolactin levels are normal, further tests are guided by the clinical impression. A LH:FSH ratio of greater than 2.5:1 confirms PCOD, but many PCOD patients have a lower ratio even though their FSH and LH levels are in the high normal range. Approximately 20% of PCOD patients will also have a modestly elevated prolactin. If PCOD is suspected, total and free testosterone and dehydroepiandrosterone sulfate (DHEAS) tests should be done; most patients with PCOD will have elevated levels even if they are not clinically manifest. Conversely, patients with hirsutism may have apparently normal circulating androgen levels because of differences in metabolic clearance rates, sex hormone binding globulins, and end-organ sensitivities.[21] Total testosterone levels greater than 200 ng/mL or DHEAS levels greater than 7.0 ng/dL should lead to the suspicion of an androgen-producing tumor. Testosterone is most often implicated with ovarian tumors and DHEAS with adrenal tumors, but a definitive evaluation is indicated.

By exclusion, all remaining patients with low or normal FSH and LH levels fall into the category of hypothalamic hypogonadotropic amenorrhea. With the exception of patients with anorexia nervosa, most have normal levels. The anorexia nervosa patient represents the extreme in hypothalamic dysfunction with disturbances of appetite, thirst, water conservation, temperature, sleep, autonomic balance, and endocrine secretion; as reflected in endocrine studies of these patients FSH and LH are low or undetectable. Cortisol is elevated due to decreased clearance with normal production rates. Prolactin, TSH, and T4 are normal, but T3 is low and reverse T3 is high. Many of the symptoms of anorexic nervosa patients are those of hypothyroidism: constipation, cold intolerance, bradycardia, hypotension, dry skin, and hypercarotenemia.

Treatment of Secondary Amenorrhea

After the etiology of the amenorrhea has been determined and confirmed by appropriate laboratory tests, rational and appropriate treatment can be instituted.

Uterine

Intrauterine synechiae (Asherman syndrome) accounts for virtually all causes of uterine secondary amenorrhea. Modern treatment involves hysteroscopic resection of the scar tissue.[22] Blind dilation and curettage is to be condemned because further destruction of the endometrium is likely to occur. For the same reason the best resection is by sharp dissection, avoiding the use of lasers and electrosurgery that offer no distinct advantage. The scar tissue rarely bleeds, although dissection beyond the margins of the scar into the myometrium can occur. To guide the dissection and to avoid inadvertent uterine perforation, concurrent laparoscopy is recommended when hysteroscopic resection of extensive adhesions is undertaken. Should bleeding occur, tamponade with a 3-cc balloon catheter will usually control it.

Successful results are dependent on the endometrium's amazing regenerative ability. To promote endometrial proliferation, postoperative estrogen in pharmacological doses is recommended. A typical regimen is conjugated 5 to 10 mg of estrogens daily for 25 days for two cycles. A progestin (5 mg of Provera) during the last 10 days of the estrogen can be given to ensure orderly withdrawal bleeding. As previously mentioned, to take advantage of the patient's own endogenous estrogen, it is beneficial to time the surgery during the patient's early proliferative phase as determined by basal body temperature charting. Formerly, the use of intrauterine devices as a splint to separate the uterine walls was recommended. The devices available today may in fact inhibit the growth of the endometrium. While the theory behind a splint seems logical, there is little evidence to support its use. If a splint is to be used, an inflatable balloon catheter filled with no more than 3 cc of fluid should be used. It should be removed in 3 to 5 days. Infections associated with hysteroscopic surgery are uncommon, but prophylactic antibiotics are recommended.

A follow-up hysterogram in 2 months will assess the results of the hysteroscopic resection. To remove any residual scar, especially in the most severe cases, repeat hysteroscopic procedures may be required. Normal menses will resume in approximately 90% of patients. Pregnancy may be allowed 2 months after restoration of normal menses. Of those women who do get pregnant, 80% will have term deliveries. However, there is a higher complication rate, with most complications involving the placental site and resulting in placenta previa, placenta abruptio, placenta accreta, and retained placenta.[23] All may result in maternal hemorrhage, so it would be prudent to consider having autologous blood available at the time of delivery if possible.

Ovarian Failure (Hypergonadotropic Amenorrhea)

Women under 30 years old with karyotypes revealing the presence of a Y chromosome should have the dysgenetic gonads removed to avoid the 25% incidence of cancer. If a polyglandular autoimmune disorder has been diagnosed as the etiology to the ovarian failure, appropriate replacement therapy should be instituted. Occasionally women will begin to menstruate normally as soon as other endocrine function has been restored.

Unfortunately, the majority of women with premature ovarian failure have irreversibly lost both their reproductive capability and their ovarian endocrine function. Even in the woman with apparent intermittent ovarian function, hormone replacement therapy is indicated. It is important to prevent the consequences of estrogen deprivation such as osteoporosis and premature heart disease as well as to treat the immediate signs and symptoms of hypoestrogenism. Many different regimens for hormone replacement therapy exist, but to avoid endometrial hyperplasia, which is associated with the use of continuous unopposed estrogen, a progestin agent must be used for at least 12 days each month.[24] To control their menopausal symptoms, younger patients might also require higher doses of estrogen than the normal menopausal woman.

Almost all women who have conceived have done so while on estrogen replacement treatment. Perhaps the estrogen suppresses the high FSH and stimulates the restoration of receptors in the few remaining responsive follicles. Pregnancies with human menopausal gonadotropin (hMG)–human chorionic gonadotropin (hCG) therapy have also been reported, as well as successful pregnancies with hMG-hCG following suppression of the elevated FSH and LH with gonadotropin hormone-releasing hormone agonists (GnRH agonists).[25] The latter is a very costly therapy with a very low success rate. In the woman desiring to carry a pregnancy, oocyte donation with embryo transfer to her uterus offers the best opportunity.

Contraception should be used in patients who do not desire pregnancy because pregnancy has occurred in up to 10% of individuals diagnosed with premature ovarian failure. Birth control pills can be used as a satisfactory form of hormone replacement and may be more acceptable psychologically to the younger patient.

Hypothalamic Amenorrhea (Hypogonadotropic Anovulation)

Hypothalamic amenorrhea has its origin from multiple factors; the most common are emotional or physical stress and low body weight. Patients with hypothalamic amenorrhea all have low or normal gonadotropins and are hypoestrogenic. Their ovaries are capable of normal ovulation and estrogen production. Treatment should be directed at correcting the underlying problem and may require psychological counseling. In the extreme case of anorexia nervosa, inpatient psychiatric care and nutritional support may be needed; a 5%

to 15% mortality rate has been reported in anorexia nervosa patients, and relapses are frequent.

The critical weight theory states that the onset of menarche and maintenance of regular menses depends on maintaining a critical weight with a critical amount of body fat.[26] The minimum body fat at age 13 to initiate menarche is 17%, and to maintain regular menses at age 16, it is 22%. A loss of body weight in the range of 10% to 15% results in a loss of one-third of the total body fat and may result in amenorrhea. The competitive female athlete has about 50% less body fat than the noncompetitor; this may place an athlete under the 22% critical level. Because the fat is replaced with lean muscle mass, this may occur without a change in total body weight.

The critical weight theory does not explain all cases of amenorrhea; therefore other factors may be additive to the hypothalamic suppression. It is probable that opioids from central nervous system (and perhaps other distant organs) may act to suppress GnRH release. After exercise β-endorphin levels are increased, which may explain the suppression of GnRH, and prolactin levels are also increased, which may be explained by the β-endorphin suppression of dopamine that inhibits the release of prolactin.

Treatment directed at educating the patient about the effect of weight, diet, and exercise may allow her to change her lifestyle stresses and to restore normal menstruation. In the interim she needs to be treated with hormone replacement therapy similar to the menopausal patient with ovarian failure. The patient who is sexually active but does not desire pregnancy should be counseled that ovulation may resume spontaneously and therefore should use contraception. Oral contraceptive agents can be used for hormone replacement. Patients should also be informed that the amenorrhea may persist after therapy is discontinued. To prevent osteoporosis, additional calcium supplementation should also be recommended, even in the physically active individual. If osteoporosis is a concern, bone density can be evaluated by computerized tomography (CT) scan or dual-photon absorptiometry.

For the patient desiring pregnancy, initial attempts at stimulating ovulation should utilize clomiphene citrate (CC). While only a few patients will respond to CC therapy, the low risk and low cost justify its trial. Those patients who fail CC treatment will readily respond to hMG-hCG therapy and need to be monitored carefully to avoid ovarian hyperstimulation syndrome. An alternative approach is to use pulsatile gonadotropin-releasing hormone (GnRH) therapy.[27]

Pituitary Amenorrhea

The most common pituitary tumor is the prolactin-secreting adenoma which will be discussed under hyperprolactinemic amenorrhea. Other pituitary tumors are rare and are investigated according to the symptoms that they cause. Amenorrhea of pituitary origin may also result from pituitary ablation by previous surgery or radiation. Fortunately, pituitary infarction secondary to hypovolemic shock related to obstetric hemorrhage (Sheehan's syndrome) is rarely seen today. All result in panhypopituitarism and require total replacement therapy. Replacement of the gonadotropins is not necessary to sustain life, and therefore only menopausal hormone replacement therapy is required. Should pregnancy be desired, the ovaries can be stimulated in the same manner as detailed with the hypothalamic patient.

Hyperprolactinemic Amenorrhea

Treatment of hyperprolactinemia not associated with a demonstrable pituitary tumor is with bromocriptine, a long-acting dopamine agonist that mimics the dopamine inhibition of prolactin secretion. It is effective in restoring menses in 80% to 90% of hyperprolactinemia patients. The side effects of bromocriptine are headache, nausea, and faintness due to orthostatic hypotension. The hypotension has both peripheral and central origins; relaxation of the vascular smooth muscle and decreased sympathetic autonomic nervous activity result. Side effects can be minimized by starting with low doses given at bedtime and gradually increasing to therapeutic levels. The usual starting dose is one tablet (2.5 mg), and the usual therapeutic level is achieved at 5 to 7.5 mg per day in divided doses. For patients who are unusually sensitive, intravaginal absorption following the same dosages may be tried to avoid the nausea and vomiting or a starting dose of 1.25 mg may be tried. Response is usually seen with normalization of the prolactin level in 2 to 4 weeks and return of menses in 6 to 8 weeks.

The treatment of prolactin-secreting adenomas is somewhat controversial.[28] Most microadenomas (less than 10 mm in diameter) are treated with bromocriptine as described above with essentially the same results. The risk of a microadenoma progressing to a macroadenoma is less than 5%. Therefore the patient with idiopathic hyperprolactinemia or a microadenoma has no urgent need for any treatment unless pregnancy is desired. If observation is elected, her prolactin levels should be periodically evaluated and if there is a distinct rise in the levels, repeat radiographic studies should be performed. Because the patient is likely to be hypoestrogenic, hormone replacement therapy should be encouraged.[29] Surgical treatment of microadenomas has an initial cure rate of 80% to 85% but has a late recurrence rate of 20%.

The treatment of macroadenomas is a challenging dilemma. The surgical cure rates are less than 50%, with recurrence rates of 20% to 50%. Bromocriptine is very effective in shrinking macroadenomas; evidence of size reduction occurs by the sixth week. Size reduction does not correlate with pre- or post-treatment prolactin levels or with the percent reduction in the prolactin levels. When bromocriptine is discontinued, most adenomas will reexpand to pretreatment size. The dilemma is whether to operate on the shrunken tumor, remove the bulk of the tumor, and then treat with bromocriptine or to keep patients on bromocriptine indefinitely. Regardless of the treatment, most patients will normalize their prolactin level, resume normal menses, and get pregnant if they desire.

Amenorrhea with Androgen or Estrogen Excess (Polycystic Ovaries)

Androgen- or estrogen-producing tumors of ovarian or adrenal origin are treated surgically. Most steroid-producing neoplasms are benign and can be associated with restoration of normal ovarian function after extirpation.

Treatment of women with secondary amenorrhea due to PCOD must take into account the patient's needs and complaints. The major concern is to prevent endometrial hyperplasia from unopposed estrogen production. The hirsute patient's primary concern is to control her abnormal hair growth.

In the amenorrheic patient with PCOD who does not desire pregnancy and who is not hirsute, therapy with an intermittent progestin (5 to 10 mg of medroxyprogesterone acetate orally for 12 to 14 days each month) or oral contraceptives if she is sexually active will prevent endometrial hyperplasia. Because ovulation can occasionally occur, those women electing progestin therapy should be counseled to use barrier contraception.

Therapy for women with hirsutism must be individualized. In the patient with only fine, scattered terminal hair, birth control pills may act to inhibit the LH secretion by the pituitary and therefore the excess androgen production by the ovaries. In addition, oral contraceptive agents may work by increasing the sex hormone binding globulins, thus reducing the amount of free androgen that reaches the end organ. Oral contraceptives may also slow the progression of hair growth in the patient with significant hirsutism, but the effect may not be very apparent to her. Spironolactone (100 to 200 mg per day) has also been shown to be effective. Spironolactone probably works by not only decreasing androgen synthesis by the ovary but also by blocking its action on the hair follicle and sebaceous gland. Patients must realize that established hair will not disappear, but the texture may be finer and the growth slower. The effect or lack of effect will not be apparent in most patients sooner than 6 months after therapy is initiated. Additional cosmetic treatments such as bleaching, electrolysis, plucking, waxing, and depilation should be encouraged.[30]

Clomiphene citrate is the initial treatment for patients with PCOD who desire pregnancy. Relatively high doses may be required in the obese patient. Ovulation occurs in 75% to 80% of patients who take clomiphene citrate; however, the pregnancy rate is only 35% to 40%. This may reflect the adverse effect that high intrafollicular androgens have on the oocyte. Failure with clomiphene citrate warrants the use of hMG-hCG. Encouraging results with human purified FSH have also been achieved in patients with elevated basal levels of LH. Occasionally, shutting down the endogenous hypothalamic input with a long-acting GnRH agonist allows for more predictable control of hMG-hCG stimulation.[31]

Classical ovarian wedge resection of the ovaries should be performed only rarely today. It is effective in only about 50% of patients, and the effect may only be temporary. Pelvic adhesions following this surgery may also adversely affect future fertility. Laparoscopic follicular puncture with electrocautery or laser ablation of surrounding theca interna may be as effective as wedge resection and less risky. Following this procedure certain patients who were resistant to ovulation induction may respond.[32] Unless there is another reason to perform a laparoscopy, medical treatment should always be attempted first.

References

1. Treloar AE, Boynton RE, Benn BG, et al. Variation of the human menstrual cycle through reproductive life. *Int J Fert*. 1975; 12:77–126.
2. Knobil E. The neuroendocrine control of the menstrual cycle. *Recent Prog Horm Res*. 1980;36:53–88.
3. Reid RL, Van Vergt DA. Neuroendocrine events that regulate the menstrual cycle. *Contemp OB/GYN*. 1987;28:147–155.
4. Gindoff PR, Ferin M. Brain opioid peptides and menstrual cyclicity. *Semin Reprod Endocrin*. 1987;5:125–133.
5. Yen SSC. Chronic anovulation due to CNS-Hypothalamic-Pituitary dysfunction. In: Yen SSC, Jaffee RB, eds. *Reproductive Endocrinology*. Philadelphia; Pa: WB Saunders Company; 1991:631–688.
6. Yen SSC. The human menstrual cycle: neuroendocrine regulation. In: Yen SSC, Jaffee RB, eds. *Reproductive Endocrinology*. Philadelphia; Pa: WB Saunders Company; 1991:273–308.
7. Speroff L, Glass RH, Kase NG. Regulation of the menstrual cycle. In: Speroff L, Glass RH, Kase NG, eds. *Clinical Gynecologic Endocrinology and Infertility*. Baltimore; Md: Williams & Wilkins Co; 1983:75–100.
8. Strauss JF, Gurpide E. The endometrium: regulation and dysfunction. In: Yen SSC, Jaffee RB, eds. *Reproductive Endocrinology*. Philadelphia; Pa: WB Saunders Company, 1991:309–356.
9. American College of Obstetricians and Gynecologists: Amenorrhea (ACOG Technical Bulletin 128). Washington DC, 1989.
10. March CM. Intrauterine and cervical pathology. In: Buttrum VC, Reiter RC, eds. *Surgical Treatment of the Infertile Female*. Baltimore; Md: Williams & Wilkins Co; 1985:249–287.
11. Jacobs HS, Knuth UA, Hull MGR, et al. Postpill amenorrhea: cause or coincidence? *Br Med J*. 1977;ii:940–942.
12. Coulam CB. Premature gonadal failure. In: Wallach EE, Kempers RD, eds. *Modern Trends in Infertility and Conception Control*. Chicago: Year Book Publishers; 1985:84–94.
13. Coulam CB. Autoimmune ovarian failure. *Semin Reprod Endocrin*. 1983;1:161–167.
14. Yamamora DLR, Reid RL. Psychological stress and the reproductive system. *Semin Reprod Endocrin*. 1990;8:65–73.
15. Fayez JA, Mutie G, Schneider PJ. The diagnostic value of hysterosalpingography and hysteroscopy in infertility investigation. *Am J Obstet Gynecol*. 1987;156:558–560.
16. Thomas R, Reid RL. Thyroid disease and reproductive dysfunction: a review. *Obstet Gynecol*. 1987;70:789–798.
17. Molitch ME. Management of prolactinomas. *Ann Rev Med*. 1989;40:225–232.
18. Averette HE, Boike GM, Jarrell MA. Effects of cancer chemotherapy on gonadal function and reproductive capacity. *CA*. 1990;40:199–209.

19. Maxson WS, Wentz AC. The gonadotropin resistant ovary syndrome. *Semin Reprod Endocrin*. 1983;1:147–160.
20. Ho PC, Tang GWK, Fu K, et al. Immunologic studies in patients with premature ovarian failure. *Obstet Gynecol*. 1988;71:622–626.
21. Kazer RR, Rebar RW. Polycystic ovary syndrome. In: Sciarra JJ, ed. *Gynecology and Obstetrics*. Philadelphia; Pa: Harper & Row, Publishers; 1988:1–10.
22. March CM. Hysteroscopy. *J Reprod Med*. 1992;37:293–312.
23. Friedman A, DeFazio J, DeCherney A. Severe obstetric complications after aggressive treatment of Asherman syndrome. *Obstet Gynecol*. 1986;67:864–867.
24. Gambrell RD Jr. Prevention of endometrial cancer with progestins. *Maturitas*. 1986;8:159–168.
25. Kreiner D, Droesch K, Navot D, et al. Spontaneous and pharmacologically induced remission in patients with premature ovarian failure. *Obstet Gynecol*. 1988;72:926–928.
26. Frisch RE. Food intake, fatness, and reproductive ability. In: Vigersky RA, ed. *Anorexia Nervosa*. New York: Raven Press; 1977:149–160.
27. American College of Obstetrician and Gynecologists. *Medical Induction of Ovulation* (ACOG Technical Bulletin 120). Washington, DC; 1988.
28. Vance ML, Thorner MO. Prolactinomas. In: Molitch, ed. *Pituitary Tumors: Diagnosis and Management*. Endocrinology and Metabolism Clinics of North America. 1987;16:731–753.
29. Klibanski A, Biller BMK, Rosenthal DI, et al. Effects of prolactin and estrogen deficiency in amenorrheic bone loss. *J Clin Endocrinol Metab*. 1988;67:124–130.
30. American College of Obstetricians and Gynecologists. *Evaluation and Treatment of Hirsute Women* (ACOG Technical Bulletin 103). Washington, DC; 1987.
31. Filicori M, Campaniello E, Michaelacci L, et al. Gonadotropin-releasing hormone (GnRH) analog suppression renders polycystic ovarian disease patients more susceptible to ovulation induction with pulsatile GnRH. *J Clin Endocrinol Metab*. 1988;66:327–333.
32. Greenblatt E, Casper RF. Endocrine changes after laparoscopic ovarian cautery in polycystic ovarian syndrome. *Am J Obstet Gynecol*. 1987;156:279–285.

7
Pelvic Inflammatory Disease

M. LeRoy Sprang

Pelvic inflammatory disease (PID) is a general term connoting inflammation caused by an infection in the upper genital tract. It is a disease that has reached epidemic proportions in the 1990s and may include infection of any or all the following: endometrium, myometrium, oviducts, ovaries, uterine serosa, parametrium, or pelvic peritoneum. Pelvic inflammatory disease is inarguably an extremely serious reproductive health problem. Each year, more than 1 million U.S. women experience an episode of PID,[1,2] with at least one-fourth of them suffering one or more serious, long-term sequelae.[3-5]

Pelvic inflammatory disease generates approximately 2.5 million outpatient visits each year. Approximately 200,000 women are hospitalized annually for PID, with over 100,000 surgical procedures performed annually.[1,2] Washington et al estimate that the direct cost for PID and PID-associated ectopic pregnancy and infertility in the United States in 1990 was $2.7 billion, and indirect costs were estimated to be $1.5 billion, for a total cost of $4.2 billion (Table 7.1).[6] If the current PID incidence persists, in the year 2000, costs associated with PID are projected to approach $10 billion (Fig. 7.1).[6] Better understanding, prevention, early aggressive diagnosis, and treatment of PID are all needed both to reduce human suffering and to contain the rising costs of PID.

TABLE 7.1. Estimated indirect cost of PID and sequelae, 1990.*

Outcome	Lost wages, $**	Lost value of household management, $**	Total, $**
Pelvic inflammatory disease			
Outpatient ($n = 1,248,000$)	314.77	441.60	756.37
Hospitalized ($n = 174,100$)	105.18	147.16	252.34
Ectopic pregnancy ($n = 44,000$)	33.81	44.39	78.20
Infertility ($n = 249,600$)	87.92	125.47	213.39
Subtotal	541.68	758.62	1,300.30
Deaths†			208.10
Total			1,508.40

* Adapted from *JAMA*, November 13, 1991; 266:2565–2569.
Copyright 1991, American Medical Association.
* All figures are millions of dollars
† Total mortality costs are due to 675 deaths from pelvic inflammatory disease and 15 from ectopic pregnancy associated with pelvic inflammatory disease (cost based on estimate of lost wages and cost of household management).

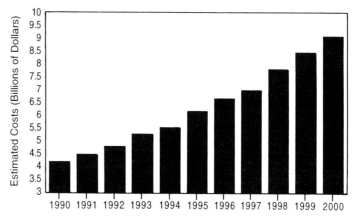

FIGURE 7.1. Projected total cost of PID to the year 2000.

Adapted from JAMA, November 13, 1991 266:2565-2569. Copyright 1991, American Medical Association.

Etiology

The majority of cases of PID are caused by *Neisseria gonorrhoeae* or *Chlamydia trachomatis*.[3,7-9] Coinfection with *N gonorrhoeae* and *C trachomatis* also occurs with PID.[8] In many series 25% to 50% of cases do not have detectable chlamydial or gonococcal infection[10-13]; other organisms contribute to PID. PID is usually a polymicrobial infection, and wide variety of anaerobic and aerobic (facultative) bacteria have been isolated from the upper genital tracts of women with PID.[10-13] Common anaerobic isolates include bacteroides, peptostreptococcus, and peptococcus species, while *Gardenerella vaginalis*, Streptococcus species, *Escherichia coli*, and hemophilus influenzas are the most common facultative bacteria found (Table 7.2). Although they are often considered part of the normal flora of the vagina, perhaps a change in the quantity of normal flora or failure of a barrier function at the cervical-vaginal interface allows the ascendance of lower genital tract flora into the endometrial cavity.

The role of genital mycoplasmas in the etiology of PID remains elusive. Cervical cultures positive for both *Mycoplasma hominis* and ureaplasma urealyticum may be obtained from the majority of young sexually active women. The rate of isolation of genital mycoplasmas from the cervix is approximately 75%; this rate is similar in both sexually active women with PID and those without PID.[14] However, serologic studies in women with acute PID have demonstrated that approximately one in four develops a significant rise in antibodies to these organisms.[14] Nevertheless, histologically mycoplasma does not appear to produce damage to the tubal mucosa. Overall, studies suggest that mycoplasma may be a commensal bacteria rather than a pathogen in the oviducts.

Neisseria gonorrhoeae and *Chlamydia*

TABLE 7.2. Bacteriology of pelvic inflammatory disease and pelvic abscess.

Aerobic Genital Bacteria
  Group B streptococci
  Other streptococci
  Coagulase-negative staphylococci
  Escherichia coli
  Gardnerella vaginalis
  Other facultative gram-negative organisms

Anaerobic Genital Bacteria
  Peptostreptococci
  Peptococci
  Bacteroides bivius
  Black-pigmented Bacteroides
  Other Bacteroides spp.

Sexually Transmitted Organisms
  Neisseria gonorrhoeae
  Chlamydia trachomatis
  Genital mycoplasmas (?)

trachomatis are considered primary pathogens of the endocervical canal and cause mucopurulent endocervicitis, which is diagnosed by the presence of mucopus within the endocervical canal or by the presence of large numbers of leukocytes on a Gram stain of cervical mucus. The endocervical canal and mucus plug within the endocervix represent the major barriers that protect the endometrium and the rest of the upper genital tract from the vaginal flora. Infection with chlamydia or gonococci may damage the endocervical canal, break down the barriers, and permit ascending infection. While chlamydia and gonococci are primary pathogens and may cause ascending infection by themselves, the damage they cause may allow the other flora to breach the barrier and to ascend to infect the endometrium, which is thought usually to be sterile.

Several features of anatomy and physiology may combine to facilitate ascendance of infection. Chlamydia grows in columnar cells, not in squamous cells. Extension of endocervical columnar epithelium outward on the surface of the cervix, or "cervical ectopy," produces a large area of columnar epithelium that may be more susceptible to infection by chlamydia. Cervical ectopy occurs in teenagers and young adults, in whom the prevalence of PID is the highest.

Other factors play a role in determining whether lower genital tract infections persist, ascend, or are cleared. Only approximately 15% of women with *C trachomatis* or *N gonorrhoeae*, endocervicitis progress to PID. Cervical mucus changes during the menstrual cycle and may permit passage of organisms, particularly at midcycle, when the mucus is thinner and more penetrable.[15] At other times during the menstrual cycle, the mucus is more viscous and acts as a plug to seal off the uterine cavity. In addition, organisms may gain access to the uterine cavity during menses after the mucus plug has been expelled with the menstrual blood. As a result of uterine contractions during the sloughing process, organisms that reach the endometrium may be propelled into the fallopian tubes at the time of menses (retrograde menstruation).[15]

Approximately 60% of *C trachomatis* and *N gonorrhoeae* PID occur within 1 week from the first day of the last menses.

Another explanation for ascendance of cervical vaginal flora into the upper genitive tract may be negative intraperitoneal pressure ("insuck" mechanism). Other possibilities include bacteria piggybacking on motile spermatozoa[16,17] or on trichomonades.[18]

Most patients (approximately 85%) develop PID through the spontaneous mechanisms described above. However, approximately 15% of PID cases may be the result of instrumentation of the female genital tract. Procedures such as endometrial biopsy, dilation and curettage, hysterosalpingography, insertion of an intrauterine device, or abortion may introduce normal cervical-vaginal flora into the uterine cavity. Although the rate of infection from such procedures is quite low, because of the large number of these procedures performed annually, a significant number of PID cases develop through this mechanism.

Pathophysiology

It is generally recognized that *Chlamydia trachomatis* infection may progress to a severe chronic inflammatory state. When it invades the upper genital tract, infertility may result. It appears that the damage and scarring are not always a direct effect of chlamydia, but rather from the host response to the infection.[19-21] This response usually peaks by 2 weeks and returns to normal in 5 weeks, thus producing a self-limited disease process.[19] The response may be mediated by antibody and complement formation. After repeated inoculations of *C trachomatis* into the fallopian tubes, salpingitis and distal tube obstruction occur.[20] In contrast to acute infection, mononuclear cells characterize the predominant inflammatory response seen in repeated infections. Repeated inoculations cause a delayed hypersensitivity that produces extensive tubal damage and paratubal adhesions.

Gonococci, on the other hand, cause more direct damage. Gonococci are capable of invading cells that line the fallopian tube. They selectively adhere to nonciliated mucus-secreting cells, and after traversing the cell they are released from the basal surface and are thus able to invade the subepithelial space.[22] Gonococci produce a number of extracellular products, including enzymes such as phospholyipase and peptidase, that might damage host cells. Despite invasion into the nonciliated epithelial cells by gonococci, the major damage is to the uninfected ciliated epithelial cells.[22] Tissue damage to these cells may be due to at least two of the structural components of the gonococcal surface, namely lipooligosaccharides[23] and peptidoglycan.[24] During growth, the gonococcus sheds outer membrane blebs that contain the lipooligosaccharide (LOS) molecules. Possibly, the LOS molecules bind to nonciliated cells and without actually invading the cells cause damage to them. Molecules of LOS from different gonococci vary in their ability to generate complement-derived neutrophil chemoattractant (C5A).[25] This may, in fact, explain why some gonococcal organisms are more virulent than others. The strains that generate C5A signal the host to mount a PMN response at the site of infection. In the upper genital tract the influx of PMNS is the hallmark of PID.

The intense inflammatory response leads to edema and damage in the tubes. Inflamed tubal surfaces may adhere to each other and may cause occlusion or the formation of blind pouches; the result can be infertility and/or increased risk of ectopic pregnancy. If the fimbriae of the tube become agglutinated and closed, a pyosalpinx may result. After treatment, sterile debris liquefies, and as a result a hydrosalpinx is formed. If the fimbriae attach to the ovary, a tubo-ovarian abscess may form. When the patient ovulates, the follicular cyst ruptures, and organisms then have access to the ovarian stroma. A gradually enlarging abscess is formed in the ovary and may result in total destruction of the ovary. Rupture or leaking of the pus-filled structures can disseminate infection intraperitoneally. Several animal studies have shown that when *E coli*, *B fragilis*, *E faecalis*, and bacteroides bivius are grown together they act synergistically to produce abscesses. Abscesses can form in the anterior or posterior cul-de-sac, between loops of intestine, or in the perihepatic areas. Sometimes the purulent exudate migrates to the right colic gutter, up to the perihepatic area, and causes perihepatitis with adhesion formation between the anterior liver capsule and the peritoneum (banjo strings). This entity, which can present with right upper quadrant pain, is referred to as Fitz-Hugh-Curtis syndrome.

Epidemiology

To avoid PID and its sequelae it is essential to identify risk factors for sexually transmitted diseases (Table 7.3). When the clinical picture is unclear, knowledge of risk indicators may increase or decrease suspicion of PID. Risk indicators also may guide the ordering and interpretation of diagnostic tests. In addition, to reduce the risk of upper tract infection, women who are at high risk should be screened routinely for lower genital tract infections.

Age in part predicts the risk of PID. Approximately 75% of patients with PID are under 25 years of age.[14] Studies have shown that sexually active teenagers are three times more likely to acquire PID than 25- to 29- year-old women.[26] This may be accounted for by both their biological and behavioral characteristics. Compared with older women, teenagers have larger zones of cervical ectopy and, during the perimenarchal period, more penetrable cervical mucus. Behavioral differences include patterns of greater numbers of sexual partners among sexually active teenagers than among older women. Teenagers tend to have one monogamous partnership at a time but in a sequential fashion. A published survey of the Centers for Disease Control (CDC) indicates that in 1970, 29% of females aged 15 to 19 had had sex, but in 1990 the number was 54%. The 1990 survey reported that 19% of U.S. high school students has had four or more sex partners.

Low levels of education[27,28] and unemployment[27] and low income[29] are all markers for increased risk of PID. In addition,

TABLE 7.3. Risk factors for STD, PID and PID sequelae.*

Risk factor	Acquisition of STD	Development of PID	Development of PID sequelae
Demographic/Social Indicators			
Age (<20 years of age)	+	+	−
Low socioeconomic standing	+	+	*
Unmarried sexually active women	+	+	*
Urban residence	+	*	*
Individual Behavior/Practices			
Sexual behavior			
Young age at first sexual intercourse	+	*	*
Multiple sex partners	+	*	*
High frequency of sexual intercourse	+	*	*
High rate of acquiring new partners	+	*	*
Contraceptive practice			
Use of a barrier method	−	−	−
Use of hormonal contraceptives	+	−	*
Use of an IUD	*	+	+
Health-care behavior			
Late evaluation of symptoms	+	+	+
Lack of compliance with STD treatment	+	+	+
Lack of notification and treatment of partner	+	+	+
Others			
Douching	*	+	*
Smoking	+	+	*
Substance abuse	+	*	*
Menses	+	+	*

Key: + Increased Risk   − Decreased Risk   * Undetermined Association

*Adapted from *JAMA*, Nov. 13, 1991; 266:2581–2586. Copyright 1991, American Medical Association.

unmarried sexually active women are at increased risk, because they are more likely to have multiple sex partners.[28]

The precise role of sexual behavior in the development of PID remains unclear. However, several dimensions of sexual behavior have been associated with increased risk. The sexual parameter that has the highest predictive value for later PID is the age of coital debut. Women with coital debut before the 16th year had twice as many PIDs as women with sexual debut at age 18 or later.[30] A CDC survey indicated that among 15-year-old females 1 in 20 was sexually active in 1970, but 1 in 4 was sexually active in 1988. Factors such as multiple sex partners or change of sex partners within the last 30 days also increase the risk of PID. The frequency of sexual intercourse also appears to be a risk factor for PID. In one study, married women with one recent sexual partner having intercourse six or more times per week had a risk of PID of 3:2 compared to similar women having intercourse less than one time per week.[31] Intercourse during menses may also be linked to an increased risk of PID.[27]

The choice of contraception appears to affect the risk of PID as well as STD and tubal infertility. Mechanical and chemical barriers such as condoms, diaphragms, and vaginal spermicides appear to decrease risk when used correctly and consistently throughout sexual activity. Latex condoms offer greater protection than do natural membrane condoms.[32]

The relationship between oral contraceptives (OCs) and the risk of PID is complex. The association appears to be different for each link in the PID risk chain. Oral contraceptives increase the risk of *C trachomatis* infection of the endocervix and decrease the risk of PID or tubal infertility. Most studies have demonstrated a twofold to threefold increase in the prevalence of cervical *C trachomatis* infections in OC users.[33-35] Several studies, however, have shown that the rate of PID decreases by as much as 50% in OC users compared to sexually active women who do not use any contraceptive method.[27,36]

Despite 25 studies conducted worldwide, the link between the use of intrauterine devices (IUDs) and PID remains one of the most controversial topics in contemporary contraception. Most studies have found an increased risk of PID and its sequelae among IUD users, with the increase ranging from twofold to ninefold. Given the data, IUD use appears to increase the risk of PID, but because of methodological flaws in early studies, the magnitude of this association seems to have been overestimated.[37] The increased risk of PID associated with IUD use centers on the time of IUD insertion. The relative risk of PID associated with IUDs (other than the Dalkon shield) appears to be highest in the first 4 months of IUD use, and is not significantly elevated above baseline at 5 months and beyond.[38]

Vaginal douching, cigarette smoking, and menses also appear to influence the risk of PID. Vaginal douching has recently received considerable attention as a practice that may be associated with PID and ectopic pregnancy. One multivariate analysis that adjusted for cofounding variables, found that douching during the past 2 months was associated with a relative risk for PID of 1.7, with the risk being positively correlated with frequency of douching.[39] Douching may alter the vaginal environment by decreasing the body's defense mechanisms, or it may flush vaginal and cervical microorganisms into the uterine cavity, thereby increasing the risk of upper genital tract infection.

In a study that used logistic regression to control for confounding variables, current cigarette smokers had a significantly elevated relative risk for PID of 1.7 compared with women who never smoked. Similarly, former cigarette smokers had a significantly elevated risk for PID of 2.3. However, a dose-response relationship was not observed.[40] Cigarette smoking has also been associated with an increased risk of ectopic pregnancy and infertility.

A relationship between the menstrual cycle and onset of acute PID has been demonstrated. Approximately 60% of cases of gonococcal and/or chlamydial salpingitis have been reported within 1 week from the first day of the last menstrual period. Only 14% of nongonococcal nonchlamydia salpingitis patients reported onset of symptoms within 1 week from the first day of the last menses.[41] This relationship may be due to the increased penetrability of the mucus plug discussed earlier.

A thorough understanding of epidemiology will help practitioners identify patients who are at risk for PID and will allow early STD testing and early intervention. Risk assessment may also increase timely diagnosis and early intervention of patients with PID. Early treatment should result in better long-term results. Risk assessment should also lead to patient counseling about changing sexual behavior to decrease the risk of infection; here is an opportunity to educate patients on STDs, PID, and PID sequelae (Table 7.4). Finally, risk assessment can help determine the need to counsel sex partners and to facilitate the treatment of sex partners when indicated. Antibiotics should be administered presumptuously to sex partners of those patients with confirmed chlamydial or gonococcal infections. This approach interrupts the chain of transmission in a community, reduces the risk of reinfection, and reduces the risk of the development of complications.

Diagnosis

Pelvic inflammatory disease involves a great diversity of diseases, diversity of etiology, and diversity of presentations. Because of the variation in symptoms and signs, PID is difficult to diagnose. In fact, PID may stand for "pretty inexact diagnosis." In a classic study by Jacobson and Westrom acute PID was laparoscopically confirmed in only 65% of 814 patients with a clinical diagnosis of PID.[42] In 12%, other pathology (appendicitis, ectopic

TABLE 7.4. Recommendations for individuals to prevent STD/PID.*

General preventive measures	Specific recommendations	Quality of evidence supporting effectiveness of intervention
Maintain healthy sexual behavior	1. Postpone initiation of sexual intercourse until at least 2 to 3 years following menarche	III
	2. Limit number of sex partners	II
	3. Avoid casual sex and sex with high-risk partners	III
	4. Question potential sex partners about STD and inspect their genitals for lesions or discharge	III
	5. Abstain from sex if STD symptoms appear	III
Use barrier methods	Use condoms, diaphragms, and/or vaginal spermicides consistently and correctly throughout all sex for protection against STD, even if contraception is not needed	II
Adopt healthy medical care-seeking behavior	1. Seek medical treatment promptly after having unprotected sex (intercourse without a condom) with someone who is suspected of having an STD	III
	2. Seek medical care immediately when genital lesions or dischange appear	III
	3. Seek routine checkups for STD if in nonmutually monogamous relationship(s) even if symptoms are not present	III
Comply with management instructions	1. Take all medications as directed, regardless of symptoms	I
	2. Ruturn for follow-up evaluation as instructed	III
	3. Abstain from sex until symptoms disappear and appropriate treatment is completed	III
Ensure examination of sex partners	1. When diagnosed as having an STD, notify all sex partners in need of medical assessment	III
	2. If preferred, assist health providers in identifying and notifying sex partners	III

* STD represents sexually transmitted disease, and PID, pelvic inflammatory disease. The grading scheme was adopted from US Preventive Services Task Force. I indicates evidence obtained from at least one properly randomized controlled trial; II, evidence obtained from well-designed cohort or case-control analytic studies; and III, opinions of respected authorities, based on clinical experience, descriptive studies, or reports of expert committees.
From *JAMA* Nov. 13 1991. Copyright 1991, American Medical Association.

pregnancy, ruptured ovarian cysts) was present. However, in 23% of patients with a diagnosis of acute PID, the pelvis was completely normal. Therefore as many as one-third of the patients may have a false-positive diagnosis of acute PID . Moreover, 91 patients who underwent laparoscopy for other diagnosis were found to have acute PID. This large study clearly demonstrates that false-positive and false-negative diagnoses are quite common using classic clinical parameters for acute PID.

Laparoscopy studies have found that only 17% of patients with proven tubal infection exhibit the classic textbook signs and symptoms of PID, including fever, elevated white blood cell count, and adnexal masses palpated on pelvic examination. No single finding or combination of historical, physical, or laboratory findings is both sensitive and specific for the diagnosis of PID. Although the importance of the diagnosis of PID is indisputable, few studies have systematically assessed its sensitivity and specificity. Despite this, sensitive criteria for the diagnosis need to be used because early, accurate diagnosis of PID is paramount for effective management of acute disease and prevention of sequelae.

The only widely disseminated criteria for PID diagnosis in the United States were formulated in 1983 and were endorsed by the Infectious Disease Society for Obstetricians and Gynecologists in the same year.[43] Clinical and laboratory information from women with laparoscopically proven PID was used to devise these guidelines. Three physical examination findings were all required: abdominal direct tenderness, tenderness with motion of the cervix, and adnexal tenderness. In addition, the patient had to have at least one of the following criteria:

1. Confirmation of *Chlamydia trachomatis* or gonococcal infection
2. Temperature greater than 38°C (100.4°F)
3. Leukocytosis greater than 10,000 WBC per cubic millimeter
4. Purulent material in the peritoneal cavity obtained by culdocentesis or laparoscopy
5. Pelvic abscess or inflammatory complex done by manual examination or sonography
6. Sexual contact with a person known to have gonorrhea, chlamydia, or nongonococcal urethritis

To make the diagnosis of acute PID, only one of the criteria on the above list need be present in conjunction with all three of the clinical criteria mentioned previously. These guidelines represent a consensus and have become the U.S. Department of Health and Human Services' case definition of PID (See Table 7.5).

PID remains a serious problem for millions of women. In this section the available data on the respective usefulness of the history, the physical exam, and the laboratory studies on the accurate diagnosis of PID are reviewed. A new approach to diagnosis of PID that emphasizes a high index of suspicion and separates patients into those who present with mild symptoms and those who present with severe symptoms are discussed. This new approach will be very useful in the early diagnosis of mild PID and can reduce the sequelae of this disease.

TABLE 7.5. Clinical criteria for the diagnosis of acute PID.*

All of the following signs should be present:
  Lower abdominal tenderness
  Cervical motion tenderness
  Adnexal tenderness
In addition, one or more of the following should be present:
  Temperature ≥38°C (100.4°F)
  White blood cell count ≥10,500/mm³
  Purulent material obtained by culdocentesis
  Inflammatory mass by bimanual examination or ultrasound
  Erythrocyte sedimentation rate >15 mm/hour
  Evidence of cervical colonization with Gonococcus or Chlamydia (via Gram stain or rapid slide test)
  Presence of >5 white blood cells per oil-immersion field on Gram stain of endocervical discharge

* Adapted from: Sweet RL, Pelvic inflammatory disease and infertility in women. *Infect Dis Clin North Am.* 1987; 1:199–215.

In most patients with PID, the history will reveal pain in the lower abdomen and pelvis. More than 90% of women with PID present with diffuse bilateral lower abdominal pain.[42,44] The pain is classically described as consistent and dull, is usually accentuated by motion or sexual activity, and is generally of less than 7 days' duration. Approximately 75% of patients with acute PID have an endocervicitis and coexisting purulent vaginal discharge. Fever and chills were reported in about 40% of patients with PID.[42] Irregular menses is also noted in approximately 40% of patients. Occasionally patients complain of urethritis or proctitis as well. Nausea and vomiting are late symptoms in the normal course of acute PID. Acute PID often occurs with minimal symptoms; in fact, 50% of women with tubal obstruction do not give a history of having PID and probably have had "silent PID."

The physical examination is critical in the diagnosis of acute PID and is more sensitive than the history. All patients with suspected PID should undergo a complete physical. Most findings will be located in the pelvic area, but a complete physical examination is needed so that subtle clues are not overlooked. Gonococcal infection can also cause pharyngitis, arthritis, proctitis, and septicemia. From 5% to 10% of women with acute PID develop symptoms of perihepatic inflammation, the Fitz-Hugh-Curtis syndrome. Symptoms and signs include right upper quadrant pain, pleuritic pain, and tenderness in the right upper quadrant. Either *N gonorrhoeae* or *C trachomatis* can produce perihepatic inflammation.

The pelvic examination will yield the most helpful information and must be performed in a gentle yet thorough manner. The pelvic examination begins with an inspection of the genitalia for lesions, including ulcerations, blisters and purulent areas. Any abnormal sites should be appropriately cultured. Bartholin's glands, Skene's glands, and the urethra should be palpated and compressed to detect the presence of purulent discharge; if present, the discharge should be cultured. A speculum, lubricated with only a small amount of water, should be carefully inserted into the vagina. Commonly used lubricants may be toxic to fastidious organisms, especially *N gonorrhoeae*. The vagina and cervix should be thoroughly inspected because the presence of other sexually transmitted diseases will increase the index of suspicion. Multiple STDs in one individual are the rule rather than the exception.

N gonorrhoeae and *C trachomatis* often produce purulent endocervicitis, that is, mucopus. The columnar epithelium may be elevated above the plane of the squamous epithelium, creating a hypertrophic cervicitis. Approximately 87% of patients with hypertrophic cervicitis harbor *C trachomatis*. In patients with cervicitis, 50% with simple ectropion harbor *C trachomatis*.

As previously stated, *N gonorrhoeae* is a fastidious organism and great care is needed in culturing it. In taking a culture, first remove the excess cervical mucus with a cotton ball or a large cotton-tipped swab. Second, place a small cotton-tipped swab or, better, a Dacron or calcium alginate swab into the endocervix and gently rotate it a complete 360 degrees for 15 to 20 seconds. Third, remove the swab and roll it directly on modified Thayer Martin or a similar medium. The medium should be at room temperature; a medium just removed from a refrigerator has little chance of recovery. Ideally, the culture should be transported to the lab within minutes or it should be incubated at 36°C for 12 to 18 hours and then transported. A single endocervical culture will detect approximately 85% of gonococcal cervical infection. Because other bacteria may appear similar, gram stain of cervical mucus for *N gonorrhoeae* is not considered specific.

C trachomatis is identified by placing a Dacron swab in the endocervical canal and rotating it for 15 to 20 seconds to abrade the columnar epithelium. The swab is placed in an antibiotic-containing transport medium and stored at 4°C. It is important to perform the definitive culture within 24 hours. Cultures for *C trachomatis* are difficult to perform in the lab, and many labs have switched to other simple, rapid, and sensitive methods to detect chlamydial antigens. One test uses fluorescein-conjugated monoclonal antibodies to detect elementary bodies of *C trachomatis* (microtrak, Syva). In performing this test remember that *C trachomatis* is an intracellular organism that resides in the columnar cells. The sensitivity of the test depends on the adequacy of the sample of the endocervical canal with a cytobrush. Before the endocervical sample is obtained, external mucus and debris should be carefully wiped away with a cotton swab. The microtrak has a sensitivity of 85% to 93% and specificity of 99% for women.

Additional information can be obtained from an endometrial biopsy. The specimen should be taken with a curette and then divided into two parts. One portion is sent for histologic evaluation, and the other is placed in an anaerobic transfer vial for the isolation of *N gonorrhoeae*, *C trachomatis*, and other aerobic and anaerobic bacteria. Two studies reported that endometrial inflammation on a biopsy significantly correlated with PID, with good sensitivity (70% to 89%), and specificity (67% to 87%).[45,46]

After the speculum is removed, a bimanual examination should be expertly performed. The vagina should be palpated for the presence of masses or pain. The cervix also should be palpated gently to discern pain, an indicator of direct cervical infection. Next, the cervix should be moved gently from side to side to determine whether there is pain or tenderness, which would suggest inflammation of the lower genital tract. The uterus should be palpated for tenderness and pain and should be moved from side to side as well as anteriorly and posteriorly. The adnexa should be gently palpated to assess size and discern pain.

Cervical motion tenderness and adnexa tenderness are present in most women with symptomatic PID. Pelvic inflammatory disease patients generally have bilateral pain and tenderness, although occasionally (5% to 10%) these symptoms may be unilateral. However, other conditions that mimic PID—such as appendicitis, ectopic pregnancy, ruptured ovarian cysts, torsion of the adnexa, and endometriosis—need to be considered. A rectovaginal examination should be done to complete the exam and to determine whether the cul-de-sac is clear. If the cul-de-sac is clear, a culdocentesis may prove helpful, and ultrasonography can be used to help make the culdocentesis successful. Culdocentesis is sometimes used to help make the diagnosis of acute PID and to help rule out other significant pelvic pathology. When purulent peritoneal fluid is aspirated, one has evidence of inflammatory process in the pelvis. In patients with acute PID or other inflammatory processes (appendicitis or pancreatitis), the white blood cell count of the peritoneal fluid is greater than 30,000 cells/mL.[45,46] The white blood cell count of peritoneal fluid in women without peritoneal inflammation is less than 1,000 cells/mL.[45,46] If serous fluid is aspirated from the cul-de-sac, it should be Gram stained and cultured for aerobes, anaerobes, chlamydia, and gonorrhoeae. The fluid should also be tested for amalse. Positive gonococcal cultures and gonococcus/chlamydia tests are significant indicators of acute PID. If the testing is performed on the fallopian tubes and/or the endometrium, the sensitivity and specificity are greater.

If the aspirated fluid is nonclotting blood, one would consider an ectopic pregnancy, hemorrhagic corpus luteum, or retrograde menstruation. Aspiration of blood that clots would suggest that the bleeding is extremely recent or that a vessel has been injured.

Several laboratory tests show good correlation with acute PID, with high sensitivity and specificity. For example, an elevated erythrocyte sedimentation rate is a significant indicator of acute PID. Sensitivity was 64% and specificity 69% for a sedimentation rate greater than 20 mm/hr.[42,44] C-reactive protein is also a significant predictor of acute PID. In several studies C-reactive protein showed high sensitivity (74% to 93%) and a wider range of specificity (50% to 90%).[47] An elevated peripheral white blood cell count does not appear to help distinguish PID from competing diagnoses. In patients with laparoscopy-confirmed PID, only 50% to 60% demonstrated white blood cell counts greater than 10,000 WBC/mL.[42] A recent finding is that serum CA-125 concentrations of greater than 16 units/mL (with a range of 20 to 1,300) correlated with the extent of inflammatory peritoneal involvement in patient with clinically diagnosed PID.[48] CA-125 is a surface antigen produced by the peritoneum in response to various stimulants. Acute PID may be the most common reason for elevated CA-125 in young women.

In conjunction with the physical examination, ultrasonography may be useful. It is especially helpful in cases where the patient has extreme tenderness, is uncooperative, or is extremely obese. It can also be used as an adjunct to surgical procedures such as culdocentesis or percutaneous draining of an abscess or in documentation of resolution or enlargement of adnexal masses during treatment. Other imaging techniques, such as CT scan or MRI, are rarely indicated in patients with acute PID.

The "gold standard" in the diagnosis of PID is laparoscopy. Direct visualization of the pelvic organs is the most accurate method of diagnosis. Tubal edema, erythema, pockets of purulent exudate, or a tubal ovarian abscess confirms the diagnosis. Not every patient with PID should undergo laparoscopy. The cost and the risk of the procedure need to be weighed against the potential benefit. When one is relatively confident of the diagnosis, there is no need to perform laparoscopy. If signs and symptoms are mild, it is probably wiser to treat and observe the patient's response than to subject the patient to laparoscopy. If, however, the differential diagnosis includes other serious conditions and the patient's presentation is severe, then laparoscopy may be indicated. Because of all the confounding variables and the seriousness of acute PID and its sequelae, a new approach to the diagnosis of PID has been suggested. The diagnosis of PID is approached by classifying the presentation as either mild or severe.[49] Patients with severe symptoms require more elaborate evaluation and diagnosis to rule out more serious disease than do patients with mild symptoms. See the table and figure adopted from *JAMA*. The definition of severe given by an expert panel includes patients who exhibit one or more of the following characteristics: malaise, pallor, diaphoresis, facial expression of distress, altered mental status, minimal spontaneous movement, nausea and vomiting, or abnormally elevated or diminished vital signs.[49] Laparoscopy is also indicated in those patients who fail to respond to appropriate medical therapy, where it can be used for both diagnostic and therapeutic purposes.

Patients with mild presentations need only exhibit the minimum criteria for the clinical diagnosis of PID, namely, lower abdominal tenderness, bilateral adnexal tenderness, and cervical motion tenderness (Table 7.6). Patients who present with only this minimal clinical criteria may be treated for PID unless there are other competing diagnoses, for example, a positive pregnancy test or concern about acute appendicitis.

This diagnostic model reflects the inherent difficulty of PID diagnosis and emphasizes diagnostic sensitivity for mild disease and thorough, accurate diagnosis for severe disease. Using a lower diagnostic threshold for mild disease is extremely important because it will allow practitioners to treat patients with mild presentations on the basis of minimal clinical criteria. The need to treat this milder PID, however, is great; severity of clinical presentations corresponds poorly with damage to the fallopian tubes and with the likelihood of developing serious long-term sequelae. The new criteria from the Centers for Disease Control are based on the principle of "lower diagnostic threshold" to detect and treat more PID.[50] At the same time, women who present with severe clinical signs will have a more elaborate diagnostic evaluation because incorrect diagnosis and treatment may cause unnecessary morbidity. In these women additional criteria therefore need to be met to confirm the diagnosis of PID (Figure 7.2). All patients suspected of having PID should have a cervical culture for *N gonorrhoeae* and a cervical culture or nonculture test for *C trachomatis*, although these are not needed to justify treatment. Although biologic testing is not necessary to justify initial treatment, it does provide diagnostic confirmation and serves as a baseline for tests for cure at follow-up visits.

TABLE 7.6. Criteria for clinical diagnosis of pelvic inflammatory disease.

Minimum Criteria for Clinical Diagnosis of PID
  Lower abdominal tenderness
  Bilateral adnexal tenderness
  Cervical motion tenderness

Additional Criteria Useful in Diagnosing PID

Routine	Elaborate
Oral temperature >38.3 C	Histopathologic evidence on endometrial biopsy
Abnormal cervical or vaginal discharge	Tuboovarian abscess on sonography
Elevated erythrocyte sedimentation rate and/or C-reactive protein	Laparoscopy
Culture or nonculture evidence of cervical infection with *N gonorrhoeae* or *C trachomatis*	

Treatment

The ultimate goals of treatment are the elimination of signs and symptoms of infection,

FIGURE 7.2. PID diagnostic algorithm.

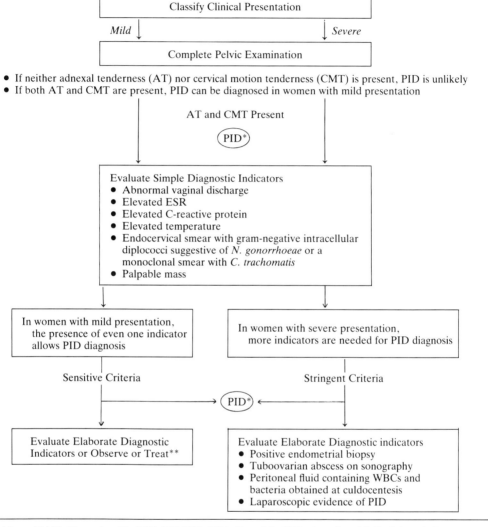

Adapted from *JAMA*: November 31, 1991; 266:2594–2604. Copyright 1991, American Medical Association.
* PID was diagnosed in the absence of strong evidence for a competing diagnosis.
** PID management depends on available tests, compliance and patient preference.

eradication of pathogens, and reduction of tubal damage. Reduction of tubal damage should help prevent the long-term sequelae such as recurrent PID, ectopic pregnancy, chronic pelvic pain, and infertility. Treatment is based on a consensus that PID is polymicrobial in etiology.[51-53] Any treatment regimen needs to include coverage for a wide range of bacteria including *N gonorrhoeae*, *C trachomatis*, and a variety of endogenous anaerobic and aerobic bacteria. In addition, coverage should be provided for penicillinase-producing *N gonorrhoeae* or chromosomally mediated resistant organisms. For many infections the choice of antibiotic therapy is usually based on culture and sensitivity of bacteria obtained directly from the site of the infection. In PID this is difficult and often impractical. It could be accomplished by culturing the fallopian tubes via the laparoscope, but that would be

TABLE 7.7. Inpatient treatment (currently recommended regimens providing sufficient antibacterial coverage).

Regimen A:	Regimen B:
Cefoxitin IV 2G every 6 hours,	Clindamycin IV 900 mg every 8 hours
or	plus
Cefotetan IV 2 g every 12 hours	Gentamicin loading dose IV or IM (2 mg/kg) followed by a maintenance dose (1.5 mg/kg) every 8 hours
plus	
Doxycycline 100 mg every 12 hours orally or IV	

Continue regimen for at least 48 hours after clinical improvement. Following hospital discharge, continue doxycycline 100 mg orally 2 times a day for 10–14 days. Clindamycin, 450 mg orally 4 times a day, for 10–14 days may be an acceptable alternative in cases in which anaerobic coverage is of primary concern.

* Centers for Disease Control, 1990.

TABLE 7.8. Outpatient treatment (currently recommended regimens providing sufficient antibacterial coverage).*

Cefoxitin 2 g IM
plus
Probenecid, 1 g orally, concurrently
or
Ceftriaxone 250 mg IM (or equivalent cephalsoporin)
plus
Doxycycline 100 mg orally 2 times a day for 10–14 days
or
Tetracycline 500 mg orally 4 times a day for 10–14 days

For patients who cannot tolerate doxycycline: substitute erythromycin, 500 mg orally 4 times a day for 10–14 days.

* Centers for Disease Control, 1990.

done at great expense both in financial terms and in potential harm to the patient. If every American woman with acute PID had a laparoscopy performed, there would be 14 deaths directly related to the procedure each year. In addition, treatment should be started as early as possible, usually before the microbial cause is established, and therefore treatment is usually emperic.

Despite general agreement that broad spectrum therapy is appropriate, questions persist regarding optimal treatment regimens. Most studies have evaluated combinations of antimicrobials that on a theoretical basis could be expected to provide broad-spectrum coverage for likely pelvic pathogens. In 1989 the CDC convened a panel of experts to develop guidelines for both the inpatient and outpatient treatment of PID.[54] These experts reviewed theoretical considerations, available data, numerous studies, and other factors, including cost. On the basis of this information the regimens recommended for inpatient treatment were similar to those recommended in 1985[55] and have been continued in the 1990–91 recommendations (Table 7.7). In regimen A (a combination of cefoxitin and doxycycline), cefoxitin was selected for coverage of *N gonorrhoeae*, enterobacteriaceae, and anaerobes. Doxycycline was chosen primarily to cover *C trachomatis*. Cefotetan was mentioned as an alternative to cefoxitin. Also recommended in appropriate doses were other cephalophorins (ceftizoxime, cefotaxime, and ceftriaxone) that provide adequate coverage. In regimen A, doxycycline was given orally or intravenously every 12 hours; the bioavailability is similar either way. Giving doxycycline orally to patients with normal gastrointestinal function is preferred.

In regimen B (a combination of clindamycin and gentimycin), clindamycin was selected to cover anaerobes and Gram-positive aerobes. The amnioglycides were chosen to cover Gram-negative aerobes including *N gonorrhoeae*. Regimen B was thought to be more useful than regimen A for patients with pelvic abscesses but provides less optimal coverage for *C trachomatis*. In both regimens the intravenous antibiotics are given for at least 48 hours after the patient clinically improves. This is followed by 100 mg of doxycycline orally twice daily to complete 10 to 14 days of total therapy. Shorter courses of therapy or lower doses may result in therapeutic failures.

Compared to 1985 recommendations, the 1989 CDC recommendations for outpatient treatment were substantially modified (Table 7.8). Recommendations are 2 grams of cefoxitin IM concurrent with probenecide orally; 250 gm of ceftriaxon IM, or the equivalent cephalosporin, plus 100 mg of doxycycline orally two times a day for 10 to 14 days; or 500 mg of tetracycline orally four times a day for 10 to 14 days. Because of the ease of remembering and therefore better patient compliance, the doxycycline dose is preferable. For patients who cannot take doxycycline, substitute 500 mg of erythromycin orally, four times a day for 10 to 14 days. It must be pointed out that little information is available on the clinical effectiveness for the ambulatory treatment of PID. Additionally, parenteral beta lactam antibiotics are recommended in all outpatient regimens.

Patients who receive the outpatient regimen need to be monitored closely and re-evaluated in 72 hours.[54,55] Within 72 hours of treatment, patients should have subjective amelioration of symptoms, be afebrile, and have significant reduction in pain and tenderness. Patients who do not appropriately respond within 72 hours should be hospitalized for parenteral therapy.[50]

Many new antibiotics have become available for the treatment of PID, including combinations of penicillins with beta lactamase inhibitors, quinolones, a monobactam, and a carbapenem. Ampicillin-sulbactam, amoxicilin-clavulanic acid, and ticarcillin-clavulanic acid have been studied for the treatment of PID. Most studies report clinical cure rates in the range of 95%.[56–58] Several studies of ampicillin-sulbactam for PID treatment have shown high rates of clinical cure. One study[59] showed that severe adhesions were absent in 11 women treated with ampicillin-sulbactam plus doxycycline but present in 4 (36%) of 11 women treated with cefoxitin plus doxycycline. The use of combinations of penicillins with beta lactamase inhibitors appears to have promise, but the use in treating cases of PID that involve *C trachomatis* remains in question.

In treating PID not complicated by tuboovarian abscess, quinolones such as ciprofloxacin and ofloxacin have cure rates in the 90% to 95% range.[60,61] Further studies are needed to document their effectiveness. Women with tuboovarian abscess have not been studied, and the limited in vitro activity of quinolones against anaerobes would suggest that these agents may not effectively treat anaerobic infections.[62] Because the quinolones are effective orally, they have a potential of being used in the ambulatory management of PID in combination with other oral antibiotics effective against anaerobes.

The monobactam aztreonam is active only against Gram-negative aerobes. As such it is an alternative to amnioglycocides in regimen B and may be considered as an alternative to amnioglycocides in patients who are at increased risk for ototoxicity or nephrotoxicity. Aztreonam is significantly more expensive than the amnioglycocides[63,64]; however, it does not cause ototoxicity or nephrotoxicity.

Imipenem, a carbapenem that is combined with cilastatin, is a potent antibiotic. It has the widest antimicrobial range of activity of all currently available beta lactam antibiotics; however, it is not expected to be active against *C trachomatis*.[65] Although it is highly effective in treating intraabdominal infections, data on its effectiveness in treating PID are not available.[63] Because of its potential to cause seizures in approximately 1% of patients and its importance in treating nosocomial infections caused by multidrug-resistant organisms, imipenem should be used judiciously. It is not generally considered a first-line antibiotic for the treatment of PID.[66]

In the United States approximately three out of four women with acute PID are currently being treated as outpatients. However, there are no data on the risk, benefits, and costs of treatment of inpatient versus outpatient treatment of PID. Oral absorption of antibiotics is generally considered too variable for reliable treatment of serious infections. Moreover, the utility of a single IM injection of a cephalosporin in eradicating organisms from the fallopian tubes has not been reported and can only be inferred. Therefore, in treating PID, hospitalization should be considered whenever possible.[62] The recent, more widespread use of home infusion teams may represent an option for outpatient management of PID with parenteral antibiotics, and this area needs further investigation. Many experts recommend that all patients with PID be

TABLE 7.9. Criteria for hospitalization of outpatients with pelvic inflammatory disease.*

The diagnosis is uncertain.
Surgical emergencies such as appendicitis and ectopic pregnancy cannot be excluded.
A pelvic abscess is suspected.
The patient is pregnant.
The patient is an adolescent (adolescent patients' compliance with therapy is unpredictable, and the long-term sequelae of PID may be particularly severe for members of this group).
Severe illness precludes outpatient management.
The patient is unable to tolerate an outpatient regimen.
The patient has failed to respond to outpatient therapy.
Clinical follow-up within 72 hours of starting antibiotic treatment cannot be arranged.

---

* Centers for Disease Control, 1991.

hospitalized.[50] Because of practical considerations, costs, inconvenience, patient desires, and so forth, all patients are not hospitalized, but it is important to be aware of situations in which hospitalization is particularly recommended. The CDC in conjunction with a group of outside experts published a list of such situations in 1991 (Table 7.9).[56] Stating that hospitalization is particularly recommended in the following situations: when the diagnosis is uncertain; for surgical emergencies (for example, when appendicitis and ectopic pregnancy cannot be excluded); when a pelvic abscess is suspected; if the patient is pregnant; if the patient is an adolescent; if severe illness precludes outpatient management; if the patient is unable to tolerate an outpatient regimen; if the patient has failed to respond to outpatient therapy; or if clinical follow-up within 72 hours of starting antibiotic treatment cannot be arranged. To demonstrate the importance of hospitalizing patients with an uncertain diagnosis, in many large series of 100 consecutive patients suspected of a diagnosis of PID, 3 to 4 will have an ectopic pregnancy and 3 to 4 will have acute appendicitis. In addition, other diagnoses will often be present, for example, torsion of the adnexa, ruptured ovarian cysts, or endometriosis.

Acute PID associated with use of an IUD is usually more advanced. It is common practice to remove and culture the IUD as soon as appropriate levels of intravenous antibiotics have been achieved.[67] Women with PID are at high risk for recurrent PID; therefore the IUD is not generally an appropriate contraceptive for women with an IUD-associated infection who remain interested in childbearing. When a tuboovarian abscess is present, it is especially important to achieve therapeutic levels of appropriate parenteral antibiotics before the IUD is removed, because septic shock has been observed in women treated otherwise.[67]

HIV-infected women with PID, or those immunocompromised for any reason, may have a more clinically severe disease and may be more refractory to medical management. Immunocompromised women are at increased risk for a complicated clinical course, are more likely to have tuboovarian abscesses, and are more likely to require operative intervention. If PID is suspected, such women should be hospitalized immediately and treated with appropriate parenteral antibiotics.

It is imperative that the sex partners of patients with PID be treated. Up to 80% of partners of women with PID attributable to *C trachomatis* or *N gonorrhoeae* are themselves infected with one or both infections. Many of these partners are asymptomatic. Failure to treat the partner places the woman at risk for recurring infection. The community is also at risk because the sex partners often unknowingly continue to transmit the infection. The CDC recommends that sex partner(s) be evaluated and, after evaluation, be empirically treated with regimens effective against both *C trachomatis* and *N gonorrhoeae*.[50] Some physicians treat the partners presumptively.

A full discussion of the surgical management of PID is not appropriate here, but surgical management has decreased markedly in the last 15 years. Operations are restricted to life-threatening infections, ruptured tuboovarian abscesses, failure to respond to appropriate medical management, and removal of persistent masses. Approximately 70% of patients with tuboovarian abscesses respond to appropriate medical management (Fig. 7.3).[68] Patients should be followed carefully, and if there is any positive response, it is reasonable

FIGURE 7.3.[68]

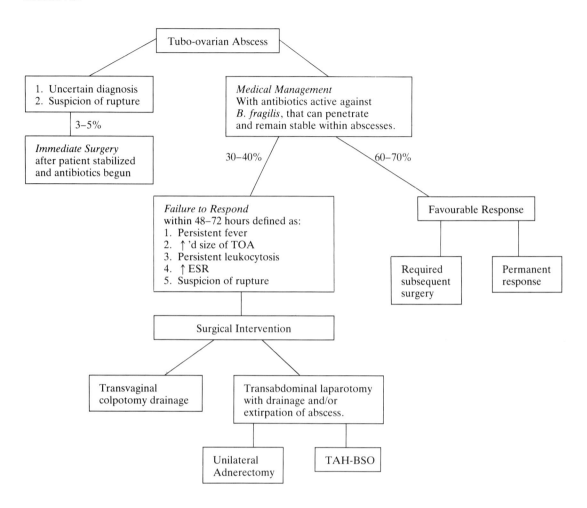

to continue medical management. An exception is a patient with a clinical picture of a ruptured tuboovarian abscess and possible sepsis or shock. Clinical evidence of rupture includes diffuse peritonitis, sudden change in the character and intensity of pain, tachacardia out of proportion to fever, critically ill appearance, or the development of shock in a patient with a tuboovarian complex. These patients should be started on triple antibiotic therapy without delay. The patient is stabilized and taken to the operating room for immediate surgical intervention. Delayed or inappropriate response in these patients could result in significant morbidity or mortality. In treating a patient with ruptured TOA, the physician must be aware of the potential for extremely serious complications such as ARDS, septic shock, acute tubular necrosis, septic thrombophlebitis, local or distant abscesses, and wound infection and/or wound dehiscence.

If the patient does not respond to medical management, the physician must carefully discuss all the options with her. Thorough counseling and informed consent are imperative. An abscess can occasionally be drained through a percutaneous-directed catheter with or without laparoscopy. Ultrasonography or CT scan is sometimes used to perform a percutaneous-directed catheter drainage. Occasionally, for a very low, dependent abscess pointing in the cul-de-sac, culpotomy can be used. In some patients, laparoscopy can be used both to confirm diagnosis and for surgical management, such as lysis of adhesions and draining abscesses. If this is not feasible, these patients will require a laparotomy, which is generally performed through a vertical incision. The entire abdomen is thoroughly explored for abscesses, and the anatomy is restored to normal as much as possible. Careful dissection of distorted tissue planes is essential. Based on earlier discussions with the patient and on individual evaluation of the extent of disease, an adenectomy or hysterectomy is performed. If a hysterectomy is performed, the vaginal cuff is left open. With the advent of IVF the preservation of a single ovary and the uterus may allow the patient the possibility of future childbearing. All reasonable options should be considered. The abdomen is then copiously irrigated with 3 or 4 liters of ringers lactate. Malecot or multiple Jackson-Pratt drains through the cuff or a posterior colpotomy to closed suction are put in place. Drains are usually left in place until there is less than 30 mL of drainage during a 24-hour period. Fascia is then usually closed with either permanent or delayed absorbable sutures. In vertical incisions, the Smeed-Jones technique is preferred. A delayed primary closure of the skin and subcutaneous tissue is the procedure of choice. A pack of moist gauze is placed in the subcutaneous layer and covered with a sterile dressing for 3 to 4 days; then the incision is reinspected and, if clean, closed with steristrips or simple sutures put in place while the patient is in her room.

Follow-Up

Patients treated for PID should have a follow-up visit 7 to 14 days following completion of therapy to ensure an appropriate clinical response. Patients with a positive culture for *N gonorrhoeae* or *C trachomatis* should be recultured for test of cure. The follow-up visit is another excellent opportunity for patient education. It gives the practitioner time to counsel the patient regarding STDs, to discuss high-risk sexual behavior, and to provide strategies to help prevent future infections. It is important to inform the patient that barrier methods of contraception can be used for personal protection against STDs. Also, remind the patient of the risk of PID recurrence and schedule another follow-up visit in 4 to 6 months to again educate, evaluate, and culture for possible recurrence.

Sequelae

Pelvic inflammatory disease has four major sequelae: recurrent upper genital tract infection, chronic pelvic pain, ectopic pregnancy, and infertility. These sequelae cause significant human suffering and morbidity, and even today PID still has mortality associated with it. Grimes has estimated that in the United States there is presently one death every other day directly related to PID. Most of these deaths occur in patients with a ruptured tuboovarian abscess. In addition, approximately 40 women die of ectopic pregnancies each year in the United States. At least half of these ectopic pregnancies occur as a result of tubal damage secondary to PID.

After a woman has had one episode of PID, she is two to three times more likely to have at least one subsequent episode of PID than a never-infected woman.[69] Twenty-five percent of women with one episode of PID develop a subsequent episode.[69,70] Approximately 20% of women with PID develop chronic pelvic pain, and it may be difficult to distinguish chronic pelvic pain from recurrence of PID. This pain is usually related to the formation of adhesions, and there is often increased pain with menses and with intercourse. Anatomic distortion and residual damage from infectious disease processes may also become components of a chronic pelvic pain syndrome.

The number of ectopic pregnancies in the United States has more than doubled in the past 10 years. The increased rate is directly proportional to the increased rate of STDs. The risk of ectopic pregnancy is 7 to 10 times greater in women who have had PID compared to women who have never had PID.[70] Approximately 25,000 women annually have ectopic pregnancies that are attributable to

PID, and half of these women will become infertile.[70]

Infertility resulting from tubal occlusion occurs in 11% of women with one episode of PID, 23% of women with two episodes, and 54% of women with three episodes.[3,69] With each successive episode the risk of infertility doubles. Another factor of infertility is laparoscopically documented severity of PID. After a mild episode of PID the risk of infertility is relatively small, that is, around 3%, whereas it is approximately 17% following severe PID.[3] Thus early intervention is extremely important in preventing infertility, and it has been shown in the animal model that early treatment decreases tubal damage. Furthermore, the specific etiologic bacteria involved affect the risk of tubal damage. For example, despite its typically mild presentation, chlamydial PID results in a threefold increase in infertility as compared to gonococcal PID.

Summary

Pelvic inflammatory disease involves a great diversity of diseases with a diversity in etiology and presentation. Generally considered to be a polymicrobial infection, PID presentation ranges from the very subtle to life threatening. Only about 40% of women with PID present with a temperature elevated greater than 100.4°F (38°C), and only about 50% to 60% present with a white blood cell count greater than 10,000. The clinical diagnosis of PID is ambiguous at best, and practitioners therefore need a very high index of suspicion. In general, PID is a condition of young women; 75% of cases occur in females younger than 25. The risk of a sexually active woman under 25 acquiring PID is 1 in 8.

Pelvic inflammatory disease has a great cost associated with it both in terms of dollars ($4.2 billion annually) and in terms of human suffering. Approximately 25% of women with PID develop significant medical sequelae. These sequelae include an increased risk of recurrent PID, chronic pelvic pain, ectopic pregnancy, and infertility.

In the 1990s pelvic inflammatory disease should be thought of as a totally new and different disease than the previous presentation of acute pain, fever, and other severe symptoms that have characterized PID in the past. The new diagnostic approach classifies patients according to their presentation, mild or severe. Patients with mild presentations can be diagnosed on the basis of minimum criteria (lower abdominal tenderness, bilateral adnexal tenderness, and cervical motion tenderness) and treatment should be instituted on the basis of these minimal criteria in the absence of competing diagnoses. For women with more severe clinical signs, more elaborate diagnostic evaluation is warranted; additional criteria such as elevation of temperature or white blood cell count, positive culture, more elaborate histologic evidence, or ultrasonography are indicated. By considering patients with a mild presentation separately, diagnostic sensitivity is emphasized, thereby assuring that fewer cases of PID will be missed.

Because PID is usually caused by a variety of organisms the use of broad-spectrum antibiotics is needed. No single agent that provides sufficient coverage is currently available. In addition to selecting appropriate antibiotics, management should emphasize compliance with a full course of therapy and also include education, risk-reduction counseling, and a follow-up evaluation. Practitioners also need to ensure that there is adequate evaluation and treatment of all sex partners.

Finally, pending better data, hospitalization where feasible, should be strongly considered, particularly for those women desiring to maintain their fertility. Hospitalization is particularly recommended in cases of uncertain diagnosis, where concomitant problems exist, for pregnant or adolescent patients, for patients unable to follow or tolerate an outpatient regimen, or for patients who fail to respond to initial therapy.

References

1. Rolfs RT, Galaid EL, Zaidi AA. Epidemiology of pelvic inflammatory disease: trends in hospitalizations and office visits, 1979–1988. In: *Proceedings of the Centers for Disease Control*

and National Institutes of Health joint meeting, Pelvic Inflammatory Disease: Prevention, Management, and Research in the 1990s. Bethesda, Md; September 4–5, 1990.
2. Washington AE, Cates W, Zaidi AA. Hospitalizations for pelvic inflammatory disease: epidemiology and trends in the United States, 1975 to 1982. *JAMA.* 1984;251:2529–2533.
3. Westrom L. Incidence, prevalence, and trends of acute pelvic inflammatory disease and its consequences in industrialized countries. *Am J Obstet Gynecol.* 1980;138:880–892.
4. Cates W, Rolfs RT, Aral SO. Sexually transmitted diseases, pelvic inflammatory disease and infertility: an epidemiologic update. *Epidemiol Rev.* 1990;12:199–220.
5. Sweet RL. Pelvic inflammatory disease. In: Sweet RL, Gibbs RS, eds. *Infectious Diseases of the Female Genital Tract.* Baltimore, Md: Williams & Wilkings; 1990:241–266.
6. Washington AE, Katz P. Cost of the payment source for pelvic inflammatory disease: trends and Projections, 1983 through 2000. *JAMA.* 1991;266:2565–2569.
7. Westrom L, Svensson L, Wolner-Hanssen P, Mardh P-A. Chlamydial and gonococcal infection in a defined population of women. *Scand J Infect Dis.* 1982;32(suppl):157–162.
8. Mardh P-A. An overview of infectious agents in salpingitis: their biology, and recent advances in methods of detection. *Am J Obstet Gynecol.* 1980;138:933–951.
9. Sweet RL, Draper DL, Hadley WK. Etiology of acute salpingitis: influence of episode number and duration of symptoms. *Obstet Gynecol.* 1981;58:62–68.
10. Eschenbach DA, Harnisch JP, Holmes KK. Pathogenesis of acute pelvic inflammatory disease: role of contraception and other risk factors. *Am J Obstet Gynecol.* 1977;128:838–850.
11. Hadley WK, Brooks GF. Microbiology and pathogenesis of acute salpingitis as determined by laparoscopy: what is the appropriate site to sample? *Am J Obstet Gynecol.* 1980;138:985–989.
12. Eschenbach DA, Buchanan TM, Pollock HM, et al. Polymicrobial etiology of acute pelvic inflammatory disease. *N Engl J Med.* 1975;293:166–172.
13. Thompson SE, Hager WD, Wong KH, et al. The microbiology and therapy of acute pelvic inflammatory disease in hospitalized patients. *Am J Obstet Gynecol.* 1980;136:179–186.
14. Droegemueller W. Upper genital tract infections. In: Droegemueller W, Herbst AL, Mishell DR Jr., Stenchever MA, eds. *Comprehensive Gynecology.* St. Louis, Mo: CV Mosby; 1987:614–642.
15. Rice PA, Schacter J. Pathogenesis of pelvic inflammatory disease. *JAMA.* 1991;266:2587–2593.
16. James AN, Knox JM, Williams RP. Attachment of gonococci to sperm: influence of physical and chemical factors. *Br J Vener Dis.* 1975;52:128–135.
17. Wolner-Hanssen P, Mardh P-A. In vitro tests of the adherence of *Chlamydia trachomatis* to human spermatozoa. *Fertil Steril.* 1984;42:102–107.
18. Keith LG, Berger GS, Edelman DA, et al. On the causation of pelvic inflammatory disease. *Am J Obstet Gynecol.* 1984;149:215–224.
19. Patton DL, Halbert SA, Kuo CC, Wang SP, Holmes KK. Host response to primary *Chlamydia trachomatis* infection of the fallopian tube in pigtailed monkeys. *Fertil Steril.* 1983;40:829–840.
20. Patton DL, Kuo CC, Wang SP, Halbert SA. Distal obstruction induced by repeated *Chlamydia trachomatis* salpingeal infections in pigtailed macaques. *J Infect Dis.* 1987;155:1292–1299.
21. Patton DL, Landers DV, Schacter J. Experimental *Chlamydia trachomatis* salpingitis in mice: initial studies on the characterization of the leukocyte response to chlamydial infection. *J Infect Dis.* 1989;159:1105–1110.
22. McGeer ZA, Johnson AP, Taylor-Robinson D. Pathogenic mechanisms of Neisseria gonorrhoeae: observations on damage of human fallopian tubes in organ culture by gonococci of colony type 1 or type 4. *J Infect Dis.* 1981;143:413–422.
23. Gregg CR, Melly MA, Hellerqvist CG, Coniglio JG, McGee ZA. Toxic activity of purified lipopolysaccharide of *Neisseria gonorrhoeae* for human fallopian tube mucosa. *J Infect Dis.* 1981;143:432–439.
24. Melly MA, McGee ZA, Rosenthal RS. Ability of monomeric peptidoglycan fragments from *Neisseria gonorrhoeae* to damage human fallopian-tube mucosa. *J Infect Dis.* 1984;149:378–386.
25. Densen P, MacKeen LA, Clark RA. Dissemination of gonococcal infection is associated with delayed stimulation of complement-dependent neurophil chemotaxis in vitro. *Infect Immun.* 1982;80:78–87.

26. Bell TA, Holmes KK. Age-specific risks of syphilis, gonorrhea, and hospitalized pelvic inflammatory disease in sexually experienced U.S. women. *Sex Transm Dis*. 1984;11:291–295.
27. Wolner-Hanssen P, Eschenbach DA, Paavonen J, et al. Decreased risk of symptomatic chlamydial pelvic inflammatory disease associated with oral contraceptive use. *JAMA*. 1990;263:54–59.
28. Marchbanks PA, Lee NC, Peterson HB. Cigarette smoking as a risk factor for pelvic inflammatory disease. *Am J Obstet Gynecol*. 1990;162:639–644.
29. Aral SO, Mosher WD, Cates W Jr. Self-reported pelvic inflammatory disease in the United States, 1988. *JAMA*. 1991;266:2570–2573.
30. Ledegaard 0, Helm P. Pelvic inflammatory disease: the influence of contraceptive, sexual and social life events. *Contraception*. 1990;41:475–483.
31. Lee NC, Rubin GL, Grimes DA. Measures of sexual behavior and the risk of pelvic inflammatory disease. *Obstetrics and Gynecology*. 1991;77:424–430.
32. Centers for Disease Control. Condoms for prevention of sexually transmitted diseases. *MMWR*. 1988;37:133–137.
33. Cromer BA, Heald FP. Pelvic inflammatory disease associated with *Neisseria gonorrhoeae* and *Chlamydia trachomatis*: clinical correlates. *Sex Transm Dis*. 1987;14:125–129.
34. Harrison HR, Costin M, Meder JB, et al. Cervical *Chlamydia trachomatis* infection in university women: relationship to history, contraception, ectopy, and cervicitis. *Am J Obstet Gynecol*. 1985;153:244–251.
35. Washington AE, Gove S, Schachter J, Sweet RL. Oral contraceptive, *Chlamydia trachomatis* infection, and pelvic inflammatory disease: a word of caution about protection. *JAMA*. 1985;253:2246–2250.
36. Senanayake P, Kramer DG. Contraception and the etiology of pelvic inflammatory disease: new perspectives. *Am J Obstet Gynecol*. 1980;138:852–860.
37. World Health Organization. *Mechanism of Action, Safety and Efficacy of Intrauterine Devices*. Geneva, Switzerland: World Health Organization; 1987; Technical report series 753.
38. Lee NC, Rubin GL, Ory HW, Burkman RT. Type of intrauterine device and the risk of pelvic inflammatory disease. *Obstet Gynecol*. 1983;62:1–6.
39. Wolner-Hanssen P, Eschenbach DA, Paavonen J, et al. Association between vaginal douching and acute pelvic inflammatory disease. *JAMA*. 1990;263:1936–1941.
40. Marchbanks PA., Lee NC, Peterson HB. Cigarette smoking as a risk factor for pelvic inflammatory disease. *Am J Obstet Gynecol*. 1990;162:639–644.
41. Sweet RL, Blankfort-Doyle M, Robbie MO, Schachter J. The occurrence of chlamydial and gonococcal salpingitis during the menstrual cycle. *JAMA*. 1986;255:2062–2064.
42. Jacobson L, Westrom L. Objectivized diagnosis of acute pelvic inflammatory disease: diagnostic and prognostic value of routine laparoscopy. *Am J Obstet Gynecol*. 1969;105:1088–1098.
43. Hager WD, Eschenbach DA, Spence MR, Sweet RL. Criteria for diagnosis and grading of salpingitis. *Obstet Gynecol*. 1983;61:113–114.
44. Hadgu A, Westrom L, Brooks CA, Reynolds GH, Thompson SE. Predicting acute pelvic inflammatory disease: a multivariate analysis. *Am J Obstet Gynecol*. 1986;155:954–960.
45. Paavonen J, Aine R, Teisala K, Heinonen PK, Punnonen R. Comparison of endometrial biopsy and peritoneal fluid cytologic testing with laparoscopy in the diagnosis of acute pelvic inflammatory disease. *Am J Obstet Gynecol*. 1985;151:645–650.
46. Bongard F, Landers DV, Lewis F. Differential diagnosis of appendicitis and pelvic inflammatory disease: a prospective analysis. *Am J Surg*. 1985;150:90–96.
47. Lehtinen M, Laine S. Neinonen PK, et al. Serum C-reactive protein determination in acute pelvic inflammatory disease. *Am J Obstet Gynecol*. 1986;154:158–159.
48. Dick JM, Kauer FM, Fleuren GJ, de Bruijin HW. Serum CA-125 levels in patients with provisional diagnosis of pelvic inflammatory disease: clinical and theoretical implications. *Acta Obstet Et Gynecol Siandi*. 1989;68:637–641.
49. Kahn JG, Walker CK, Washington AE, Landers DV, Sweet RL. Diagnosing pelvic inflammatory disease: a comprehensive analysis and considerations for developing a new model. *JAMA*. 1991;266:2594–2604.
50. Centers for Disease Control. Pelvic inflammatory disease: policy guidelines for prevention and management. *MMWR*. 1991;40(RR-5):1–25.

51. Brook I. Bacterial synergy in pelvic inflammatory disease. *Arch Gynecol Obstet.* 1987;241:133–143.
52. Westrom L. Pelvic inflammatory disease: bacteriology and sequelae. *Contraception.* 1987;36:111–128.
53. Paavonen J, Teisala K, Heinonen PK, et al. Microbiological and histopathological findings in acute pelvic inflammatory disease. *Br J Obstet Gynecol.* 1987;94:454–460.
54. Centers for Disease Control. 1989 Sexually transmitted diseases treatment guidelines. *MMWR.* 1989:38(No. S-8):31–33.
55. Centers for Disease Control. 1985 Sexually transmitted diseases treatment guidelines. *MMWR.* 1985;34(Suppl 4S):19–21.
56. Obwegeser J, Kunz, J, Wust J, Schar G, Steiner R, Buchi W. Clinical efficacy of amoxycillin/clavulanate in laparoscopically confirmed salpingitis. *J Antimicrob Chemother.* 1989;24(suppl B):165–176.
57. McGregor JA, Christiansen FB. Treatment of obstetric and gynecologic infections, with an emphasis on beta-lactamase-producing organisms. *J Reprod Med.* 1988;33(suppl 6):595–597.
58. Crombleholme WR, Landers D, Ohm-Smith M, et al. Sulbactam/ampicillin versus metronidazole/gentamicin in the treatment of severe pelvic infections. *Drugs.* 1986;3(suppl 2):11–13.
59. Wolner-Hanssen P, Paavonen J, Kiviat N, Young M, Eschenbach DA, Holmes KK. Outpatient treatment of pelvic inflammatory disease with cefoxtin and doxycycline. *Obstet Gynecol.* 1988;71:595–600.
60. Apuzzio JJ, Stankiewicz R, Jain S, Kaminski Z, Louria D. Comparison of parenteral ciprofloxacin with clindamycin-gentamicin in the treatment of pelvic infection. *Am J Med.* 1989;87(suppl 5A):148S–151S.
61. Crombleholme WR, Schachter J, Ohm-Smith M, Luft J, Whidden R, Sweet RL. Efficacy of single-agent therapy for the treatment of acute pelvic inflammatory disease with ciprofloxacin. *Am J Med.* 1989;87(suppl 5A):142S–147S.
62. Peterson HB, Walker CK, Kahn JG, Washington AE, Eschenback DA, Paro S. Pelvic inflammatory disease: treatment issues and options. *JAMA.* 1991;266:2605–2611.
63. Dodson MG, Faro S, Gentry LO. Treatment of acute pelvic inflammatory disease with aztreonam, a new monocyclic B-lactam antibiotic and clindamycin. *Obstet Gynecol.* 1986;67:657–662.
64. Pastorek JG II, Cole Cm, Aldridge KE, Crapanzano JC. Aztreonam plus clindamycin as therapy for pelvic infections in women. *Am J Med.* 1985;78(suppl 2A):47–50.
65. Jones RN. Review of the in vitro spectrum of activity of imipenem. *Am J Med.* 1985;78(supple 6A):22–32.
66. Duff P. *Obstetrics and Gynecology Clinics of North America: Antibiotic Use in Obstetrics and Gynecology.* Philadelphia, Pa:WB Saunders; 1992:Vol 19, No. 3.
67. American College of Obstetricians and Gynecologists. *Precis III: An Update in Obstetrics and Gynecology.* Washington, DC: The American College of Obstetricians and Gynecologists; 1986:17.
68. Landers DV, Sweet RL. Current trends in the diagnosis and treatment of tuboovarian abscess. *Am J Obstet Gynecol.* 1985;151:1098–1110.
69. Sweet RL. Pelvic inflammatory disease: prevention and treatment. *Modern Medicine.* 1987;55:64–70.
70. Esenbach DA. Acute pelvic inflammatory disease. In: Schiarra JJ, Droegmeuller W, eds. *Gynecology and Obstetrics*, Rev ed. Philadelphia, Pa:JP Lippincott; 1990:(1)44:1–21.

8
Human Immunodeficiency Virus (HIV) and Hepatitis

Patricia Herrera and J. Paul O'Keefe

Viral hepatitis is caused by five known enterically or parenterally transmitted viruses. The disease can be severe, especially in pregnancy, and chronic liver disease results from infection with two of the five agents. In addition, congenital infection may occur. Likewise, human immunodeficiency virus (HIV) infection and acquired immunodeficiency syndrome (AIDS) are becoming increasingly common and are also transmitted vertically. Unique manifestations of HIV infection in women are only recently receiving emphasis in the medical literature. In this chapter the epidemiology, clinical manifestations, treatment, and prevention of these important viral diseases in women and their offspring are reviewed.

Viral Hepatitis

In the 1940s, viral hepatitis, which is a common condition caused by several viral agents, was divided into type A (infectious hepatitis) and type B (serum hepatitis). Since then, three other agents have been described. Besides the hepatitis A and B viruses, the delta agent and two non-A non-B hepatitis viruses have been identified.[1] The non-A non-B viruses include a parenterally transmitted agent (hepatitis C)[2] and an enterically transmitted agent (hepatitis E) (See Table 8.1).[3]

The five hepatitis viruses and their associated epidemiology and clinical syndromes are described in the following sections. In general, management of the disease is symptomatic, but biochemical and serologic follow-up is indicated for infections with the viruses that may produce chronic infection (HBV, delta agent, and HCV). Specific management in pregnancy and risk to the fetus or newborn are also discussed.

Hepatitis A

Hepatitis A virus is caused by an RNA virus that is most often transmitted by the fecal-oral route. Person-to-person transmission among intravenous drug users has also been recognized.[4] Transmission via blood transfusion has been described but is uncommon.[5] The incubation period of hepatitis A ranges between 15 and 45 days. Patients characteristically have abrupt onset of symptoms that may include fever, anorexia, jaundice, malaise, and dark urine. Pre- and postexposure prophylaxis with immune globulin has been shown to be protective against clinical illness. Preexposure prophylaxis is recommended for international travel to developing countries. Postexposure prophylaxis is recommended if exposure has been within 2 weeks for the following exposed groups: household and sexual contacts of the index case; staff of daycare centers; residents and staff in nursing homes, and persons who have had close contact with the index case.[6]

A person with hepatitis A is infectious for the period of approximately 2 weeks immediately preceding the onset of jaundice. Serologically, one finds IgM anti-HAV in

TABLE 8.1. Summary of hepatitis viruses and their effects on pregnancy and the fetus.

Virus	Mode of transmission	Clinical course of pregnancy	Risk to fetus
Hepatitis A	Fecal-oral	Risk of premature labor and delivery	Premature labor, minmal risk of transmission
Hepatitis B	Parenteral Sexual	Risk of premature labor and delivery	Prematurity, chronic hepatitis B, hepatocellular carcinoma, cirrhosis
Hepatitis C	Parenteral Sexual	Risk of premature delivery; fulminant hepatitis in the third trimester	Prematurity, fetal death as a result of maternal death in fulminant disease
Hepatitis D	Parenteral Sexual	As in hepatitis B since may exist as a coinfection	Potential for transmission since may be cotransmitted with the hepatitis B virus
			No chronic course described is as opposed to hepatitis B
Hepatitis E	Fecal-oral	Fulminant hepatitis, particularly in the third trimester	Fetal death as a result of maternal death in fulminant hepatitis

acute disease. The presence of IgG anti-HAV indicates past infection and immunity.

Hepatitis A is self-limited and usually followed by recovery. Neither chronic carriers nor a chronic disease state have been identified.[6]

In pregnancy hepatitis A infection is associated with an increased risk of premature labor and delivery.[7] Generally it is not transmitted to infants.[8]

Hepatitis B

The hepatitis B virus is a DNA virus. It is transmitted parenterally, perinatally, or sexually. Nosocomial infections via transfusion, hemodialysis, and accidental parenteral exposure are also well documented.[9] It causes both acute and chronic hepatitis, is responsible for a significant number of cases of cirrhosis, and is associated with hepatocellular carcinoma. The incubation period averages 70 days, with a range of 30 to 180 days.

As opposed to hepatitis A, the onset of hepatitis B infection is insidious. Besides symptoms such as malaise, anorexia, low-grade fevers, and jaundice, acute hepatitis B can present with a serum sicknesslike syndrome with prominent arthralgias and rash. Aminotransferases peak at 20 to 50 times normal.[10]

Serologically, one finds hepatitis B surface antigen in the acute phase. HBsAg usually disappears with convalescence, and persistence of HBsAg for more than 6 months indicates chronic infection. Antibody to hepatitis B core appears at the onset of symptoms. It is a more reliable indicator of past infection with hepatitis B virus because it is not cleared as rapidly as HBsAg. Hepatitis B antigen indicates viral replication and is associated with a high degree of infectivity.

Hepatitis B (HBV) vaccine was initially recommended for high-risk health care workers, homosexually active men, hemodialysis patients, household and sexual contacts of HBV carriers, intravenous drug users, hemophiliacs who receive factor VIII or IX blood products, and HIV-infected persons.[6] Because of poor compliance on the part of providers in vaccinating risk groups and because of continued sexual and perinatal transmission, candidate population groups for the vaccine are being expanded to include all children.[11,12] Recombinant DNA-derived vaccines are the preferred vaccines. Primary vaccination consists of three intramuscular doses, and the second and third doses are given at 1 and 6 months after the first. A four-dose regimen has been approved for one of the recombinant vaccines.

Hepatitis B infection during pregnancy may be associated with premature labor and delivery.[7] Women who are both HBsAg and HBeAg positive at the time of delivery have a 70% to 90% chance of transmitting the infection to their newborn infants. With HBsAg alone the infection rate is 25%. Ninety percent of infected infants will become chronic HBsAg carriers, and of these, 25% will develop cirrhosis or hepatocellular carcinoma as adults.[13] Pregnant women should be screened for HBsAg. Infants born to HBsAg-positive mothers should receive HBIG and hepatitis B vaccine at birth.[14] As mentioned above, the CDC has recently recommended universal vaccination of children.

Delta Hepatitis

The delta agent is a defective RNA virus that requires HBsAg for replication.[1] It is transmitted parenterally and sexually.[15] It can occur as an acute infection in conjunction with acute hepatitis B or as a superimposed acute infection in a chronic HBsAg carrier. Infection in a chronic carrier tends to be aggressive, causing the underlying liver disease to progress rapidly. Of patients with delta infection, 60% to 70% will develop cirrhosis, and most of these die of liver disease. Patients with simultaneous acute infection with HBV and delta recover with resolution of hepatitis and clearance of HBsAg from the serum. Hepatitis D infection in pregnancy usually follows the same course as hepatitis B infection. Fulminant hepatitis D infection in neonates is uncommon.[8]

By detecting HDAg, a test available only in research laboratories, or by measuring total or IgM-specific antibody (antidelta) during or after infection, a specific diagnosis of delta infection can be made.[16]

Patients immunized with hepatitis B vaccine are also protected from infection with the delta agent.

Hepatitis C

Hepatitis C is caused by an RNA virus following an incubation period of approximately 50 days. It is transmitted parenterally and currently accounts for approximately 85% of cases of posttransfusion hepatitis. Sporadic cases have also been reported. When compared to hepatitis B infection, hepatitis C is characterized by a more indolent course. The death rate is less than 1 percent.[17] However, hepatitis C commonly progresses to chronic hepatitis and cirrhosis. Interferon alpha has recently been shown to induce clinical and biochemical remission of hepatitis caused by HCV.[18] The long-term efficacy of interferon alpha is not yet established.

Diagnosis of HCV infection is established by demonstrating antibody to HCV. Since the antibody to hepatitis C arises late in the course of the acute illness (approximately 3 months into the infection), the test is of little use in the diagnosis of acute infection.

Vertical transmission of HCV has been documented. Epidemiologic studies have shown that women who become infected with HCV in the third trimester pose the greatest risk to their infants.[8] Studies have shown that immune globulin given before blood transfusion reduces the severity and transmissibility of HCV.[19] Thus in pregnant patients requiring blood products, pretransfusion immune globulin should be considered. The CDC has not specifically recommended immune globulin for persons with percutaneous exposure to HCV-positive blood from a patient with HCV infection but says it may be reasonable to use it as soon as possible after exposure.[6]

Hepatitis E

Hepatitis E virus is an RNA virus responsible for epidemics of food or waterborne hepatitis in Asia, North and West Africa, and Mexico.[6] With the exception of immigrants and returning travelers, the disease has not been reported in the United States. Hepatitis E is also known as enterically transmitted non-A non-B hepatitis and is transmitted via the fecal-oral route. Infection usually results in a self-limited disease that typically presents with a cholestatic picture and does not progress to chronic hepatitis. In epidemics the highest prevalence has been in young adults. The highest mortality has been in pregnant women[20]; this is particularly true if infection occurs in the third trimester, when it produces a particularly fulminant infection. The best

HIV Infection and AIDS

Infection with HIV is emerging as a serious health concern for women. In New York City, AIDS is the leading cause of death among women between the ages of 25 and 34. The first female with AIDS in the United States was reported in 1981.[21] As of December 31, 1991, a cumulative total of 206,392 cases of AIDS had been reported to the Centers for Disease Control. In that same year there were 45,506 new cases in women, representing a 5% increase over 1990. The number of cases in women in 1991 increased by 15% over those in 1990, compared with a 3.6% increase in men over the same period.[22]

Although AIDS in women has increased dramatically over the years, the male-to-female ratio for adults, excluding men with homosexual contact, has remained constant at approximately 2.5:1. In 1990 intravenous drug use was listed as a risk behavior in 48% of the 4,890 reported AIDS cases in adolescent and adult women. Heterosexual contact with a man infected with HIV or at risk for HIV infection was the reported means of contact for 34% of these women, up from 30% in the previous year.[23] AIDS in women has been reported in all 50 states. The highest cumulative incidence of AIDS in women occurs in New York, Washington, D.C., Florida, Puerto Rico, and New Jersey. In these areas, most women with AIDS reside in a small number of counties with large urban populations. Racial or ethnic minority groups are disproportionately represented among HIV-infected women. Although 19% of U.S. women are black or Hispanic, 72% of U.S. women diagnosed with AIDS come from these two minority groups.[24] Since nearly 85% of women with AIDS are of reproductive age, transmission must be concomitantly considered both a medical and a social issue.

The Centers for Disease Control has also studied the epidemiology of AIDS in lesbians. Although little is known about transmission of HIV among women who have had sex only with other women, most lesbians with AIDS also report use of intravenous drugs. Although woman-to-woman transmission of HIV appears to be an extremely rare event, two reports of AIDS among lesbian and bisexual women indicate that women who engage in sex with other women can be exposed to HIV.[25,26] Case reports of female-to-female sexual transmission of HIV and the well-documented risk of female-to-male HIV transmission indicate that vaginal secretions and menstrual blood are potentially infectious and that oral vaginal exposure to these secretions can potentially lead to HIV transmission.[27]

Course of HIV Infection

Knowledge about the natural history of HIV infection is derived primarily from individuals who differ from women in gender, income, and risk factors. Specific information regarding the natural course of HIV infection in women is not available. Most of what is known about HIV in women has focused on the effect of HIV infection on pregnancy and perinatal transmission.

After infection with HIV most people remain asymptomatic for a long period. However, the infection is not entirely latent. Viral replication proceeds at low levels, and these asymptomatic individuals are infectious. Studies show that after infection the virus attaches to and is taken up by T-lymphocytes with the T4 or CD4 surface receptor. These cells are called T-helper lymphocytes and play a key role in modulating the immune response. Other cells of the monocyte/macrophage lineage also have the CD4 surface receptor and may be infected. Over time there is a gradual depletion of the T-helper population. When the number of T4 cells reaches a critical low level, the host's immune response is no longer adequate, and opportunistic infections develop.

Care of HIV-Infected Women

The impact of gender on the course and presentation of HIV disease, coupled with increasing numbers of infected women, has created a need for guidelines on the care of

HIV-infected women. It has also been shown that certain gender-specific infections such as moniliasis, pelvic inflammatory disease, and human papilloma virus and associated cervical neoplasia have a more aggressive and rapid progression in women who are HIV infected.[28] In the following sections, the clinical behavior and specific management of HIV disease in women as well as management of selected opportunistic infections and malignancies will be discussed.

Primary Care of HIV Disease

With the increasing rate of HIV infection, it is important for primary care providers to be able to recognize the unique manifestations of HIV infection in women. It is also imperative that these physicians be familiar with the complex medical, reproductive, psychosocial, and ethical issues in this population. Many women in the United States are not aware that they have been infected with HIV and remain undiagnosed until their first AIDS-defining illness or until a child, infected perinatally, becomes ill.

The reasons that HIV-infected women present at a later stage in their illness than do HIV-infected men is not understood. It may be related to poor access to health care, poor utilization of health care, a low index of suspicion among providers of health care to women, or a greater tolerance of symptoms before seeking medical care.[29]

Public health authorities and AIDS experts recommend that screening for HIV occupy a more prominent place in primary medical care. The virus and the disease should be discussed during primary care encounters, including routine prenatal and gynecological office visits. Individuals whose sexual or social history suggests that they may be at risk should be counseled and tested for the virus. Counseling and HIV testing procedures include basic education about the biology and epidemiology of HIV and its mechanism of transmission, about the disease AIDS, and about the implications of testing. Counseling should also include recommendations for testing of sex partners, discussion of safer sex practices, and counseling in regard to childbearing.[29] Finally, referral for emotional counseling or for help with social or economic problems is mandatory.

Written informed consent for HIV testing is required in many states. Testing must be carried out in a manner that will protect the confidentiality of the patient. Alternatively, the patient may prefer to be tested at a designated anonymous testing center if one is available.

Primary care of HIV-infected persons does not require special training and can be provided by obstetrician-gynecologists. After the diagnosis is established, a careful review of systems, physical examination, and a baseline evaluation of the immune status are performed. During the examination careful attention should be paid to the skin, oral cavity, and lymph nodes. Laboratory screening should include the following: hemoglobin, white blood cell count with differential, platelet count; chemistry screen including glucose, creatinine, cholesterol, liver function tests, albumin, and globulin; RPR; toxoplasmosis titer; skin testing with PPD and an anergy battery; and enumeration of CD4+ T-lymphocytes.[30]

Healthy patients are seen at 6 month intervals. Repeat physical examination including Pap smear and repeat CD4 count is performed. Antiretroviral therapy with zidovudine (AZT) is begun when the CD4 count reaches 500/mm^3. To monitor for toxicity, patients are seen monthly or bimonthly, and a complete blood count is obtained. When the CD4 count reaches 200/mm^3, the risk for opportunistic infections increases. At that time prophylaxis for *Pneumocystis carinii* pneumonia is begun with trimethoprim sulfamethoxazole. The recommended dose of trimethoprim sulfamethoxazole is one double-strength tablet daily, but many clinicians follow a schedule of one double-strength tablet three times weekly.[31]

Management of Specific Problems in Women with HIV Infection

Candidiasis

The most commonly observed presentation of AIDS in women is mucocutaneous candidiasis. Vaginal candidiasis might be considered an indicator of possible HIV infection. Mucocutaneous infections are known to occur in a hierarchical pattern.[32] Recurrent and often severe vaginal candidiasis can occur when there is no significant reduction in CD4 lymphocyte counts, whereas oropharyngeal candidiasis is associated with highly significant reductions in CD4 lymphocyte counts. Esophageal candidiasis occurs with advanced immunodeficiency. Although vaginal candidiasis often occurs with normal CD4 lymphocyte counts, a few studies have shown that chronic unexplained vaginal candidiasis may be the presenting symptom in women with severe T-cell depletion.[33] In these patients chronic yeast vaginitis can predate oral thrush and is the only clinical indication of a severe underlying immunodeficiency. Because vaginal candidiasis is a common clinical entity in women without an underlying immunodeficiency, health providers may not recognize it as a sign of HIV infection. Therefore all women with vaginal candidiasis should be thoroughly questioned about the number of sexual contacts and *their* habits, blood transfusions, and injectable drug use. They should also undergo a careful physical examination for oral thrush and lymphadenopathy to help the physician identify patients who require further evaluation for HIV disease.

Initial therapy of symptomatic vaginal candidiasis in HIV-infected women consists of a standard topical antifungal medication. If a follow-up examination indicates a treatment failure or frequent recurrences in the absence of correctable causes such as oral contraceptives or antibiotic therapy, oral preparations such as ketoconazole or fluconazole should be prescribed.

Pelvic Inflammatory Disease

Studies have shown that PID may manifest differently in patients with HIV infection. At presentation there may be minimal leukocytosis. Pelvic abscesses with associated requirement for surgical intervention have been noted more frequently in patients infected with HIV than in those not infected.[28] Up to 17% of HIV-infected women have pelvic inflammatory disease. Detailed studies documenting the etiologic organisms in these cases of PID have not been reported, but both *Chlamydia trachomatis* and *Neisseria gonorrhoeae* are seen. Information regarding the natural history of PID caused by specific pathogens in HIV-infected patients is also lacking. Furthermore, it is not known whether infection with HIV increases the likelihood that lower genital tract pathogens will invade the uterus and fallopian tubes.[34,35]

Treatment of PID should consist of a broad-spectrum antibiotic regimen that would cover *N gonorrhoeae*, *C trachomatis*, facultative Gram-negative rods, genital anaerobes, and group B streptococcus. Suggested regimens include cefoxitin and doxycycline or clindamycin and gentamicin.[36] Experts recommend that immunocompromised patients with PID including HIV infection be managed in the hospital.[28,37]

Genital Ulcer Disease

Sexually transmitted diseases associated with genital ulcers are considered as possible cofactors that might increase transmission of HIV. The denuded cutaneous or mucosal surface is thought to represent a damaged mechanical barrier. This increases the susceptibility to infection of the HIV-exposed partner as well as increases the infectiousness of the HIV-infected individual.[38] Indeed, epidemiological studies done in Nairobi, Kenya, identified genital ulcers as independent risk factors for HIV infection.[39]

The differential diagnosis of infectious genital ulcer disease includes herpes simplex, chancroid, and syphilis. In addition, just as ulcerative lesions in the esophagus and other parts of the gastrointestinal tract have been convincingly attributed to HIV itself, recent reports of culture negative genital ulcers responding only to AZT suggest a role for HIV in this location as well.[40]

Syphilis

There has been a dramatic increase in the incidence of syphilis associated with the HIV epidemic. This increase is occurring at a time when studies indicate that patients with syphilis are more likely to be infected with HIV.[41] Furthermore, HIV-infected patients with syphilis are more likely to progress to clinical neurosyphilis and are more likely to fail treatment with standard drug regimens for syphilis.[42] The CDC now recommends screening with RPR in all HIV-infected persons and HIV testing of all patients with syphilis. In addition, all HIV-infected patients with syphilis should undergo CSF evaluation as part of the routine diagnostic work-up.[36] Standard treatment regimens for syphilis are used in patients who are also infected with HIV, but more rigorous follow-up is recommended. HIV-infected patients treated for syphilis should be followed clinically and with quantitative nontreponemal serologic tests at 1, 2, 3, 6, 9, and 12 months. Patients treated for early syphilis whose nontreponemal test either increases or fails to demonstrate a fourfold decline within 6 months should undergo CSF examination and be retreated.

Treatment of syphilis in pregnant women should be with penicillin only.[36] Erythromycin does not reliably treat the fetus, and tetracycline is toxic to the fetus. Recommended preparations and doses of penicillin are similar to those recommended for nonpregnant patients. Pregnant women who are allergic to penicillin should be skin tested and desensitized with penicillin.[43] Posttreatment follow-up should include nontreponemal testing monthly so that retreatment can be given if needed.

Herpes Simplex

Genital herpes infections in HIV-infected women behave more aggressively than in the general population. More symptomatic at presentation and often displaying a more protracted course in HIV-infected women, HSV infections should be treated with acyclovir. Response to treatment is not universal. Failure to respond may be a manifestation of viral resistance to acyclovir, probably resulting from the long-term use of the drug in these patients. Higher doses of oral acyclovir or intravenous therapy have been successful, but the highly resistant virus does not respond.[44] Alternative therapy with foscarnet or ganciclovir has been successful in treating resistant HSV.[45,46] Unfortunately, both drugs must be given intravenously and are more toxic than acyclovir. Frequent recurrences of genital HSV infection can be suppressed with daily oral acyclovir.

Cervical Neoplasia

Cervical neoplasia has many characteristics of a sexually transmitted disease; thus an association with HIV can be anticipated on the basis of common behavioral risk factors. It is known that women with impaired cellular immunity are susceptible to genital papilloma virus and neoplasia.[47] Since immunosuppressed women are known to be at risk for lower genital tract neoplasia, immunodeficient HIV-infected women are predictably at high risk. CIN is often higher grade and more extensive among HIV-infected women than noninfected women. Multiple sites on the cervix, other genital sites, and even perianal neoplastic lesions have been reported.[44] As a result of published studies documenting greater frequency of cervical neoplasia in HIV-infected patients compared with noninfected patients, some authors recommend HIV testing of young women with cervical neoplasia in areas of high HIV prevalence.[48]

Human papilloma virus infection is a common problem in HIV-infected women. Condylomata acuminata occurs more frequently and may be difficult to manage.[44] In studies of HIV-infected women categorized by status of HIV disease, HPV infection was detected in 67% of symptomatic HIV-positive women, 31% of asymptomatic HIV-positive women, and 27% of HIV-negative women.[49] The HIV-infected women in whom HPV was not detected did not demonstrate an increased incidence of CIN.[48] Results of this investigation suggest that HIV-induced immunosuppression may predispose to HPV-mediated cervical abnormalities.

With recognition of the increased incidence of HPV-associated cervical neoplasia in HIV-infected women, the CDC has published guidelines for Pap smears.[50] Annual tests are recommended for women whose results are normal and indicate that the specimen is adequate. Given the risk in this population, other experts have recommended testing more frequently and lowering the threshold for performing colposcopy.[51]

Abnormalities of Menses

Menstrual disorders have been noted anecdotally in HIV-infected women. The most common abnormalities are amenorrhea and oligomenorrhea. This may cause anxiety in patients because of difficulty in predicting times of fertility.

Contraception

Family planning in HIV-infected women has been a difficult problem.[28] What prevents conception most effectively may not be the best method for preventing transmission of HIV. Although a diaphragm and sponge or an IUD may be used, each has potential for causing abrasions and thus may provide a portal of entry for HIV infection.[52] Women may stop using oral contraceptives and start using condoms to prevent HIV transmission; however, the risk of conception increases with condoms. On the other hand, oral contraceptives alone do not prevent HIV transmission. Because of this dilemma, women may find themselves using two methods, one to prevent pregnancy and the other to prevent HIV transmission, for example, birth control pills and condoms. Using two methods requires high motivation.

HIV Infection and Pregnancy

HIV Infection

Eighty percent of reported AIDS in women has been diagnosed during the reproductive years. Data regarding outcome of pregnancy in HIV-infected women have been conflicting. Retrospective studies are faulted by including patients who present very late in pregnancy or even in labor. Thus factors other than HIV that adversely affect the pregnancy could not be studied.

In 1987 Minkoff et al published a retrospective analysis of pregnancy outcome in women whose children subsequently developed HIV.[53] In the study one-third developed premature rupture of membranes and preterm birth. However, there was no control group, and the majority of women studied were injectable drug users, a lifestyle known to affect pregnancy adversely. Prospective controlled studies have shown no differences between seropositive and seronegative groups in the frequency of elective or spontaneous abortions, preterm delivery, still birth, or low birth-weight infants.[54] While asymptomatic HIV infection may not affect pregnancy outcome, other studies have reported that children of HIV-infected women who are symptomatic during pregnancy have significantly higher mortality rates.[55]

Both HIV and pregnancy are associated with alterations in cell-mediated immunity. In pregnancy this is seen in the second and third trimesters. In HIV-infected patients CD4 percentages have been noted to increase early in pregnancy and then to decline. Because of this, authorities recommend measuring the CD4 count during each trimester so that appropriate therapy can be instituted when the count declines.

Opportunistic Infections

Opportunistic infections in HIV-infected pregnant women also present a problem. Not only is the mother's health jeopardized by the infection, but the fetus may be infected with CMV, HSV, or toxoplasmosis. In addition, the fetus may also be subjected to the adverse effects of treatment for the opportunistic infection in the mother.

Use of Medications for HIV Disease and Opportunistic Infections in Pregnancy

Recommendations for AZT and PCP prophylaxis were based on large, multicenter studies conducted in men and women. However, evaluations in pregnant HIV-infected women were not extensive. There is scant information regarding the effect of AZT on pregnancy, on the fetus, or on reproductive capacity. AZT has not been shown to concentrate in the placenta or fetal tissue. But there are differences of opinion among high-risk perinatologists regarding the point at which AZT should be started in pregnancy. Most agree that AZT should be instituted as soon as the CD4 count reaches $200/mm^3$.[28,31,56,57]

Prospective studies on PCP prophylaxis in pregnancy are lacking. Most recommendations are based on inference. Although trimethoprim sulfamethoxazole has been contraindicated in the last trimester of pregnancy,[57a] it is the preferred prophylactic agent throughout pregnancy in many centers.[28,57] Aerosolized pentamidine has the advantage of little systemic absorption.[58] There may, however, be altered distribution when lungs and diaphragm are compressed by the gravid uterus.[57] Dapsone, an effective alternative to trimethoprim sulfamethoxazole, has been used in dermatitis herpetiformis and leprosy in pregnant women.[28]

Pyrimethamine, used for toxoplasmosis, appears to be safe in pregnancy.[59] However, care should be taken since it is a folate antagonist. One study in mice has shown that ganciclovir, used for the treatment of symptomatic infections with CMV, caused embryonic retardation and congenital abnormalities.[45]

A discussion of treatment of the infectious complications of AIDS during pregnancy is beyond the scope of this chapter. In general, studies on treatment of opportunistic infections in HIV-infected pregnant women are lacking. Reviews on this topic have been published elsewhere.[28,55]

Perinatal Transmission of HIV

Although children make up only a small percentage of reported AIDS cases in the United States, only crude estimates of the prevalence of HIV infection in children are available.[60] Throughout the epidemic the incidence of HIV-infected women has increased. Since most HIV-infected women are of childbearing age, the number of infected children is expected to increase. A greater portion of minority populations are injectable drug users, and therefore up to 75% of HIV-infected children are African-American or Hispanic.[61]

There are three possible means of viral spread from mother to child: intrauterine; intrapartum (through the birth canal); and postpartum (through breastfeeding).

Intrauterine transmission across the placenta has been shown by the identification of virus in fetuses as early as the 15th week of gestation.[62,63]

After early studies noted HIV in both cervical and vaginal secretions, intrapartum transmission initially became a concern. Some authors contend that intrapartum transmission is frequent. However, studies showing no difference in infection rates in full-term infants delivered vaginally versus abdominally indicate that intrauterine transmission is more likely.[59] Premature infants, on the other hand, appear to have a higher risk of becoming infected with HIV than full-term infants. This suggests either that prematurity may result from early in utero HIV infection or that prematurity per se places the infant at higher risk of becoming infected during labor, delivery, or both.

Most studies suggest that perinatal transmission occurs in approximately 30% of infants born to seropositive mothers.[59] However, the reported range is great.[64] Some studies have suggested perinatal transmission may be more common in women at a later stage of their illness.[65] However, other studies have shown that neither CD4 count at time of delivery nor diagnostic category, that is, asymptomatic or symptomatic HIV infection or AIDS, are associated with frequency of transmission.[60]

The third potential method of vertical transmission, which occurs postpartum, is breastfeeding. HIV has been isolated from cells in human milk. Breastfeeding was established as a mechanism of transmission when accounts of HIV-infected children, whose sole exposure was breastmilk, were reported.[66,67,68] Mothers who are recently infected and have viremia appear to be at higher risk of infecting their children by breastfeeding.[69] Because of the potential risk of transmission, authorities recommend that women who are HIV infected or at high risk of being HIV-infected should not breastfeed their children.[69,70,71] Formula, where available and affordable, should be used. Conversely, in developing countries the advantages of breastfeeding in preventing death from infantile diarrheal disease in the newborn may outweigh the risk of HIV transmission in breastmilk.[69]

All three methods of perinatal transmission appear to occur. Precautionary measures can be taken in an attempt to decrease the risk of transmission. While HIV infection is not a contraindication to vaginal delivery, invasive procedures such as amniocentesis, chorionic villous sampling, and scalp monitoring during delivery are relatively contraindicated.[72]

Diagnosis of Neonatal Infection

Transplacental transfer of maternal antibody to the fetus renders early diagnosis of postnatal HIV infection problematic. Maternal IgG can be found in children up to 15 months of age. Consequently, children cannot be reliably tested for HIV antibody with conventional ELISA or Western blot testing until they are 15 months of age. If treatment is to be introduced before symptoms of HIV infection appear, availability of a laboratory test that will detect infection regardless of maternal antibodies is imperative.

The recent development of the polymerase chain reaction, (PCR), a technique that amplifies sequences of proviral HIV DNA, is an exquisitely sensitive method for the diagnosis of HIV infection.[73,74] PCR sensitivity for detecting HIV in a newborn can be as high as 95%. However, the test is not readily available. Detection of IgA antibodies to HIV is another method that has been developed for early diagnosis of perinatally acquired HIV infection.[75] Because IgA does not cross the placenta, finding these antibodies would indicate the presence of infection. The sensitivity is greatest after the third month of age.[76] Although still considered investigational, one or both of these tests should make accurate, early diagnosis of HIV infection in neonates possible and should open up opportunities for early intervention.

Occupational Exposure to Hepatitis and HIV

Transmission of HBV and HIV to health care workers is well documented.[77,78] Transmission of HCV also occurs but is less common than hepatitis B virus.[79] HBV and HIV infections of health care workers occur with parenteral or mucous membrane contact with infected blood or other virus-containing body fluids. Studies suggest that injuries by needle sticks or other sharp instruments pose a far greater risk of infection than exposure to a mucosal surface or intact skin.[80]

Studies have shown that the risk of hepatitis B infection after needle stick exposure to HBsAg-positive blood is as high as 30%.[77] Similar studies have shown that risk of HIV is 0.4%, and thus HIV is less infectious than hepatitis B.[78] Risk after mucous membrane exposure or cutaneous exposure is not quantitated, but the importance of nonintact skin (that is, abrasions or dermatitis) in the exposed area as an additional risk factor has been emphasized.[81]

In 1987 the Centers for Disease Control recommended that universal precautions be adopted to protect health care workers from infection with these agents.[82] In 1992 the Occupational Safety and Health Administration published regulations that mandated the Centers for Disease Control recommendations.[83] With universal precautions, blood and body fluid precautions are applied to all patients rather than only to those known or suspected of being infected with HIV or other blood-borne viruses. The application of universal precautions requires that barrier precautions to cover the hands, mucous membranes of the eyes and mouth, and clothing be employed when exposure to blood or body fluids is anticipated. Thus gloves are required during procedures such as phlebotomy, when placing vascular access catheters, or when dressing wounds in which hand exposure is likely. Goggles, mask, and impermeable gowns must be worn during procedures characterized by splashing or aerosolization of fluids such as delivery.

The second component of the OSHA regulations is sharp instrument safety. Sharp instrument safety programs include the use of disposable instruments whenever possible and the widespread availability of puncture-proof containers for their disposal. Risky procedures such as blind surgery, use of the opposite hand to guide sharp instruments, and even instrument passing should be minimized or avoided. Educational programs emphasizing sharp instrument safety as well as the elimination of careless disposal of sharp instruments are required.

A protocol for management of accidental exposures should be established, which should include testing of known source patients for HBsAg and HIV. The exposed employee should receive first aid for sharp instrument injury and should undergo baseline testing for immunity to HBV and HIV. Depending on the status of source and recipient, the employee should receive postexposure prophylactic treatment for HBV infection and/or prophylactic AZT. The staff of a private physician's office may develop agreements with the Occupational Health Service of a local hospital for the management of exposures in the office.

Hepatitis B vaccine must be provided to all health care workers whose occupation might place them in contact with blood or body fluids. Previous infection or vaccination confers immunity indicated by the presence of antibody to HBsAg (anti-HBs). Nonimmune employees may decline to receive the vaccine but must sign a waiver indicating that they are aware of the benefits of vaccination.

Much has been written or spoken recently in the media as well as professional literature about health providers infected with HIV or HBV. Twenty outbreaks of hepatitis B in patients of infected physicians, including obstetrician gynecologists, have been well studied and reported.[84] Up to the present only one outbreak of HIV infection in five patients of a dentist with AIDS has been documented.[85] The Centers for Disease Control issued guidelines for exposure-prone invasive procedures in July 1991.[84] The guidelines stated that practitioners who perform these procedures were required to "know their HIV status" and, if positive, to inform patients before performing the procedure. Extensive debate ensued, and, although still supported by the Centers for Disease Control, the mechanism for implementation, a listing of the procedures, and a schedule for testing have not been published. Whatever becomes of these guidelines, HIV-infected physicians who know their status should establish a relationship with their own doctor through which both personal health can be optimally maintained and a program for management of professional matters can be developed.

Summary

Viral hepatitis and HIV infections are common, important problems seen in routine practice of obstetrics and gynecology. Practitioners should be familiar with the epidemiology and clinical syndromes so that affected patients can be identified and managed.

References

1. Centers for Disease Control. Delta hepatitis. *MMWR*. 1984;33:493.
2. Choo Q-L, Kuo G, Weiner AJ, et al. Isolation of a cDNA clone derived from a blood-borne non-A, non-B hepatitis genome. *Science*. 1989;224:359–362.
3. Ramalingaswami V, Purcell RH. Waterborne non-A, non-B hepatitis. *Lancet*. 1988;1:571–3.
4. Centers for Disease Control. Hepatitis A among drug users. *MMWR*. 1988;37:297–300,305.
5. Nobel RL, Kane MA, Reeves SA, et al. Post transfusion hepatitis A in a neonatal intensive care unit. *JAMA*. 1984;252:2711-S.
6. Centers for Disease Control. Protection against viral hepatitis. *MMWR*. 1990;39:1–26.
7. James S, Disaia P, Hammond CP, et al. General medical and surgical diseases in pregnancy. *Danforth Obstetrics and Gynecology*. 1990;508–509,540–541.
8. Snydman D. Hepatitis in pregnancy. *N Engl J Med*. 1985;313:1398–1401.
9. Oren I, Hershow R, Ben-Porath E, et al. A common source outbreak of fulminant hepatitis B in a hospital. *Ann Intern Med*. 1989;110:691–698.

10. Sherlock S. Predicting progression of acute type B hepatitis to chronicity. *Lancet.* 1976; 2:354–356.
11. Centers for Disease Control. Hepatitis B virus: a comprehensive strategy for eliminating transmission in the United States through universal childhood vaccination. *MMWR.* 1991;40 (RR13):1–19.
12. Koretz R. Universal prenatal hepatitis B screening. *Obstet Gynecol.* 1989;74;808–814.
13. Reeink HW, Cafever, Scishut, et al. Prevention of chronic HBsAg carrier state in infants of HBsAg positive mothers, by hepatitis B immunoglobulin. *Lancet.* 1979;1:436–438.
14. Kane M, Hadler S, Margolis H. Routine prenatal screening for hepatitis B surface antigen. *JAMA.* 1988;259:408–409.
15. Rocca G, Poli G, Gerard P, et al. Family clustering of delta infection. Viral hepatitis and delta infection. New York: Alan R. Liss; 1984:133–137.
16. Rizzetto M, Verma G, Recchia 5, et al. Chronic HBsAg hepatitis with intrahepatic expression of delta antigen. *Ann Intern Med.* 1983;98:437–441.
17. Dienstag JL. Non-A non-B hepatitis: recognition, epidemiology, and clinical features. Gastroenterology. *N Engl J Med.* 1983;84:133–137.
18. Davis G, Balart LA, Schiff ER, et al. Treatment of chronic hepatitis C with recombinant interferon alpha. *N Engl J Med.* 1989;321: 1501–1506.
19. Davis GL, Balart LA, Schiff ER, et al. Treatment of chronic hepatitis C with recombinant interferon alpha: a multicenter randomized, controlled trial. *N Engl J Med.* 1989; 321(22):1501–1506.
20. Gust I, Purcell R. Report of a workshop: waterborne non-A, non B hepatitis. *J Infect Dis.* 1987;156:630–635.
21. Ellerbrock T, Rogers M. Epidemiology of human immunodeficiency virus infection in women in the United States. *Obstet Gynecol Clin North Am.* 1990;17:523–541.
22. Centers for Disease Control. The second 100,000 cases of acquired immunodeficiency syndrome—United States, June 1981–December 1991. *MMWR.* 1992;41:28–29.
23. Centers for Disease Control. Update: acquired Immunodeficiency syndrome—United States, 1981–1990. *MMWR.* 1991;40:358–363,369.
24. Centers for Disease Control. AIDS in women—United States. *MMWR.* 1990;39:845–846.
25. Marmor M, Weiss LR, Leyden M, et al. Possible female-to-female transmission of human immunodeficiency virus (letter). *Ann Intern Med.* 1986;105:969.
26. Monzon OT, Capelian JMB. Female-to-female transmission of HIV (letter). *Lancet.* 1987; 2:40–41.
27. Chu S, Buehler J, Fleming P, et al. Epidemiology of reported cases of AIDS in lesbians, United States; 1980–1989. *Am J Public Health.* 1990;80:1380–1381.
28. Minkoff H, DeHovitz J. Care of women infected with the human immunodeficiency virus. *JAMA.* 1991;266:2253–2258.
29. Allen M. Primary care of women infected with the human immunodeficiency virus. *Obstet Gynecol Clin North Am.* 1990;17:557–567.
30. Kess S, Bresolin L, Henning J. HIV early care. *AMA Physician Guidelines.* 1990;1–15.
31. Centers for Disease Control. Recommendations for prophylaxis against pneumocystis carinii pneumonia for adults and adolescents infected with human immunodeficiency virus. *MMWR.* 1992;41:1–11.
32. Iman N, Carpenter C, Mayer K, et al. Hierarchical pattern of mucosal candida infections in HIV-seropositive women. *Am J Med.* 1990;89:142–146.
33. Rhoads J, Wright C, Redfield R, et al. Chronic vaginal candidiasis in women with human immunodeficiency virus infection. *JAMA.* 1987;257(2):3105–3107.
34. Hoegsberg B, Abulafia 0, Sedlis A, et al. Sexually transmitted diseases and HIV infection among women with pelvic inflammatory disease. *Amer J Obstet Gynecol.* 1990;163:1135–1139.
35. Safrin S, Dattel B, Hauer L, et al. Seroprevalence and epidemiologic correlates of human immunodeficiency virus in women with acute pelvic inflammatory disease. *Obstet Gynecol.* 1990;75:666–670.
36. Centers for Disease Control. Sexually transmitted diseases: therapy guidelines. *MMWR.* 1989;38:5–15.
37. Peterson HB, Galaid EI. Pelvic inflammatory disease: review of therapeutic options. *Rev Infect Dis.* 1990;12:S656–664.
38. Holmes K, Kreiss J. Heterosexual transmission of human immunodeficiency virus: overview of a neglected aspect of the AIDS epidemic. *J Acq Immun Def Synd.* 1988;1(6):602–610.
39. *Abstracts of The Third International Conference in AIDS.* Washington, DC; 1987:25.
40. Pomerantz R, DeLaMonte S, Donegan P, et al. Human immunodeficiency virus infection of

the uterine cervix. *Ann Intern Med.* 1988; 108(3):321–327.
41. Tramont EC. Syphilis in the AIDS era. *N Engl Med.* 1987;316:1600–1601.
42. Musher DM, Hamill RJ, Baughn RE. Effect of human immunodeficiency virus in the course of syphilis and on the response to treatment. *Ann Intern Med.* 1990;113:872–881.
43. Saxon A. Immediate hypersensitivity reactions to beta lactam antibiotics. *Ann Intern Med.* 1987;107:204–215.
44. Allen M. Primary care of women infected with the human immunodeficiency virus. *Obstet Gynecol Clinic North Am.* 1990;17(3):557–567.
45. Erlich KS, Jacobson MA, Kochler JE, et al. Foscarnet therapy for severe acyclovir-resistant herpes simplex virus type-Z infections in patients with the acquired immunodeficiency syndrome. *Ann Intern Med.* 1989;110:710–713.
46. Klug S, Lewandowski C, Merker H, et al. In-vitro and invivo studies on the prenatal toxicity of five virustatic nucleoside analogues in comparison to acyclovir. *Arch Toxicol.* 1991; 65:283–291.
47. Sillman F, Stanek A, Sedlis A. The relationship between human papilloma virus and lower genital neoplasia in immunosuppressed women. *Am J Obstet Gynecol.* 1984;150:300–308.
48. Maiman M, Fruchter R, Serure, et al. Human immunodeficiency virus infection and cervical neoplasia. *Gynecol Oncol.* 1990;38:377–382.
49. Carpenter C, Mayer K, Stein M, et al. Human immunodeficiency virus infection in North American women: experience with 200 cases and a review of the literature. *Medicine.* 1991;70(5):307–325.
50. Center for Disease Control. Risk for cervical disease in HIV-infected women—New York City. *MMWR.* 1990;39:846–849.
51. Maiman M, Tarricone N, Vieira J, et al. Colposcopic evaluation of human immunodeficiency virus-seropositive women. *Obstet Gynecol.* 1991;78(1):84–88.
52. Senanayake P, Kramer DG. Contraception and the etiology of pelvic inflammatory disease: new perspectives. *Am J Obstet Gynecol.* 1980;138:852–860.
53. Minkoff H, Nanda D, Menez R. Pregnancies resulting in infants with acquired immunodeficiency syndrome or AIDS-related complex: follow-up of mothers, children, and subsequently born children. *Obstet Gynecol.* 1987; 69:285–291.
54. Selwyn P, Schoenbaum E, Davenny K. Prospective study of human immunodeficiency virus infection and pregnancy outcomes in intravenous drug users. *JAMA.* 1989;261(9): 1289–1294.
55. Minkoff H, Willoughby A, Mendez H. Serious infections during pregnancy among women with advanced human immunodeficiency virus infection. *Am J Obstet Gynecol.* 1990;162:30–34.
56. Sperling R, Stratton P, O'Sullivan M. A survey of zidovudine use in pregnant women with human immunodeficiency virus infection. *N Engl J Med.* 1992;326:857–861.
57. Nanda D. Human immunodeficiency virus infection in pregnancy. *Obstet Gynecol Clin North Am.* 1990;17(3):617–626.
57a. Baskin CG, Law S, Wenger NK. Sulfadiazine rheumatic fever prophylaxis during pregnancy: does it increase the risk of kernicterus in the newborn? *Cardiology.* 1980;65:222–225.
58. Montgomery AB, Luce JM, Turner J, et al. Aerosolized pentamidine as sole therapy for pneumocystis carinii pneumonia in patients with acquired immunodeficiency syndrome. *Lancet.* 1987;1:480–482.
59. Minkoff H. Care of pregnant women infected with human immunodeficiency virus. *JAMA.* 1987;258(19):2714–2712.
60. MacGregor S. Human immunodeficiency virus infection in pregnancy. *Clin Perinatol.* 1991; 18(1):33–50.
61. MacDonald, Ginzburg H, Bolan J. HIV infection in pregnancy: epidemiology and clinical management. *J Acq Immun Def Synd.* 1991;4(2):100–108.
62. Sprechers S, Soumenkoff G, Puissant F, et al. Vertical transmission of HIV in 15-week fetus (letter). *Lancet.* 1986;2:288.
63. Rubinstein A, Bernstein L. The epidemiology of pediatric acquired immunodeficiency syndrome. *Clin Immunol Immunopathol.* 1986; 40:115–121.
64. Newell ML, Dunn D, Peckham CS, et al. Risk factors for mother to child transmission of HIV-1. *Lancet.* 1992;339:1007–1012.
65. Blanche S, Rourioux C, Moscato ML, et al. A prospective study of infants born to women seropositive for human immunodeficiency virus type 1. *N Engl J Med.* 1989;320:1643–1648.
66. Lepage P, Van de Perre P, Caraal M. Transmission of HIV from mother to child. *Lancet.* 1987;2:400.
67. Colbunders RL, Kapita B, Nekwer W, et al. Breast-feeding and transmission of HIV. *Lancet.* 1988;2:1487.

68. Logan S. Newell ML, Ades T, et al. Breast-feeding and HIV infection. *Lancet.* 1988;1:1346.
69. Hira SK, Mangrola UG, Mwale C. Apparent vertical transmission of human immunodeficiency virus type 1 by breast feeding in Zambia. *J Pediatr.* 1990;117:421–430.
70. Mendez H, Jule J. Care of the infant born exposed to human immunodeficiency virus. *Obstet Gynecol Clin North Am.* 1990;17:637–649.
71. Oxtoby MJ. Human immunodeficiency virus and other viruses in human milk: placing the issue in broader perspective. *Pediatr J Infect Dis.* 1988;7:825.
72. Minkoff H. Care of Pregnant women infected with human immunodeficiency Virus. *JAMA.* 1987;258:2714–2717.
73. Rogers M, Ou C, Rayfield M, et al. Use of the polymerase chain reaction for early detection of the proviral sequences of human immunodeficiency virus in infants born to seropositive mothers. *N Engl J Med.* 1989;320:1649–1654.
74. Comeau AM, Harris JA, McIntosh K. Polymerase chain reaction in detecting HIV infection among seropositive infants: relation to clinical status and age and to results of other assays. *J Acq Immun Def Synd.* 1992;5:27–278.
75. Weiblen BJ, Lee FK, Cooper ER, et al. Early diagnosis of HIV infection in infants by detection of IgA HIV antibodies. *Lancet.* 1990;335:988–990.
76. Quinn T, Kline R, Halsey N, et al. Early diagnosis of perinatal HIV infection by detection of viral-specific IgA antibodies. *JAMA.* 1991;266(24):3439–3442.
77. Seiff LB, Wright EC, Zimmerman HJ. Type B hepatitis after needlestick exposure: prevention with hepatitis B immunoglobulin: final report of the Veterans Administration cooperative study. *Ann Intern Med.* 1978;88:285–93.
78. McCray E. The cooperative needlestick surveillance group. Occupational risk of acquired immunodeficiency syndrome among health care workers. *N Engl J Med.* 314:1127–1132.
79. Hepatitis C in hospital employees with needlestick injuries. *Ann Intern Med.* 1991;115:367–369.
80. Update: universal precautions for prevention of transmission of human immunodeficiency virus, hepatitis B virus and other blood-borne pathogens in health care settings. *MMWR.* 1988;37:377–388.
81. Henderson DK, Fahey BJ, Willy M, et al. Risk for occupational transmission of human immunodeficiency virus type 1 (HIV-1) associated with clinical exposures: a prospective evaluation. *Ann Intern Med.* 1990;113:740–746.
82. Recommendations for prevention of HIV transmission in health care settings. *MMWR.* 1987;36-25:15–185.
83. Department of Labor Occupational Safety and Health Administration. Occupational exposure to bloodborne pathogens. *Federal Register.* 1991;56:64175–64182.
84. Recommendations for preventing transmission of human immunodeficiency virus and hepatitis B virus to patients during exposure-prone invasive procedures. *MMWR.* 1991;40(RR-8):1–9.
85. Centers for Disease Control. Possible transmission of human immunodeficiency virus to a patient during invasive dental procedure. *MMWR.* 1990;39:489–495.

9
Ectopic Pregnancy

James V. Brasch

Introduction

An ectopic pregnancy (EP) occurs whenever implantation of the fertilized ovum takes place on any tissue other than endometrium, including abnormal intrauterine implantations (the uterine cornua, uterine cervix) as well as more obvious extrauterine implantations (fallopian tube, peritoneal cavity). The first historical reference made to this entity was by the Arabic writer Albucasis in the year A.D. 963.[1] Described in 1693 by Busiere upon the examination of the body of an executed prisoner,[2] it was finally discussed in print in the 17th century in F. Mariceau's obstetrical textbook.[3] Originally, EP was treated by observation, starvation, purgation, ergot, and electrocution and had a maternal mortality of nearly 70%. The first successful operation was performed March 1, 1883, when Lawson Tait performed a salpingectomy for the treatment of an ectopic pregnancy.[4]

Although the diagnosis and treatment of EP have progressed greatly since the days of purgation, it is still a significant entity and can have serious consequences. The Centers for Disease Control reports that from 1970 to 1987 the incidence of EP in the United States has quadrupled. A total of 78,400 case EP were diagnosed in 1987, or, put another way, 1 out of every 66 diagnosed pregnancies was an ectopic pregnancy.[5] This does not include the now-recognized cases of tubal abortions that occur in significant numbers. Although there has been a tremendous rise in the incidence of the condition, there has actually been a concomitant fall in the incidence of maternal mortality; the last reported incidence of maternal mortality was 0.042%.[6] Still, even in large centers, fully 10% of all EP present in the ruptured state, and it represents an increasing proportion of maternal deaths.[7,8] More important, regardless of outcome the mere diagnosis of EP carries with it serious ramifications with respect to future fertility. Successful intrauterine pregnancy occurs in no better than 70% of future pregnancies, and EP recurrence is as high as 12% to 15%.[9] The statistics are worse yet if there already exists a diagnosis of relative infertility[10] or if the contralateral tube is anatomically abnormal at the time of surgery.[11] Perhaps only with earlier diagnosis and conservative surgical or medical management might these consequences be altered.

Anatomy of the Fallopian Tubes

The fallopian tubes are paired structures that extend from the ovary to the uterus in the upper free margin of the broad ligament. The tubes are 10 to 12 centimeters long and are separated both anatomically and histologically into four distinct sections (see Fig. 9.1).[12] From proximal to distal these sections are (1) the interstitial segment, the segment that pierces the uterine wall; (2) the isthmus, the segment that comprises the most proximal one-third of the extrauterine tube; (3) the ampulla, the segment that comprises the distal two-thirds of the extrauterine tube; and (4) the infundibulum,

Tubal anatomy

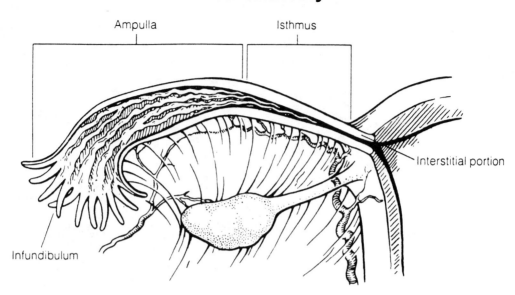

FIGURE 9.1. Most ectopic pregnancies occur in the ampulla. There is anastomosis of vessels from the ovarian, or infundibulopelvic, to the uterine portion of the tube.

the 1 to 2-centimeter portion that contains the fimbriae and the tubal ostium. Although each segment is unique with respect to its anatomic and hisotologic appearance, some general statements can be made. First, the wall of the tube thickens as one progresses toward the uterus. Second, the tubal lumen decreases in size as one moves toward the uterus (from 1 cm in the outer ampulla to less that 100 um in the interstitium). Last, the blood supply is the greatest in the area of the uterine cornua, where the branches of the uterine artery and ovarian artery meet.

Histologically, three layers exist. The innermost layer is the mucosa, a thin-walled layer designed chiefly for exchange and lubrication. This layer has ciliated cells that beat toward the uterus, interspersed among islands of lubricating cells. In the ampulla, the mucosa takes on the appearance of a wildly complex pattern of branching folds. Fertilization most commonly takes place in this part of the tube. The second layer is the muscularis, which is arranged with an inner circular layer and an outer longitudinal layer. These muscles propel the fertilized ovum toward the uterus with peristaltic waves of movement. The muscularis also increases in width as one moves toward the uterus. The outer layer is the serosa, the single-celled layer that serves mainly as a line of separation from other organs in the peritoneal cavity.

Therefore a wide variance exists in histologic appearance throughout the tube, all related to that specific location's role in fertility. A complete understanding of the pathogenesis and management of EP depends wholly on an understanding of tubal anatomy and histology.

Pathogenesis of Ectopic Pregnancy

Etiologic factors can be divided into two groups: those that cause a delay or actually prevent passage of the fertilized ovum into the endometrial cavity and those that are related to the embryo itself.

The most commonly mentioned factor is pelvic inflammatory disease (PID). The reported incidence of PID rose dramatically in the early 1960s. With the associated tissue damage and scar-tissue formation, it seemed plausible that PID and EP would be associated. Early pathologic studies at the time of salpingectomy found evidence of PID in nearly 50% of patients.[13] However, the increase in PID has not

TABLE 9.1. Histopathologic findings of salpingitis isthmica nodosa and prior tubal inflammatory disease in fallopian tubes that contained an ecyesis.[15]

		No. of cases	Inflammatory disease (%)	Salpingitis isthmica nodosa (%)
Wrork	1942	100	—	13
Persoud	1970	100	43	49
Honore	1978	35	—	37
Majmudar	1983	100/100	—	27/57
Stock	1984	58/80	37	14

kept pace with the continued rise in the frequency of EP.[14] Moreover, recent studies suggest evidence of PID in only 30% of salpingectomy specimens (Table 9.1).[15]

Congenital anomalies also play a role in the development of EP. DeCherney studied the laparoscopic appearance of the fallopian tubes in 16 infertile women known to be exposed to DES in utero. He found that their tubes were foreshortened and sacculated, and had pinpoint os with constricted fimbriae. He speculated on the known increase of ectopic pregnancy in this group of women and the associated abnormal anatomy present.[16] Salpingitis isthmica nodosa has also been implicated as an etiologic factor, as concluded by Budowick,[17] and was widely held until challenged by Stock in 1985. Stock found but one case of salpingitis isthmica nodosa on his review of over 120 salpingectomy specimens.[15]

Another more theoretical event that could contribute to the development of EP is transmigration of the fertilized ovum with its associated delay in implantation. This event could either be internal (with passage of the fertilized ovum across the uterine cavity into the opposite tube with implantation there) or external (with migration of the oocyte to the opposite tube within the peritoneal cavity and fertilization there). Evidence of this includes the presence of corpus luteum formation contralateral to the affected tube, the incidence of which has been reported as high as 50%.[18,19]

Other factors also exist and include the use of fertility medications, which can sometimes alter the motility of tubal mucosa. IUD use more effectively inhibits intrauterine gestation, and therefore the likelihood of ectopic pregnancy is more common if one conceives while using this form of contraception. Also, the increased use of conservative surgery to treat EP maintains fertility but also results in a 12% to 15% risk of recurrence.

Evidence that embryonic factors may influence implantation was first noted by Stratford, who found that 64% of 44 ectopics in his study were grossly malformed.[20] Pland found that nearly 33% (5/16) of the ectopic pregnancies in his study had gross chromosomal anomalies.[21] Many speculate that embryonic factors may give rise to ectopic pregnancies in anatomically normal tubes.

Pathology

All ectopic pregnancies begin by abnormal implantation in tubal mucosa rather than in the endometrial cavity. Originally, Budowick felt that trophoblastic tissue aggressively invaded the muscularis, leading to hematoma formation between the muscularis and serosa.[17] Bleeding from the tubal fimbriae was actually excess blood dissecting between the outer two layers of the tube. When the tubal serosa became too distended, rupture occurred. Until Stock challenged this theroy in 1985,[15] the majority of ectopic pregnancies were managed by salpingectomy, with the feeling that hemostasis could not be achieved using a more conservative approach. If Budowick were correct, one could envision the creation of an artificial space, with possible scarring and kinking of the tube. However, there existed hystersalpingographic evidence that the ectopic pregnancy was usually contained within the mucosa, and this spurred Stock to reevaluate this theory.

In his study, Stock retrospectively evaluated the histologic preparations of over 120 salpingectomies and found no cases of interposing hematoma between muscularis and serosa. Instead, he noted attenuation of the muscularis, with associated vascular compromise, leading to local hemorrhagic necrosis and tubal rupture. His work cleared the way for a more conservative surgical approach, and corroborated surgical evidence was already accumulating in

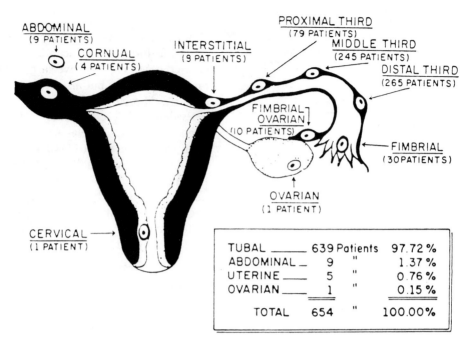

FIGURE 9.2. Anatomic site of ectopic pregnancy. (From Breen JL: *Am J Obstet Gynecol.* 106:1004, 1970.)

this regard. Later work, also done by Stock, revealed that fallopian tubes that had undergone conservative surgery to treat ectopic pregnancy (either salpingotomy or salpingostomy) were remarkably well-healed and free of scar-tissue formation. He concluded that underlying tubal disease was the more likely cause of recurrent ectopic pregnancy in the specimens he evaluated.[22]

Location

In a 21-year survey by Breen (Fig. 9.2),[23] 97.7% of all ectopic pregnancies were found to occur in the fallopian tube itself. The most common site within the tube was the ampulla, followed, in order of most to least frequent, by the isthmus, the fimbriae, and finally the interstitium. Nearly 1.3% were found to be abdominal implantations, and 0.75% were abnormal intrauterine implantations (cervical or interstitial). Only 0.15% of all EPs are ovarian, and these must pass the scrutiny of Spiegelberg's classification to be diagnosed as such. The criteria, recounted here for the sake of completeness, are (1) the fallopian tube must be intact and clearly separate from the ovary, (2) the gestational sac must occupy the normal position of the ovary, (3) the sac must be connected to the uterus by the ovarian ligament, and (4) there must be ovarian tissue demonstrated within the walls of the sac.[24]

Evaluation and Diagnosis

Introduction

Patients generally present in one of two conditions. The clinically unstable patient who is admitted to the emergency center with shock secondary to the blood loss associated with ruptured EP offers no diagnostic dilemma. A positive pregnancy test, vital signs consistent with blood loss, and a surgical abdomen are usually present. This presentation as such dictates management, that is, surgical exploration. This also represents the condition that is to be avoided if preservation of reproductive organs is desired. The more common presentation is that of the clinically stable patient in whom the diagnosis of EP is entertained. This patient can either be mildly symptomatic, with mild lower pelvic discomfort associated with a positive pregnancy and irregular vaginal bleeding, or be totally asymptomatic but at high risk for the development of ectopic pregnancy. Although this discussion can be related to both patient presentations, in the latter condition early diagnosis is especially important and difficult.

History

A complete history is essential, and knowledge of those factors that might predispose the patient to EP could raise or lower suspicion. Therefore every evaluation must begin with a complete history, including a thorough review of pertinent medical history. The most frequently associated historical event is that of previous EP. As noted, the recurrence risk is as high as 10% to 15%. Other conditions associated with EP include DES exposure in utero, IUD use (past or present), and a history of PID. Past emergency center admissions associated with fever, pelvic discomfort, and an elevated white blood cell count, especially if the patient were treated with antibiotics, is historical evidence of a possible previous pelvic infection. Surgical history often associated with EP incudes past tubal surgery (including infertility surgery and tubal ligation) and appendectomy, especially in cases of appendiceal rupture. Finally, a history of infertility in general should alert one as to the possible diagnosis of this condition.

The most common group of presenting symptoms is the triad to pain and vaginal bleeding in a woman who has had a preceding period of amenorrhea (Table 9.2).[25] Fully 90% to 100% of patients present with some form of pelvic pain, usually starting as vague lower pelvic discomfort, then increasing to sharp, colicky pain. With tubal leaking and the resultant mild hemoperitoneum, there is often vague nausea. When tubal rupture occurs and more serious hemoperitoneum ensues, the diaphragm becomes irritated and referred pain to the shoulder and neck takes place. Bleeding is the second most common presenting symptom (50% to 80% of cases) and is usually most

TABLE 9.2. Symptoms of ectopic pregnancy.[25]

Symptom	% of patients with symptoms
Abdominal pain	90–100
Amenorrhea	75–95
Vaginal bleeding	50–80
Dizziness, fainting	20–35
Urge to defecate	5–15
Pregnancy symptoms	10–25
Passage of tissue	5–10

consistent with the bleeding of a light period or that seen in early threatened abortion. The bleeding can also be quite heavy and is often associated with the loss of tissue. This tissue is frequently the decidual cast that forms in response to the hormones produced in early pregnancy. It also confuses the examiner and can be a source of a false sense of security that a threatened or complete miscarriage is taking place. Finally, the bleeding can also take the form of light spotting at or near the time of expected menses, further clouding the diagnosis. Fully 25% of patients present without a prior history of amenorrhea. Most early ectopic pregnancies would not be missed if the diagnosis were entertained at all, and a good rule of thumb is that all threatened abortions are ectopics until ruled otherwise. This thought alone would lead to an earlier diagnosis.

Physical Findings

The most common physical finding is that of adnexal tenderness, often accompanied by abdominal pain. (Table 9.3).[25] Interestingly, this pain can be unilateral, bilateral, and even contralateral to the affected tube and is due either to the condition of the tube at the time of diagnosis (ruptured or not) or to additional factors that may also be present. A palpable adnexal mass is far less common (roughly 50%) and also can be contralateral due to the presence of a corpus luteum cyst on the side opposite the ectopic (as mentioned earlier). A physical finding often seen is that of cervical motion tenderness. This nonspecific finding

TABLE 9.3. Signs of ectopic pregnancy.[25]

Sign	% of patients with sign
Adnexal tenderness	75–90
Abdominal tenderness	80–95
Adnexal mass	50
(20% present on side opposite ectopic)	
Uterine enlargement	20–30
Orthostatic changes	10–15
Fever	5–10

occurs in most cases of adnexal irritation, whether it be from infection, cyst formation, or torsion. However, because it is only rarely seen with threatened abortion, it can help greatly in differentiating between the two. Finally, physical findings such as orthostatic changes and rebound tenderness can be found, depending on the point at which the diagnosis is made.

Differential Diagnosis

Obviously, given the variety of presenting features, the differential diagnosis is rather large and must include threatened abortion and corpus luteum cyst (ruptured or unruptured). Other causes of pelvic pain not related to pregnancy, including salpingitis, appendicitis, torsion of the adnexa, a degenerating fibroid, urinary tract infection, and endometriosis, must also be considered. The key issue in this diagnostic nightmare is the presence or absence of pregnancy or, rather, the ability to diagnose pregnancy accurately. Full knowledge as to the sensitivity of the urinary B-hCG testing employed in one's unique setting is of utmost importance. Currently, most emergency centers employ techniques that allow for diagnosis to be made 7 to 10 days after ovulation, or before the missed period. Additionally, there has been movement toward a clarification in the reference standard, and in 1974, the International Reference Preparation (IRP) was introduced. Once the diagnosis of pregnancy is made, the remainder of the evaluation depends on the particular presentation at hand. Very little information is gained from quantitative hCG results when the presenting symptom is that of a surgical abdomen associated with a positive pregnancy test. On the other hand, the true diagnostic dilemma is the clinically stable patient who might have an EP. It is in this patient that additional diagnostic tests are necessary.

Ancillary Diagnostic Aids

Perhaps the most significant development in the last 10 years was the elucidation of hCG curves in normal and abnormal pregnancy by Kadar et al in 1981.[26] Their work described a "discriminatory zone," the level of hCG hormone (6,000 to 6,500 mIU/mL) at which transabdominal ultrasound (TAUS) should be able to delineate the presence of an intrauterine gestation and therefore rule out EP. Conversely, the presence of an intrauterine sac found by ultrasound associated with a quantitative level of hCG hormone less than 6,000 mIU/mL suggested the presence of an abnormal pregnancy (either a missed abortion or an ectopic pregnancy). Finally, the absence of an intrauterine gestational sac associated with a level greater than 6,500 mIU/mL was always associated with the diagnosis of EP. Kadar et al also delineated a production curve for hCG and stated that the average normal doubling time was 1.98 days. If the quantitative hCG level rose less than 66% over the course of 2 days, there was a high probability (positive predictive value of 86%) of an abnormal gestation (either missed AB or ectopic pregnancy).

These concepts are quite useful in following the clinically stable patient. A baseline quantitative hCG can be drawn at the time of presentation, and if a value is in the discriminatory zone, it can be compared with the expected TAUS data. If no gestational sac is seen, a high probability of an EP exists, and one can proceed with operative intervention. However, in a later prospective study, only 30% of clinically stable patients had levels in the discriminatory zone at the time of presentation.[27] The concept of the "20-day window" refers to those days that TAUS cannot reliably demonstrate a gestational sac (roughly 28 days after conception),

TABLE 9.4. Relationship of gestational age, hCG levels, and transvaginal ultrasound findings.[30]

Ultrasound findings	Days from last menstrual period	B-hCG (mIU/ml) IRP
Sac	34.8 +/− 2.2	1,398 +/− 155
Fetal pole	40.3 +/− 3.4	5,113 +/− 298
Fetal heart motion	46.9 +/− 6.0	17,208 +/− 3,772

a time critical to the early diagnosis and management of EP. Because of this, Kadar et al's concept of doubling time became as important as their original assertions. One can now follow the stable Patient with hCG levels every 48 hours to check for appropriate doubling. If the level shows less than a 66% rise, then one's suspicion is raised. By multiple regression analysis using more than two hCG levels on the same patient, Pittaway and Wentz corroborated the work of Kadar et al.[28] Despite later reports that demonstrated a 36% sensitivity and a 70% specificity using the 66% rule,[29] the concept of doubling time is still valid and is a standard approach to the management of the stable patient.

Another way to deal with the "20-day window" is to reduce it. If the presence of a gestational sac could be demonstrated at an earlier time, and therefore at lower hCG levels, the discriminatory zone could be shifted downward. In 1988 Fossum et al described transvaginal ultrasound (TVUS) findings associated with gestational age and hCG levels normal pregnancies (Table 9.4).[30] Basically, a true gestational sac could be visualized at hCG levels of roughly 1,400 mIU/mL (IRP) or a menstrual age of 35 days. This is well within the time during which early intervention can reduce morbidity and have a positive effect on future fertility. Again, each center must demonstrate its own discriminatory zone, and in some centers TVUS is still unavailable. Other useful radiographic findings reported include the combination of cul-de-sac fluid and a nonhomogeneous mass, which had a positive predictive value of 94% for the diagnosis of ectopic pregnancy when studied by Hobbins and Romero in 1988.[31] Their study used TAUS as opposed to TVUS and was done on patients whose presenting hCG level was below the discriminatory zone. Similarly, these findings using TVUS have now been described and at correspondingly lower hCG levels.[32]

An important point to remember, however, is that the combination of demonstrated intrauterine pregnancy and normal hCG levels does not necessarily rule out an ectopic gestation. The above concepts rely on early data with respect to heterotopic pregnancy (simultaneous intra- and extrauterine pregnancy), with an incidence in 1948 of roughly 1/30,000.[33] This may not be as valid today given the increased incidence of EP as well as the increased rate of dizygous twins (associated with ovulation induction). In fact, some in vitro fertilization (IVF) programs report a rate of heterotopic pregnancy of nearly 1/100, and others report the incidence anywhere from this to 1/4,000.[34-36]

The latest diagnostic aid is the use of a single serum progesterone level. That abnormal pregnancies produced lower progesterone levels than normal pregnancies was first reported by Mildwidsky et al in 1977.[37] The unique property of progesterone production that allows a single sample to become predictive is that, after the luteal phase, progesterone has relatively stable serum values during the first 8 weeks of pregnancy. Therefore an absolute knowledge of the gestational age of the pregnancy is not necessary. This was later confirmed[38] and tested retrospectively.[39,40] Yeko et al[40] would have successfully diagnosed 28 out of 28 EPs and 17 out of 18 otherwise abnormal gestations using levels of below 15 ng/mL, and all 24 normal pregnancies had levels greater than this. Stovall et al attempted to approach this prospectively and, although not quite as successful, noted that 81% of diagnosed ectopic pregnancies had levels below 15 ng/mL.[41] Only 2% of ectopic pregnancies had levels greater than 25 ng/mL. The addition of single serum progesterone to the diagnostic profile might greatly improve the chances for early intervention in that the level of suspicion would increase before receiving a second hCG level when following the stable patient into the discriminatory zone.

The clinician is often faced with the diagnostic dilemma of the patient who has regular periods, and therefore should have good menstrual dating, but who has an ultrasound inconsistent with such dates. The possibility of an ectopic pregnancy must be entertained at this point, especially in the high-risk individual. An approach using the above information would be to obtain a serum progesterone and a quantitative hCG. The patient could then undergo TVUS. If the patient is in the discriminatory zone but ultrasound is negative for an intrauterine gestation, the solution is simple. However, if the patient is not in the discriminatory zone, she must be followed until she is. Shepherd et al. found that the first two hCG levels might indeed demonstrate appropriate doubling time despite the presence of an ectopic pregnancy, but eventually all ectopic pregnancies in their prospective study dropped off the normal production curve (less than 66% over 48 hours).[29] The serum progesterone levels can be used to raise or lower suspicion or can be another piece of supporting evidence.

Follow-Up

To allow for earlier intervention and therefore for more conservative surgical approaches through earlier diagnosis is a major goal in dealing with EP. There are some complications with respect to such conservative measures. Most important is the problem of persistent trophoblastic tissue, which has a reported incidence as high as 5%.[42] Therefore it is important to follow patients who have had a conservative surgical approach with weekly quantitative hCG levels until the levels reach zero. Treatment for persistent trophoblastic tissue is similar to that of molar pregnancy and consists of single-agent chemotherapy (methotrexate). Moreover, there are numerous reports demonstrating the usefulness of methotrexate as a primary treatment for small unruptured ectopics.[43,44]

Summary

In summary, EP is still occurring at near epidemic proportions, with the incidence nearly quadrupling over the last 17 years for which data is available. Although case fatality has decreased, EP continues to represent a fairly high percentage of maternal mortality in the United States. Moreover, EP carries with it substantial risk with respect to future reproducve integrity. Short of prevention, early diagnosis and conservative management hold the most promise in decreasing maternal morbidity and in ensuring future fertility. Using the combination of newer diagnostic methods, such as hCG curves, TVUS, and serum progesterone, might greatly aid the clinician in coming to earlier diagnosis. Finally, the concept of nonsurgical care in the earliest of ectopic pregnancies through the use of methotrexate might further reduce scarring and produce better results than currently seen.

References

1. Albucasis. *Altrasrif* (11th century). English translation: Oxford, 1778.
2. Ectopic pregnancy. In: Mattingly RF, Thompson JD, eds. *TeLinde's Operative Gynecology*. Philadelphia, Pa: JB Lippincott Company; 1985:429–448.
3. Mariceau, F. *Traité des Femmes Grosses*. Paris, 1694.
4. Tait, RL. *Ectopic Pregnancy and Pelvic Hematocoel* (lecture). Birmingham, England: Journal Printing Works, 1888.
5. Ectopic pregnancy surveillance, United States, 1970–1987. *MMWR*. 1988;39(SS-4):9–17.
6. Ectopic pregnancy surveillance, United States, 1984 and 1985. *MMWR* 1988;7:637.
7. DeCherney A, Kase N. The conservative surgical management of unruptured ectopic pregnancy. *Obstet Gynecol.* 1979;54:451.
8. Barnes AB, Wennberg CN, Barnes BA. Ectopic pregnancy: incidence and review of determinant factors. *Obstet Gynecol Surv.* 1983;38:345–356.
9. Novy M. Surgical alternatives for ectopics: is conservative treatment best? *Contemp Obstet Gynecol.* 1983;21:91.

10. Pouly JL, Mahnes H, Mage G, et al. Conservative laparoscopic treatment of 321 ectopic pregnancies. *Fertil Steril.* 1986;46:1093–1097.
11. Langer R, Bukovsky I, Herman A, et al. Conservative surgery for tubal pregnancy. *Fertil Steril.* 1982;38:427–430.
12. Altchek A. Ectopic pregnancy in young patients. *Contemp Obstet Gynecol.* 1984;22:133.
13. Persoud V. Etiology of tubal pregnancy. *Obstet Gynecol.* 1970;36:257.
14. Ory S. New options for diagnosis and treatment of ectopic pregnancy. *JAMA.* 1992;267:534–537.
15. Stock RJ. Histopathologic changes in tubal pregnancy. *J Reprod Med.* 1985;30:923–928.
16. DeCherney, AH. Structure and function of the fallopian tubes following exposure to DES during gestation. *Fertil Steril.* 1981;36:741.
17. Budowick M, Johnson TRB, Genadry R, et al. The histopathology of the developing tubal ectopic pregnancy. *Fertil Steril.* 1980;34:169.
18. Berlind M. The contralateral corpus luteum—an important factor in ectopic pregnancies. *Obstet Gynecol.* 1960;16:51.
19. Insunza A, DePablo F, Croxatto HD, et al. On the rate of tubal pregnancy contralateral to the corpus luteum. *Acta Obstet Gynecol Scand.* 1988;67:433–436.
20. Stratford BF. Abnormalities in early human development. *Am J Obstet Gynecol.* 1970;107:1223.
21. Pland BJ, Dill FJ, Stylblo C. Embryonic development in ectopic human pregnancy. *Teratology.* 1976;14:315.
22. Stock, RJ. Histopathology of fallopian tubes with recurrent tubal pregnancy. *Obstet Gynecol.* 1990;75:9–14.
23. Breen JL. A 21-year survey of 654 ectopic pregnancies. *Am J Obstet Gynecol.* 1970;106:1004.
24. Spiegelberg O. Zur casuistik den ovarialschwangenschaft. *Arch Gynaekol.* 1878;13:73.
25. Wekstein LN. Current perspectives on ectopic pregnancy. *Obstet Gynecol Surv.* 1985;40:259–272.
26. Kadar N, Devore G, Romero R. Discriminatory hCG zone: its use in the sonographic evaluation for ectopic pregnancy. *Obstet Gynecol.* 1981;58:156–161.
27. Holman JF, Tyrey EL, Hammond CB. A contemporary approach to suspected ectopic pregnancy with use of quantitative and qualitative assays for B-subunit of human chorionic gonadotropin and sonography. *Am J Obstet Gynecol.* 1984;150:151–157.
28. Pittaway DE, Wentz AC. Evaluation of early pregnancy by serial chorionic gonadotropin determinations: a comparison of methods by receiver operating characteristic curve analysis. *Fertil Steril.* 1985;43:529–533.
29. Shepherd RW, Patton PE, Novy MJ, et al. Serial B-hCG measurements in the early detection of ectopic pregnancy. *Obstet Gynecol.* 1990;75:417–420.
30. Fossum GT, Davajan V, Kletzky OA. Early detection of pregnancy with transvaginal ultrasound. *Fertil Steril.* 1988;49:788–791.
31. Romero R, Kadar N, Castro D, et al. The value of adnexal sonographic findings in the diagnosis of ectopic pregnancy. *Am J Obstet Gynecol.* 1988;158:52–55.
32. Bateman BG, Nunley WC, Kolp LA, et al. Vaginal sonography findings and hCG dynamics of early intrauterine and tubal pregnancy. *Obstet Gynecol.* 1990;75:421–427.
33. DeVoe R, Pratt J. Simultaneous intrauterine and extrauterine pregnancy. *Am J Obstet Gynecol.* 1948;56:1119.
34. Molloy D, Deambrosis W, Keeping D, et al. Multiple-sited (heterotopic) pregnancy after in vitro fertilization and gamete intrafallopian transfer. *Fertil Steril.* 1990;53:1068–1071.
35. Dimitry ES, Subak-Sharpe R, Mills M, et al. Nine cases of heterotopic pregnancies in four years of in vitro fertilization. *Fertil Steril.* 1990;53:107–110.
36. Bello GV, Schonholz D, Mashirpur J, et al. Combined pregnancy: the Mount Sinai experience. *Obstet Gynecol Surv.* 1986;41:603–613.
37. Mildwidsky A, Adoni A, Dalti Z. Chorionic gonadotropin and progesterone levels in ectopic pregnancy. *Obstet Gynecol.* 1977;50:145–147.
38. Radwanska E, Frankenberg J, Allen EI. Plasma progesterone levels in normal and abnormal early human pregnancy. *Fertil Steril.* 1978;30:398–402.
39. Matthews CP, Coulson PB, Wild RA. Serum progesterone levels as an aid in the diagnosis of ectopic pregnancy. *Obstet Gynecol.* 1986;68:390–394.
40. Yeko TR, Gorrill MJ, Hughes LH, et al. Timely diagnosis of early ectopic pregnancy using a single blood progesterone measurement. *Fertil Steril.* 1987;48:1048–1050.
41. Stovall TG, Ling FW, Cope BJ, et al. Prevent-

ing ruptured ectopic pregnancy with a single serum progesterone. *Am J Obstet Gynecol.* 1989;160:1425–1431.
42. Pouly JL, Mahnes H, Mage G, et al. Conservative laparoscopic treatment of 321 ectopic pregnancies. *Fertil Steril.* 1986;46:1093–1097.
43. Leach RE, Ory SJ. Modern management of ectopic pregnancy. *J Reprod Med.* 1989;34:324–338.
44. Stovall TG, Ling FW, Buster JE. Outpatient chemotherapy of unruptured ectopic pregnancy. *Fertil Steril.* 1989;51:435–438.

10
Office Ultrasonography of the Pelvis

Steven R. Goldstein

The Rationale for Pelvic Ultrasonography

The path that diagnostic ultrasound has followed into clinical use in obstetrics and gynecology has been anything but straightforward. Many gynecologists and obstetrician are perplexed but probably do not understand why. Initially, real-time equipment had barely enough resolution to do little more than identify breech versus vertex, measure a BPD, and localize placenta. A complete exam was extremely limited and consisted of two to three Polaroid pictures. Office nurses or receptionists learned obstetrical scanning through various short courses, and the practitioner was then in the obstetrical ultrasound business. Gynecological imaging, however, with static arm scanners and later with small-headed sector scanners was seemingly more involved. The equipment was large, expensive, and for the most part remained the domain of the imager.

Over time, however, obstetrical scanning has become increasingly more sophisticated. The standard obstetrical exam now mandates an understanding of cross-sectional body anatomy (heart, brain, and so forth) that goes beyond the scope of routine training in obstetrics and gynecology. Residents in radiology spend 4 months rotating through neuroradiology. They understand intracranial anatomy, both normal and aberrant. Gynecologists, however, spend 4 months in the gyn path lab. They operate in the pelvis and routinely examine the pelvis; certainly *imaging* in the pelvis can and should be the domain of the gynecologist.

An important bridge that allows gynocologists to make the leap into gynecologic imaging has been the introduction of the vaginal probe. It is smaller, less expensive equipment, and very adaptable to being utilized by physicians themselves at the time of the bimanual exam, mainly because of the empty urinary bladder and because the procedure requires very little time once the operator is adequately trained.

Many gynecologists perform vaginal scanning only. If the clinical situation mandates doing full-bladder transabdominal views, that patient is referred to an imaging consultant. For some practitioners this will be a technician (the preferred term is *sonographer*) who functions as an "imager" elsewhere within the same practice.

Vaginal Ultrasound Equipment for the Office

Equipment Selection

Most currently available equipment was not designed specifically for vaginal probe applications. Manufacturers initially achieved vaginal probe capability merely by taking an existing full-service machine and adding a vaginal transducer to it. Almost all equipment still provides full service (obstetrical, transabdominal, and gynecologic) as well as vaginal probe applications. As a result, such equipment may be adequate for vaginal scanning in the office but will rarely be optimal.

Ideally a machine should be lightweight, easily moved, and small; bigger is not always better. Even if a machine will be placed permanently in one exam room, it is important that it be able to move 18 to 36 inches; it should be able to placed flush to the end of the exam table for vaginal scanning. Between uses it is convenient to roll the machine back near the head of the table. Thus it must be light (some small machines are still very heavy), and the cart should have easily movable wheels. The cord from the machine to the probe must be long enough so that it is easily maneuverable but not so long that it will constantly drag on the floor and be damaged by the wheels rolling over it.

One advantage of the vaginal probe is that it utilizes high-frequency equipment in close proximity to the structure being studied. This results in excellent near-field resolution in spite of high degrees of magnification. This also results in limited penetration and requires one to give up the panoramic view that can be obtained with full-bladder transabdominal scanning.

It is extremely helpful to be able to generate more than one frequency from the same probe. Most of the time the highest frequency available (7.5 mHz) will want to be used because the best resolution, and thus the clearest, sharpest, most easily recognized images of anatomic structures, result. However, there will be certain types of patients (i.e., those with obesity, enlarged fibroids, myometrial contractions) where the increased penetration provided by lower frequencies will be necessary. Remember that the entire peritoneal cavity, even when viewed from the vagina, is surrounded by fat. Fat attenuates sound, and for obese patients a 5-mHz probe will be more suitable to visualize the endometrium or to search for a tiny gestational sac, or if structures are fixed high in the pelvis secondary to adhesions. Such lower-frequency probes will also be more effective in penetrating fibroids or myometrial contractions. Because fluid is easily imaged with all types of ultrasound, stimulated ovaries with multiple follicles or intrauterine gestations beyond 7 to 8 weeks will easily be seen with virtually any type of equipment, regardless of quality. However, postmenopausal ovaries without folliculogenesis or very small early gestations (especially if they are extrauterine) will require significantly better image quality and resolution for successful scanning. The monitor screen attached to the equipment should be as large as possible. When used with a larger monitor, the same basic equipment will provide better results. The monitor screen should be movable so that the patient can be shown the findings on the screen. Thus vaginal scanning can become an excellent teaching tool and help to explain anatomy and, in fact, physiology to patients. The patients become very involved with their own anatomy, and their inclusion in the examination process is essential to successful routine use of the vaginal probe.

Transducer Types

The vaginal probe consists of a long shaft that connects the handle to the transducer located at its tip. Mechanical sector probes provide the best image quality throughout the entire field of view. They are relatively simple to build and thus usually less expensive, are capable of very high frequencies, and have a rounded small "footprint" (the area that makes contact in the vagina). The transducer is moved back and forth by means of a motor, and thus these probes are not "solid state." Some physicians have had long experience with such devices and have never had any breakdown in the moving parts of a mechanical sector probe. Such equipment, however, cannot share electronics and thus will not be capable of Doppler assessment.

Curvilinear probes represent a solid-state extension of linear array technology in which a variable depth beam is focused by geometry. The transducers are fixed and fired in sequences that produce the "real time" images. They are also capable of variable depth focusing, but such a feature will be less important with vaginal probes because of the short field of view that makes variable depth focusing not as critical. The footprint of such devices may be large and irregular in shape and may yield inconsistent contact along the entire edge. Such transducers can accommodate a simultaneous Doppler signal.

Phased-array transducers have the beam electronically steered, but this factor may not be especially important in vaginal scanning. Such transducers are more costly than mechanical ones. Phased-array transducers can also have large irregular footprints, and adequate contact in the fornix of the vagina may be suboptimal. They are capable of simultaneous Doppler signal.

The angle at which the sound emanates from the transducer is another extremely important consideration. With an endfire probe the sound emanates in parallel with the probe shaft. Once a structure in one plane is located simply turn the handle 90 degrees to find a concordant right-angle view of the same structure. If, however, the beam of sound is off-axis (usually between 10 and 30 degrees), when the handle is turned 90 degrees one is no longer in the corresponding right-angle plane of the structure viewed previously.

Field of view is yet another important factor. Because vaginal scanning affords highly magnified targeted organ views, initially the orientation may be confusing (see below). There is no panoramic view such as can be obtained with full-bladder transabdominal scanning. However, resolution diminishes as the field of view broadens. The number of lines of information remains constant; thus the image quality is less clear. In addition, if the perspective is wide enough to display multiple structures, such as the uterus and ovary, in one scanning field, the magnification employed will be obviously minimal. One benefit of vaginal scanning is that in spite of high degrees of magnification, resolution is not lost.

In terms of the shape of the probe itself, remember that the narrowest portion of the vagina is the introitus. The thickness of the shaft at the introitus will determine the probe's degree of maneuverability.

Probe Preparation

The vaginal probe is inserted into the patient's vagina. Between usages the probe should be disinfected with a commercially available disinfectant. The probe should be covered with a condom, a finger of an examining glove, or a specially fitted sterile sheath. Condoms are not sterile; in fact, they can be bought in bulk. However, having individually foil-wrapped condoms that can be torn open under the patient's watchful eye may help alleviate any patient anxiety concerning transmission of disease. If condoms are employed, it is recommended to use nonlubricated condoms without a tip. Prelubricated condoms are messy to work with and should be avoided.

Whether a sheath, condom, or glove is used, the standard ultrasound coupling gel should be applied. Care should be taken to avoid trapping any air bubble at the tip of the probe where it interfaces with the covering sheath. A small amount of coupling gel or lubricating gel can be applied to the outside of the probe for its insertion into the vagina. Note that if an insemination of sperm is to follow the vaginal ultrasound exam, some coupling gels may in fact interfere with sperm motility.

Patient Preparation

The procedure should be explained to the patient. Often, if the patient sees the transducer, some anxiety may ensue. It may be helpful to explain that only a small portion of the probe is inserted into the vagina. Phrases such as "smaller than a speculum" or "like a tampon on the end of a handle" can be helpful. Vaginal scanning should be performed on a gynecologic table that allows lithotomy position. Thus the patient can move her perineum down to the edge of the table, and there will be full range of motion of the transducer's handle, even below the plane of the horizontal. A flat examining table would not allow movement of the handle below the horizontal plane. The probe is inserted much like a speculum. A finger is placed in the posterior fourchette of the vagina and gentle pressure is

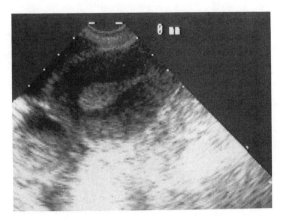

FIGURE 10.1. The uterus is seen here in its long axis. The myometrial mantle is seen to nicely surround this echogenic secretory appearing endometrium. The uterus in long axis serves as a landmark from which to begin the remainder of the pelvic vaginal probe ultrasound examination.

exerted posteriorly. The vaginal area is much more sensitive anteriorly than posteriorly; therefore depressing the posterior wall will allow for easy insertion of the probe. Realize that once the probe is inserted into the vagina it is manipulated posteriorly, anteriorly, and obliquely to the right and left to image recognizable anatomic structures.

The uterus in long axis (Fig. 10.1) serves as a basic anatomic landmark from which to orient one's eye and hand. One looks for recognizable endometrial echo surrounded by typical myometrium. The contour of the myometrial border, the homogeneity of the myometrium itself, and a characterization of the endometrium both subjectively and by measurement should be carried out. The transducer is then turned 90 degrees to view the uterus in a semicoronal plane. By moving one's hand in an anterior–posterior fashion one will see this plane from cervix to fundus. Returning to the long axis view, one then moves in a sagittal oblique fashion, looking for ovarian structures. These are recognized by their location (in premenopausal patients) immediately adjacent to the iliac vessels (Fig. 10.2). Remember that the iliac artery is smaller than the iliac vein and can be seen to pulsate; the iliac vein lies above the artery and can often be seen to have flow on real time. Premenopausal patients without significant pelvic adhesions will have ovaries that, by gravity in lithotomy position, will be located immediately adjacent to the iliac vessels. Ovaries are also recognized by their appearance. An ovary has a typical gray pattern to the stroma and contains numerous follicles at various stages of development, depending on where in the menstrual cycle the patient is or

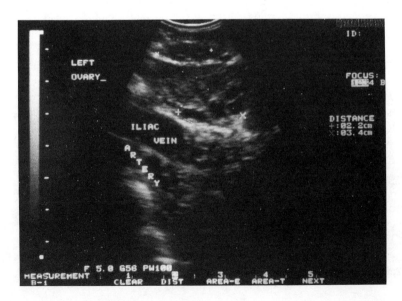

FIGURE 10.2. This normal ovary of a premenopausal patient measuring 3.4 by 2.2 cm is shown here in its anatomic relation just medial to the iliac vein and artery.

if the patient is on oral contraceptives. When either ovary is found in this long axis orientation, the transducer handle once again is turned 90 degrees. Assessing the ovary in two planes at right angles to each other allows three-dimensional assessment of the structure. The method of finding the contralateral ovary is exactly the same as finding an ovary. A complete pelvic ultrasound examination should include an assessment of the cul-de-sac and of the region anterior to the uterus where the bladder will often be seen to start to fill with a small amount of sonolucent urine. Characterization of the cervical region can be carried out as well.

Orientation

Traditional transabdominal pelvic ultrasound makes uses of multiple two-dimensional images (tomograms) to recreate three-dimensional anatomy mentally. The concept of longitudinal and transverse is a throwback to the Cartesian axis, which served as the basis for the original static arm scanners that were moved at 1/2- to 1-cm increments at perfect right angles to each other. Although scans are continued to be labeled as being "longitudinal" or "transverse," these frozen images are often made with slight angulations of the transducer. The image is frozen to depict maximal recognizable anatomy. This is especially true in obstetrical situations where one is interested in recognizable anatomy such as a fetal kidney or fetal spine rather than the axis of orientation of the transducer. With the high degree of magnification yielded in vaginal sonography and the short field of vision, concepts of longitudinal and transverse lose their meaning. Rather, what one is interested in obtaining is "anatomy-derived orientation." For example, one is much more concerned with obtaining a small gestational sac within the uterus than with determining whether the transducer is in a true longitudinal or transverse fashion. One is interested in imaging an ovary often seen overlying iliac vessels regardless of the transducer's orientation. One is anxious to observe the endometrial echo and its characterization in a relatively long-axis view of the uterus even if one's probe is slightly oblique.

Furthermore, with the vaginal probe a view of a structure and then a corresponding right-angle view to the initial view will allow a three-dimensional view of that structure (Fig. 10.3A, Fig. 10.3B). In terms of image labeling there is no absolute standard of convention. Most American physicians have chosen to display their images such that a long-axis view will show the region of the cervix in the upper right side of the screen and the fundus in the lower left portion of the screen. A small amount of urine filling the bladder will be seen in the upper left portion of the screen. Beginners often feel more secure with a small amount of urine in the bladder as an anatomic landmark. However, if the bladder is allowed to fill during the course of an exam, the pelvic structures are sometimes carried farther away from the scanning head and its optimal focal range. This can create a situation where a large portion of the field of vision is taken up by the urinary bladder. For this reason once a physician is comfortable with the anatomic-based orientation, the patient should be scanned with as empty a bladder as possible.

The beginner should be cautioned not to mistake uterine vasculature caught in a sagittal oblique section for an ovary (Fig. 10.4). In addition, with use of the vaginal probe, the bowel can be seen to fill the pelvis (Fig. 10.5). The bowel contains gas, feces, and a fluid within it that will produce bizarre, complex echo patterns that are best recognized by their motion (peristalsis). Along these lines, however, occasionally a focal paralytic ileus will be confusing because identification of the bowel usually requires its motion. With the vaginal probe physiologic findings may often be seen more readily and with more detail than one is accustomed with traditional transabdominal scans. This will be especially true of small amounts of fluid in the cul-de-sac or occasionally fluid in the endometrium of postmenopausal patients. Similarly, the corpus luteum cyst is not a smooth-walled unilocular structure; it can have an irregular thick wall or contain hemorrhage and debris. One should be cautioned against overinterpreting findings

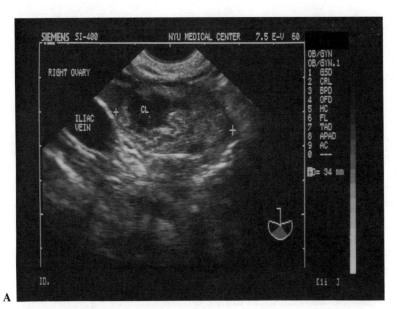

FIGURE 10.3A. View of the right ovary containing a thick-walled typical corpus luteum. The ovary in maximal dimension measures 3.4 cm (calipers). Note its location medial to the iliac vein.

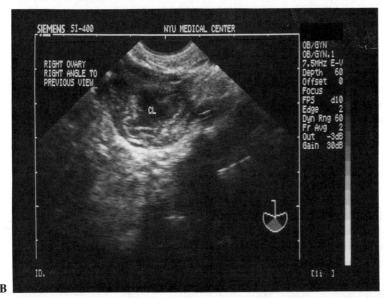

FIGURE 10.3B. The same ovary as in Figure 10.3A seen in a view at a right angle to the previous view. Once again notice the corpus luteum (labeled CL).

that may be physiologic even though they previously have not been possible to image with traditional transabdominal techniques.

The Ultrasound-Enhanced Bimanual Exam

The traditional bimanual exam actually has an *objective* component and a *subjective* component. Based on thousands of pelvic exams, many with surgical or pathologic confirmation, mental image of anatomic findings that are in fact objective (size, consistency, and so forth) is created. However, during the routine bimanual exam an attempt is also made to assess nonquantifiable findings such as mobility or tenderness. These cannot be replaced by an image; they rely on the experience of the examiner.

There is a distinction between the concept of an ultrasound examination as it is now performed by referral for indication and the con-

FIGURE 10.4. As one moves the probe in a sagittal oblique fashion from the uterus one can often produce an image such as this, which appears like an ovary containing numerous small follicles. This is uterine vasculature cut at an oblique fashion and should not be mistaken for ovary.

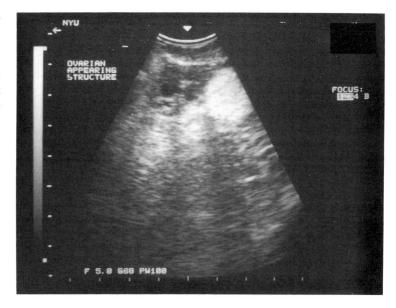

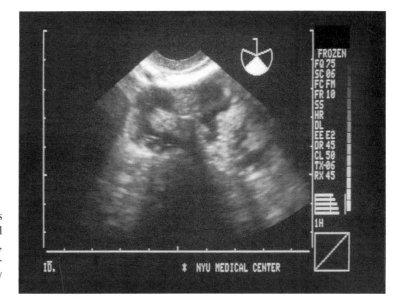

FIGURE 10.5. Bowel contains gas, liquid, and solid material that can give off confusing, bizarre, complex echo patterns. It is best recognized by its motion or peristalsis.

cept of examining one's patients with a vaginal probe to assess the objective component as part of an overall examination. Previously many of the "indications" for obtaining pelvic imaging were based on nuances appreciated at the time of the palpatory bimanual exam (tenderness, adnexal "fullness," an inadequate exam secondary to obesity, or patient guarding). Many of these so-called indications are extremely subjective and nonreproducible. For these reasons it makes sense to enhance the overall bimanual exam with vaginal probe assessment. With an empty urinary bladder, higher-frequency equipment closer to the pelvic structures results in excellent resolution. The drawbacks to the vaginal probe compared with traditional transabdominal scanning are the short field of view, the lack of sound enhancement of a filled urinary bladder, the limitations of the probe by the confines of the vagina, the new orientation for those familiar with traditional transabdominal panoramic views, and the presence of loops of bowel and the confusing echoes they generate throughout the pelvis. The advantages are that the procedure is operator friendly once the operator is adequately trained. The procedure is time efficient because it is meant to be done at the time of the bimanual exam with the patient still in lithotomy position. It adds very little time to the overall bimanual exam and has a very high patient acceptance.

Routine Office Uses

Confirm a Normal Bimanual Exam

The most common application for the vaginal probe in the hands of the office practitioner will be to confirm what appeared to be a normal palpatory exam. The uterus is seen in long axis and is normal in size, shape, and contour (Fig. 10.6). The endometrial echo is visualized and can be correlated physiologically to the findings within the ovary. This will be extremely reassuring. Remember that many healthy patients do not lend themselves to a meaningful palpatory exam, especially patients who may be tense and thus voluntarily guard or patients who are obese.

Confirm an Abnormal Finding on Bimanual Exam

Practitioners are often confronted with definitive or suspected abnormal findings on the palpatory portion of the exam. The uterine contour may be irregular in its shape, suggesting the presence of intramural/subserosal myomas (Fig. 10.7). A "fullness" in the adnexa may be seen to represent hydrosalpinx or, when imaged, may only represent a prominent loop of bowel. When one encounters a seeming enlargement in the adnexa, the vaginal probe may distinguish between a prominent ovary and actual cystic enlargement (see below).

Characterize *Consistency* of Structures Regardless of Size

On palpation two structures may have similar size and dimension but may have internal characteristics that are extremely divergent. This would true of a prominent ovary with normal sonographic texture and appearance versus a thick-walled ovarian cyst containing debris.

Identify Findings Missed on a Bimanual Exam

Often findings that were either missed at the time of the bimanual exam (such as small subserosal myomas or adnexal findings) or that might not be expected to be able to be palpated can be imaged with a vaginal probe. Included here would be abnormal endometrial findings such as a polyp (Fig. 10.8), small dermoids (Fig. 10.9), or endometriomas that are totally contained within the body of the ovary itself.

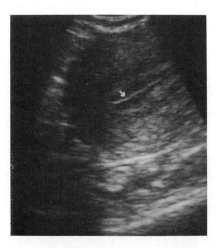

FIGURE 10.6. This is a long axis view of the uterus just prior to ovulation. Note the central endometrial echo (white arrow). On either side of this is a sonolucent "halo". This multilayered endometrium is typical of the time just prior to ovulation.

Early Pregnancy—Traditional Indications

In the past ultrasound was used in early pregnancy for "indication." Such indica-

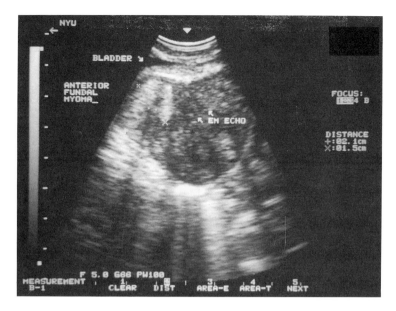

FIGURE 10.7. The uterus is seen here in long axis. An endometrial echo is labeled (white arrows). There is an anterior fundal myoma measuring 2.1 by 1.5 cm. A small amount of urine is seen to be filling anteriorly in the bladder.

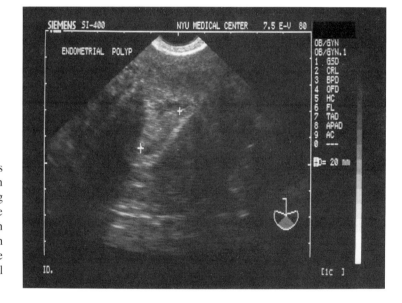

FIGURE 10.8. This long axis view of the uterus reveals an endometrial polyp measuring 2 cm in longest diameter. The uterus itself was normal in size, shape, and contour. Such a finding could not possibly be appreciated on traditional bimanual exam.

tions include bleeding, size-date discrepancies, previous ectopic pregnancies, patients requesting termination of pregnancy, and any coexisting adnexal masses. Consider the pregnant patient who is bleeding with a closed cervical os. All such patients represent "threatened abortion" but at the same time, until proven otherwise, they are "r/o ectopic." Use of the vaginal probe can help to distinguish the patient with obvious intrauterine pregnancy failure (Fig. 10.10) from the patient with normal-appearing gestation but areas of subchorionic hemorrhage or the patient with no evidence of a gestational sac. This can be especially valuable in the patient with severe abdominal pain and guarding, which prevent any meaningful traditional bimanual exam. With the vaginal probe one can differentiate the patient with an intrauterine pregnancy and pain from the patient with a ruptured corpus

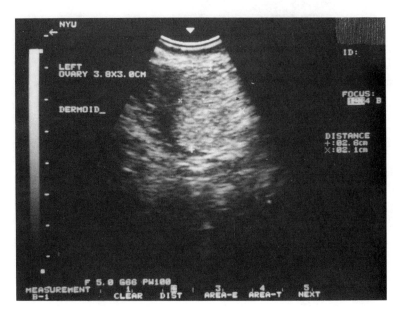

FIGURE 10.9. This represents a normal sized ovary (3.8 by 3.0 cm). Contained within it there is a 2.6 by 2.1 cm densely echogenic structure compatible with a small dermoid. Such a finding could not possibly be appreciated by palpation because it has not caused overall enlargement of the ovary.

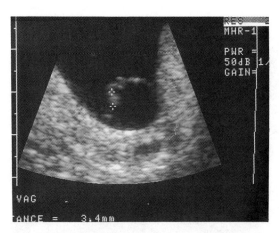

FIGURE 10.10 This shows an amorphous 3.4 mm embryonic structure contiguous with an irregularly shaped abnormal enlarged yolk sac. Such a picture is compatible with intrauterine pregnancy failure and allows for appropriate elective intervention.

Early Pregnancy—Routine Surveillance

Routine use of the vaginal probe can help to document an intrauterine location as well as to identify pregnancies that may not be developing normally. The sonographic approach to early pregnancy failure could be an entire chapter in itself. Suffice it to say here that a normal gestational sac grows approximately 1 mm per day and a normal embryo grows approximately 1 mm per day. Once ectopic pregnancy is ruled out, serial ultrasound examinations should be carried out to verify normal continued growth. By the time an embryo reaches 5 mm, cardiac activity should be seen.

Postmenopause—Traditional Indications

Patients with postmenopausal bleeding or palpable adnexal masses can certainly be evaluated with the use of the vaginal probe. In the light of postmenopausal bleeding an endometrial echo less than 5 mm (Fig. 10.12) has uniformly been shown to be associated with a lack of significant tissue. Thus, endovaginal ultrasound examination can spare such patients the discomfort, expense, and risk of

luteum cyst or from the patient with an obvious ectopic pregnancy (Fig. 10.11). Significant size-date discrepancies can often be resolved in a matter of moments with the vaginal probe. A gestational sac coexisting with fibroids can easily be distinguished from the uterine enlargement that one sees with multiple gestation in the first trimester.

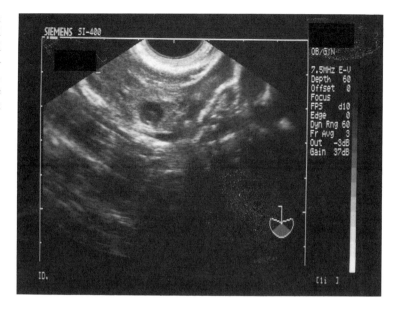

FIGURE 10.11. This shows an extrauterine gestation located in the right adnexa. Typical gestational sac is seen containing a well-formed yolk sac. The uterus is in a separate scanning plane and was seen to be empty.

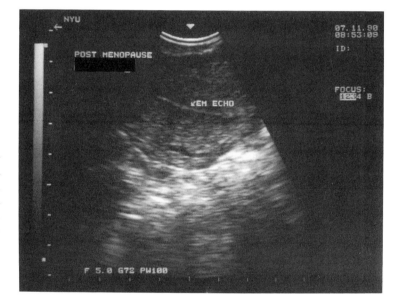

FIGURE 10.12. This "pencil-line" endometrial echo in a postmenopausal patient with vaginal bleeding is typical of a lack of significant tissue. Such a patient exhibits senile endometritis. Endometrial sampling can be bypassed when an ultrasound picture such as this is observed.

endometrial sampling. With postmenopausal bleeding an endometrial thickness greater than 6 mm has been shown to be associated with virtually all pathology tissue types, and sampling remains mandatory. Routine use of the vaginal probe may show cystic adnexal masses (Fig. 10.13) in as many as 10% to 15% of postmenopausal patients. These masses are unilocular unilateral, and they will have virtually no malignancy. However, such findings should be followed at relatively close intervals initially to be sure that there is no progression and enlargement nor development of septation or acidic fluid.

Postmenopause—Screening

Use of the vaginal probe as part of screening for ovarian cancer is another area of great interest. Great debates exist about the cost-

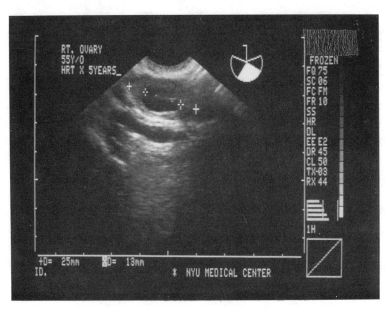

FIGURE 10.13. Right ovary of a 55-year-old patient on hormone replacement therapy for 5 years. The overall ovary is 2.5 cm in maximal diameter. Contained within it is a 13 mm unilocular cystic structure. The incidence of malignancy in such structures approaches zero.

effectiveness of such screening. It is the author's contention that patients are already screened for ovarian cancer with a totally suboptimal inadequate annual visit when palpation alone is used. Incorporation of the vaginal probe into the overall bimanual exam could allow screening of the adnexa in a much more sensible fashion than setting up separate ultrasound screening facilities. Any screening program that will look for ovarian cancer in this age group should also be looking at the size and texture of the endometrium and should be correlating this to whether or not the patient is on hormone-replacement therapy. Still, normative data for endometrial thickness (mean and two standard deviations around the mean) for various hormone-replacement regimens have not yet been established. There will often be instances where obvious abnormal endometrial findings can be discerned.

Infertility

The use of the vaginal probe for infertility problems could itself be a subject of an entire book, let alone another chapter. Suffice it to say that it is almost impossible to have an infertility practice and certainly impossible to perform various assisted reproductive technologies without the use of the vaginal probe. The probe has revolutionized follicle surveillance and oocyte retrieval for the various forms of assisted reproductive technologies.

Miscellaneous Applications

Another useful application of the vaginal probe is finding the lost IUD (Fig. 10.14), which can easily be imaged in a matter of seconds. Postoperative cervical stenosis is also easily seen with the vaginal probe.

Training and Cost

Important issues in the training of clinicians and the cost of vaginal probes must still be addressed. Once the use of vaginal probes is incorporated into residency training like fetal monitoring or laparoscopy are now, all graduating residents will ultimately become part of the solution rather than part of the problem. For physicians already in practice, postgraduate courses will have to fill the need. Such courses already exist from organizations such

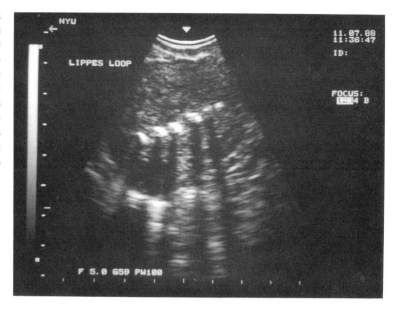

FIGURE 10.14. This shows a Lippes loop IUD seen within the uterus in long axis. Note the highly echogenic appearing Lippes loop and the acoustic shadowing that it produces. The string on this IUD was not visible transcervically. In a matter of seconds use of the vaginal probe allowed localization of it as being intrauterine.

as the American College of Obstetricians and Gynecologists, the American Institute of Ultrasound in Medicine, and various universities and institutions. Fortunately this is a *diagnostic* modality. Any physician acquiring the skill need not act on it initially. Each physician may have a personal "learning curve" that can be variable from physician to physician. During that learning curve the patient may still be triaged according to the physician's method of practice. Once that physician becomes comfortable with the information derived from the vaginal probe examination, then it can begin to be factored into the overall clinical management.

Admittedly the equipment will add cost for the practitioner. However, as the technology has become refined dedicated examination room devices with markedly decreased prices are already being seen. Soon a machine in every exam room may become a reality. But, all patients cannot be examined routinely with ultrasonography and yet charged for a pelvic ultrasound scan the way the system now charges. Ideally the global fee will be raised enough to reflect the true cost of administering such an ultrasound-enhanced bimanual exam.

Summary

The use of vaginal ultrasound by the office practitioner will have tremendous assets. Diagnoses will be able to be made on the spot more often, paving the way either for more timely therapy or, in the case of normal findings, to relieving patient anxieties. Vaginal ultrasound will save time for both the physician and the patient. With transabdominal scanning patients must often be rescheduled. Time often does not permit waiting for the bladder to fill. Vaginal scanning in the office will cut down on the need to refer patients elsewhere (except for consultation). This will reduce the number of errors that occur when information must be transmitted from one source to another. Patients uniformly prefer vaginal scanning over the discomfort of a full-bladder transabdominal approach. Patients can also better understand their care; the pictures generated on the screen are very helpful in communicating the pathology back to them.

As practiced today endovaginal ultrasound scanning is not merely a subset of conventional ultrasonography and should not be reserved for the imaging specialist. Its unique characteristics allow instant confirmation of the primary physician's findings at the time of the pelvic examination.

Suggested Readings

1. Goldstein SR, et al. Very early pregnancy detection with endovaginal ultrasound. *Obstet Gynecol*. 1988;72:200–204.
2. Goldstein SR. Early pregnancy ultrasound: a new look with the endovaginal probe. *Contemp Ob Gyn*. 1988;31:54.
3. Goldstein SR. Incorporating endovaginal ultrasonography into the overall gynecologic examination. *Am J Obstet Gynecol*. 1990;162: 625–633.
4. Goldstein SR, Nachtigall M, Snyder JR, et al. Endometrial assessment by vaginal ultrasonography before endometrial sampling in patients with postmenopausal bleeding. *Am J Obstet Gynecol*. 1990;163:114–123.
5. Letourneau J. Gynecologic sonography: report of the ultrasonography task force. *JAMA*. 1991;265:2851–2855.
6. Mendelson EB, Bohm-Velez M. Transvaginal sonography assessed early pregnancy. *Diagn Imaging*. November 1987.
7. Silva P, Platt L, Yee B. Transvaginal ultrasound in the infertility workup. *Female Patient*. 1987;12:14–24.
8. Timor-Tritsch IE, Bar-Yam Y, Egfali S, Rottem S. The technique of transvaginal sonography with the use of a 6.5 MHz probe. *Am J Obstet Gynecol*. 1988;158:1019–1024.
9. Timor-Tritsch IE, et al. Review of transvaginal ultrasonography. *Ultrasound*. 1988;6:1.
10. Timor-Tritsch IE. Is office use of vaginal sonography feasible? *Am J Obstet Gynecol*. 1990;162:983–985.
11. Weiss RR, Khulpathea T, Lau GK. Transvaginal ultrasonography screening in early gestation. Presented at the Third World Conference on Vaginosonography Gynecology, June 14–17, 1991.

11
Dysmenorrhea and Chronic Pelvic Pain

H. Jacob Saleh

Introduction

Dysmenorrhea

Dysmenorrhea, the most common gynecologic problem, means painful monthly flow. It affect between 30% and 75% of all women in the reproductive age group. Primary or physiologic dysmenorrhea is painful menses that occurs in the absence of pelvic disease. It has been shown that excessive production of prostaglandins by the endometrium can result in uterine ischemia and hypercontractibility and can cause severe dysmenorrhea. This has clarified a previous "psychosomatic" disorder, and many strides have been made in effective treatment of dysmenorrhea using nonsteroidal anti-inflammatory medications.[1,2]

Secondary or acquired dysmenorrhea always has an underlying cause (see Table 11.1). The pain pattern associated with secondary dysmenorrhea has a variable time of onset. In contrast, primary dysmenorrhea usually begins within a year of menarche. Primary dysmenorrhea is usually spasmodic and affects the lower abdominal and midline areas, with radiation to the lower back or thighs. The lateralizing, deep pain, associated with dysparunia is likely to have a preexisting cause. A complete history and careful pelvic examination are mandatory in the evaluation of patients with dysmenorrhea. Because of the variable diagnostic and treatment options available, it is important to distinguish between primary and secondary causes of dysmenorrhea.

TABLE 11.1. Causes of secondary dysmenorrhea.

Endometriosis
Pelvic adhesive disease
Fibroid uterus
Ovarian cyst
  Rupture
  Torsion
  Bleeding
Miscarriage
  Threatened
  Incomplete
  Complete
IUD use
Pelvic inflammatory disease
Endometritis
Chronic pain syndrome
Pelvic congestion
Intestinal colic
Renal colic
Adenomyosis

Chronic Pelvic Pain

Heterogenous factors, including duration of pain, anatomic location, severity, association or lack of association with tissue damage, and psychologic causes, have lead to a less than succinct definition of chronic pelvic pain. Classically it is believed that pelvic pain lasting 6 months or longer may be defined as chronic pelvic pain.[3] In many patients no specific organic cause is found. However, laparoscopy may reveal specific findings such as endometriosis or pelvic adhesive disease. Even when pelvic pathology is present, certainty that the pain is completely explained by these findings is difficult. Furthermore, psychiatric or behavioral factors may play a significant role in chronic pelvic pain; however, it is often uncertain if the emotional problems preceded the pain, developed as a result of the pain, or are a separate issue altogether.

This chapter begins with a review of the gate control theory of pain modulation, believed to be important in integrating physiologic and psychogenic processes in pain perception. Secondary dysmenorrhea will be discussed with chronic pelvic pain. Primary dysmenorrhea will be discussed separately.

Chronic Pain Perception

The Cartesian theory describes direct pain signals from the periphery to the brain as a direct circuit, where the intensity of the perceived pain is related to the severity of tissue damage. Although this may be an adequate explanation of acute pain transmission, it does not explain the transmission of chronic pain, where the extent of tissue damage is often unrelated to perceived discomfort or its associated behavioral ramifications. Several recent studies have shown the lack of association with extent of documented tissue damage and perceived pain in endometriosis.[4]

A theory that combines pain perception with behavioral integration through a network of fibers involving the periphery, spinal cord, cortex, and motor mechanisms should better explain chronic pain states. The gate control theory as proposed by Melzack[5] describes a system where incoming nocioceptive signals are filtered through the spinal cord before reaching the central control processes in the brain. Moreover, the degree to which these signals are allowed to traverse the spinal cord and ascend to the brain may be influenced by modulating signals coming down the spinal cord "gate" from high centers (Fig. 11.1). This bidirectional system may, in turn, allow for

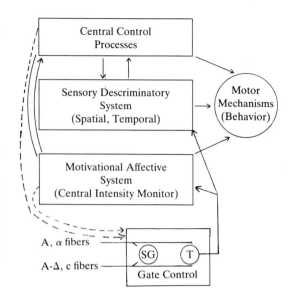

FIGURE 11.1. The gate control system of pain perception: modulating influences (scored lines) from the central control level and from the motivational affective system may alter central perception of pain by impinging on the gate control system.

psychologic processes to play a direct role in pain perception and behavioral response.

Thus a person who has had pain for 6 months or longer may appear depressed, which makes it clinically difficult to distinguish which came first, the pain or the depression. Opening the gate may allow a disproportionately intense pain experience, despite a limited amount of tissue damage. As time goes by, personality and early psychologic experience may come into play, further confounding a true cause-and-effect relationship between the pain and the actual organic disease, if any.

Because of these inconsistent pain patterns pelvic pain is therefore a diagnostic challenge. The multiple organ systems that are involved with multiple pathogenic mechanism (for examples, ischemia, distention, inflammation, muscular contraction, and psychosexual factors) may further complicate the work-up, diagnosis, and treatment of chronic pelvic pain.

Differential Diagnosis

Psychogenic Pelvic Pain

The percentage of patients with psychiatric illness underlying their pelvic pain is difficult to estimate. A recent study by Walker et al[6] demonstrated that women with chronic pelvic pain were more likely to have depression, somatization disorders, sexual dysfunction, and substance abuse. Studies by Reiter et al[7,8] show that 75% of women with chronic pelvic pain had some type of somatic diagnosis and 65% had a psychologic diagnosis, most commonly depression. Some of these women clearly have concurrent conditions, necessitating a multidisciplinary approach to diagnosis and treatment.

In such patients diagnosis is suggested from the lack of objective findings of organic disease. The pain is usually dull, nonspecific in location, and without a reasonable radiation pattern. Other clues to psychogenic etiology include aggravation of pain by stress, premenstrual onset, and occurrence of the pain as the day progresses. A history of depression, family disorders, or childhood sexual abuse should be sought. Reiter observed that in the pain-free population the rate of major childhood physical and sexual abuse was 7%, whereas in women with chronic pelvic pain the rate was over 50%. Furthermore, depression was found at a significantly higher rate in families of women with chronic pelvic pain than in the general population.

Nongynecologic Pelvic Pain

Urinary tract, bowel, or musculoskeletal origin organic pain may account for 20% to 40% of pelvic pain.[9] Interstitial cystitis has been described as a potential cause. Because of the insidious nature of this entity and the association with dysmenorrhea, diagnosis is typically delayed. Chronic posterior urethritis is another elusive urinary tract lesion causing nonspecific pelvic pain. A high index of suspicion should be used in these patients, with appropriate referral for urologic evaluation.

TABLE 11.2. Nongynecologic causes of pelvic pain.

Gastrointestinal Tract	Musculoskeletal
Constipation	Back/pelvic postural changes
Irritable bowel synchrome	Nerve entrapment
Inflammatory bowel disease	Neuropathy
Diverticulitis	Piriformis syndrome
"Smoldering appendicitis"	Ventral hernia
	Inguinal/femoral hernia
Urinary Tract	
Chronic cystitis	
Uretheral syndrome	
Interstitial cystitis	

Table 11.2 summarizes the etiologies of chronic pain related to nongynecologic disease. Irritable bowel syndrome, colitis, and vertebral disc disease should always be considered.

Gynecologic Causes

Endometriosis

The most common cause of chronic pelvic pain is endometriosis. Steege[10] recently reviewed laparoscopic findings in a total of 2,615 patients in 15 studies; endometriosis was found in up to 51% of patients. With widespread use of laparoscopy histologic confirmation has become possible, and studies using histology have confirmed endometriosis in 20% to 50% of patients who underwent laparotomies for gynecologic disease.[11]

The diagnosis of endometriosis is often made in women in their 20's and 30's who are nulliparous. Although heredity and multiple demographic factors have been implied in identification of patients with endometriosis, little scientific confirmation is available. Pregnancy may have a transient beneficial effect, but there is no evidence that pregnancy either cures or prevents later development of the disease.[12]

Pain with endometriosis may occur at any time during the menstrual cycle, but a temporal relationship with dysmenorrhea is common.

The degree of discomfort is generally not related to the extent of endometriotic lesions. A slowly expanding endometrioma may cause little or no discomfort, whereas small implants in strategic areas such as the uterosacral ligaments may cause significant symptoms. Dysparunia is generally related to stretching and irritation of fixed implants and concurrent adhesion. Dysuria and dyschezia may reflect involvement of implants and adhesions to the bladder and bowel wall, respectively.

Infertility is twice as common in patients with endometriosis as in the general population.[13] Although the fallopian tubes are generally patent, endometriosis is believed to distort tubal motility and egg transport.

Adhesive Pelvic Disease

As with endometriosis, the amount of adhesive disease is unrelated to the intensity of pain. The most common cause of pelvic adhesions is previous pelvic and abdominal surgery. Other causes include a history of pelvic inflammatory disease, endometriosis, and IUD use. Adhesion formation begins as early as several hours after the initial insult due to coagulation of exudate secondary to tissue trauma. Although adhesions are stable in their extent and location several months after injury, the pain associated with them may progress in severity. Pelvic pain due to adhesions is usually gnawing, appears more on one side than the other, and increases near or during menses. Dysparunia is a common complaint. The increase in intensity of pain progression with adhesions is related to physical, emotional, and cognitive factors as opposed to actual progression of tissue damage.

Adhesions are very common following surgery for ectopic pregnancy, but they occur less frequently after laparoscopy than after laparotomy.[14] Laparoscopic lysis of adhesions when possible may result in improved symptoms. In a unique study by Peters et al[15] the use of laparotomy for adhesiolysis did not significantly show improvement of pain symptoms in the treated patients as opposed to the control group, which was not treated.

Leiomyomata

Pelvic pain with uterine tenderness is associated with uterine fibroids. The location of the pain and associated symptoms may predict the extent of fibroid disease. A fibroid that presses on the bladder may cause urgency, urinary retention, and overflow incontinence. Degeneration and ischemia may cause severe pelvic pain, low-grade fever, and peritoneal irritation. In young patients with significant symptoms and infertility myomectomy is considered. The selection of patients for myomectomy versus hysterectomy must be individualized, but as seen in Table 11.3 some guidelines involving uterine size and symptomatology may be useful.

Because estrogen appears to play a role in the pathogenesis of myomata, using GnRH analogues to create a pseudomenopausal state in an attempt to reduce the size of these lesions has been widely investigated.[16] A decrease in the size of a myoma and in the thickness and vascularity of the surrounding myometrium have been noted in most patients after 3 to 6 months of treatment. GnRH treatment may be associated with bone density loss, which is reversible after treatment is discontinued. A reenlargement of the myoma has been noted in some patients after treatment is completed. Thus the major indications for GnRH use may be in the younger woman as premyomectomy treatment or in the perimenopausal woman to avoid a hysterectomy prior to a natural menopause.

Adenomyosis, a disease involving uterine tenderness and enlargement, may be found in up to 50% of patients with uterine leiomyomata. Hysterectomy specimen in patients with adenomyosis demonstrates heterotropic islands of endometrial tissue within the myometrium. The patients are generally middle aged and multiparous, and have uterine tenderness to palpation. Unlike leiomyomas, adenomyosis typically produces a symmetrical or globular enlargement of the uterus associated with a dysmenorrhea, menorrhagia, and dysparunia. Although definitive diagnosis and treatment may require a hysterectomy, it may

TABLE 11.3. Suggested management of patients with myomas.*

Fertility status	Asymptomatic		Symptomatic**
	< 12 wk size	> 12 wk size	
Desires pregnancy now	Conception trial	Conception trial	Myomectomy
Desires pregnancy later	Observation	Myomectomy	Myomectomy
No desire for pregnancy	Observation	Hysterectomy	Hysterectomy

* Modified from Buttram VC Jr, Reiter RC. Uterine leiomyomata: etiology, symptomatology, and management. *Fertil Steril.* 1981;36:433.
** Symptoms include bleeding, recurrent abortion, impingement, infertility and pain: after all other causes were ruled out the patient failed conservative therapy.

be feasible to manage adenomyosis by using hormonal manipulation.

Other Causes of Gynecologic Pelvic Pain

Pelvic relaxation as a source of chronic pelvic pain is more common in the postmenopausal patient. The patient usually complains of fullness, heaviness, or a dropping sensation. Lower pelvic pain can occur secondary to voluntary attempts to "hold in" prolapsing organs. Excessive mobility of the pelvic organs due to childbirth has been theorized to result in stretching and tearing of the ligaments supporting the uterus, resulting in chronic pelvic pain. Patients with cystocele may experience urine loss with intercourse, further adding to the discomfort and anxiety in the progression of chronic pelvic pain.

In the patient who has undergone hysterectomy enterocele should be considered as a source of chronic pelvic pain. Poor postoperative pelvic floor support associated with activities that increase intraabdominal pressure may aggravate symptoms of cytocele. As time passes progressive dissection of the peritoneal sac into the rectovaginal space may occur easily. It is important to distinguish between enterocele and an often coexisting rectocele. This is performed by having the patient bear down during a rectovaginal examination and palpating for an overriding enterocele. It may also be possible to palpate small bowel content of the enterocele on rectovaginal examination. This is an important differential when planning treatment options.

Intermittent pelvic aching after intercourse, usually on one side, which is exacerbated by postural changes, is associated with pelvic congestion syndrome. The typical patient is multiparous, in her 30s, and often complains of anorgasmia. In a study by Beard et al[17] clinical findings of point tenderness over the involved ovary and a history of postcoital ache was 94% sensitive and 77% specific for discriminating pelvic congestion from other causes of pelvic pain. Sonography with Doppler blood-flow studies of the pelvic vessels may provide further clues to diagnosis, although a specific treatment plan is difficult to find. It is believed that there is a high comorbidity with pelvic congestion syndrome and psychologic stress factors; as evidenced by Farquhar et al[18], while medroxyprogesterone may help the symptoms by eliminating menses, the best long-term results followed psychotherapy in conjunction with medroxyprogesterone.

Diagnostic Evaluation of Pelvic Pain

History

A large array of diagnostic aids is available to the gynecologist dealing with pelvic pain. By far the most essential aspect of the investigation is a complete patient history and physical exam. Although many strides have been made in diagnostic and operative pelviscopy as well as in radiologic imaging (for example, sonography and magnetic resonance imaging), the patient will usually tell her diagnosis if given the chance. The patient with chronic pelvic pain may have undergone previous treatments with partial or no relief of symptoms. Likewise she may suffer impaired physical function and depression out of proportion to the actual pathology. For these reasons an objective and comfortable relationship should be established.

The significance of the pain in terms of day-to-day function, missed work hours, and the effect on family dynamics should be sought. As opposed to the patient actually awakening from the physical pain, sleep disturbance may be a sign of an underlying depression, as can weight loss. As mentioned earlier, problems with childhood sexual abuse and sexuality in general may underlie a chronic pain syndrome. It is important to question the patient about her satisfaction with adult relationships as well as about previous situations that may have shaped her feminine identity.

The history should include specific descriptions of the onset, timing, character, and radiation pattern of the pain. Provocative and palliative factors should be sought. In patients with endometriosis, the pain is pronounced with menstruation or intercourse. Likewise, pain with bowel movements or urination may signal endometriotic involvement in adjacent organs. Pain that is sudden, spasmodic, stabbing, and lasting a short time per attack will rarely have an identifiable cause. Pain that is described as burning, hot, gnawing, and associated with paresthesia suggests nerve compression or injury. This type of pain is usually noticed after surgery and may be related to the surgical scar. Pain that is less difficult to localize, described as diffuse, throbbing, cramping, pulling, and lasting for hours, is more likely to be visceral in origin, involving pelvic or abdominal structures.

Questions about age of menarche, menstrual history, sexual history, parity, and contraception should be addressed. To avoid repetition of time-consuming, costly, and uncomfortable testing previous diagnostic and treatment record should be obtained and reviewed.

Physical Examination

A complete examination with a special focus on the location of the pain should be performed. Although the exact pain is often not present at the initial office visit, it is helpful to examine patients with chronic pelvic pain during an occurrence. An effort should be made to recreate the pain with palpation of the abdominal wall, uterus, adnexal regions, and vagina. A rectovaginal examination is mandatory in evaluating the uterosacral ligaments and posterior pelvic structures. During the exam position changes may be useful to delineate causes of pain due to impingement on visceral structures or neurologic causes.

To delineate uncommon causes of pelvic pain, chest and upper abdominal examination is useful. Chronic undiagnosed pneumonia or plueral effusion may underlie abdominal or pelvic pain. Upper abdominal findings such as right upper quadrant pain suggest perihepatitis associated with pelvic inflammatory disease (Fitz-Hugh-Curtis syndrome); likewise, periumbilical or abdominal firmness or mass may be a sign of a primary ovarian neoplasm with metastasis. Umbilical or abdominal wall hernias should be looked for because they may cause chronic pain. Point tenderness along the abdominal wall and inguinal area is also important in defining causes such as muscle injury or hernia. The inguinal region should be examined for lymph nodes. Enlarged lymph nodes may reflect lower genital tract tumors or infections.

The pelvic examination should begin with careful inspection of the vulva and vagina. An external vulvar lesion (ulcer or vesicle) may suggest a concomitant internal pelvic infection. Any suspicious vaginal lesions should be inspected or undergo biopsy. Pigmented lesions may be seen in 5% of patients with endometriosis.[19] A common but often overlooked cause of dysparunia is vestibulitis. With careful palpation or the use of a Q-tip it is simple to localize vestibular pain and itching to the vaginal opening below the hymeneal ring. Vaginal support is important to assess during examination. Cystocele, cystourethocele, and uterine prolapse are easily identified by asking the patient to bear down during palpation. Asking the patient to "push out the speculum" after the Pap smear is performed may demonstrate an occult uterine prolapse, sometimes in dramatic fashion. A rectocele is characterized by decreased sphincter tone, and as described previously an attenuated rectovaginal septum should be distinguished from an enterocele. The cervix should be inspected for lesions or discharge. Cervical motion tenderness is not only indicative of pelvic inflammatory disease but may be present with peritonitis of any cause. Furthermore, a cervix that appears to be normal should be carefully palpated in search of firm areas or blunting of the lateral vaginal fonices, which may suggest neoplasia.

Examination of the uterus and adnexa should be performed in systematic fashion. The size, orientation, contour, and mobility of the uterus should be noted. Uterine palpation may illicit pain over areas of fibroid degeneration or adenomyosis. An acutely retroverted uterus may cause back pain or urinary retention, especially around the time of menses or if associated with posterior uterine leiomyomata. A retroverted uterus is best appreciated on rectovaginal examination, where it may be displaced anteriorly and checked for mobility. Pain of pelvic origin is often reproducible on pelvic examination using the vaginal examining hand. If the pain is only elicited on bimanual examination with abdominal palpation, it is more likely of the abdominal wall or of nonpelvic organ origin. Likewise, back pain alone, if pelvic in origin, is usually located caudally to the ischial spines and may radiate to the buttocks and upper thighs. Pain distributed in this area may originate from the cervix, upper vagina, or lower bladder.

The adnexa should be palpated for size, position, mobility, and tenderness. Pain on adnexal palpation may reproduce the cause if endometriotic lesions or adhesions are involved. Chronic salpingitis as a result of previous pelvic infection may cause chronic pelvic pain that is indistinguishable from pain of adnexal or uterine origin. Generalized adnexal thickening or fullness, albeit nonspecific, is important to note on pelvic examination because it may direct further diagnostic testing. The rectovaginal examination may prove invaluable in assessing ovarian masses that are posterior in location. Adnexal tumors may be distinguished from posterior leiomyomata and with an adequate rectovaginal examination may be better palpated for size, consistency, and mobility.

Laboratory Studies

Relatively little information is gained using laboratory testing in chronic pelvic pain. Baseline blood count and urinalysis may be useful in patients suspected to have anemia from menorrhagia or chronic cystitis from cystocele. Erythrocyte sedimentation rate is nonspecific and essentially has no place in evaluating patients with chronic pelvic pain. In women over 30 years of age with a suspected pelvic mass or in women with a strong family history of ovarian cancer, a CA-125 titer is indicated. CA-125 assay is limited: It may not be elevated in up to 20% of ovarian malignancies, and it may be elevated in the presence of benign processes such as functional ovarian cysts or endometriosis. As opposed to first-line diagnostic tests hormonal testing (FSH, LH, testosterone, DHEAS) should be reserved to confirm physical findings in patients with suspected hormone-producing adnexal masses.

Imaging Studies

In patients with negative physical examinations, costly imaging studies such as CT scans and MRIs usually add very little to the diagnostic database. In patients who are obese or who have postoperative adhesive disease due to artifacts these studies may actually be misleading. On the other hand, ultrasound examination of the pelvis has been quite helpful in defining pelvic anatomy and the diagnoses of even unsuspected lesions in patients who are difficult to examine or in whom examination is equivocal. The dynamic ultrasound image is important in differentiating pelvic organs from the bowel, omentum, and so forth. Because of the relative cost-effectiveness and safety of ultrasound examinations, their liberal use is justified in patients with chronic pelvic pain with suspicion of pelvic disease. Cystic adnexal masses are easily seen and may be differentiated from solid tumors using ultrasound. Fluid in the cul-de-sac may result from cyst rupture or bleeding. Ultrasound is also useful to follow the size and appearance of functional ovarian cysts causing mild symptoms.

Other imaging studies may be necessary. Hysterosalpingogram (HSG) may uncover intrauterine lesions such as fibroids or adhesive disease. Tubal obstruction or defects are seen in patients with chronic or recurrent pelvic infections, and adhesive disease may be confirmed using HSG. In endometriosis where implants and adhesions cause peritubal constriction, complete obstruction is uncommon.

Laparoscopy

In addition to the diagnostic potential of laparoscopy, recent innovations and training have progressed rapidly and now incorporate a long litany of surgical procedures that can be performed through the laparoscope. In patients with suspected pelvic disease such as endometriosis or adhesions laparoscopy should be performed liberally. It is possible to carry out diagnosis and treatment in these patients by using pelviscopic techniques. Unfortunately, the patient with chronic pelvic pain is often not helped by laparoscopic surgery, even when pathology is present. This is likely explained by the complex pain perception theory in chronic pain states described earlier. Nevertheless, adding to the database by ruling in or out pelvic pathology using laparoscopy will still aid in the management of these patients. Because some patients without pelvic pathology may unjustly undergo laparoscopy this approach may appear aggressive, but perhaps more unjust may be the approach of labeling pelvic pain as psychosomatic or chronic pain syndrome in patients with actual pelvic pathology who do not undergo laparoscopy.

Primary Dysmenorrhea

Primary dysmenorrhea is pain with menses that is not associated with pelvic pathology. Usually seen in younger patients, it often begins with the onset of regular ovulating menstrual cycles and may improve spontaneously with time. The pain is typically midline and cramping in nature and may radiate to the lower back; the flanks or lower abdominal quadrants can also be involved. Prostaglandin-induced uterine contractions during menses may cause vigorous uterine spasms that may occur every 2 minutes and increase intrauterine pressure tenfold.[20] Prostaglandin usually occurs locally at the myometrial level but may cause systemic symptoms such as nausea, diarrhea, and lightheadedness. When present these symptoms raise the suspicion of primary dysmenorrhea. The pain is usually worse during the first day or two of the menstrual cycle and may be incapacitating, causing significant absenteeism and loss of work hours. Dysparunia, if present, is usually mild.

If performed in the nonmenstrual phase of the cycle, physical examination in the patient with primary dysmenorrhea is typically negative. A careful examination and an appropriate work-up are mandatory to rule out potential organic causes of dysmenorrhea. If the pain is reproducible on examination, it should be nonspecific and involve only midline structures.

Treatment for primary dysmenorrhea is discussed below. Often a trial of prostaglandin synthetase inhibitors and/or oral contraceptives maybe used as a diagnostic aid. The young patient with painful menses and without apparent organic disease who responds to a trial of medical therapy is likely to have primary dysmenorrhea. Care must be taken to explain this expectant approach to the patient and her family. Because of the interruption of school and recreational activity, attempting a several-month trial of therapy prior to ascertaining a definite diagnosis may lead to significant anxiety, both patient and parental. The usual inquiries about anticipated chronicity of pain, inability to perform physical activity, and, of course, future fertility should be expected and addressed in a thoughtful and reassuring manner. Finally, it is important to note that 15% of patients with early-onset dysmenorrhea have secondary causes; therefore a definitive diagnosis should not be postponed simply because of a patient's young age.[21]

Treatment of Chronic Pelvic Pain and Dysmenorrhea

The cornerstone of therapy for chronic pelvic pain is medical management. Widespread use of low-dose combination oral contraceptives have proven effective in the treatment of primary dysmenorrhea and secondary dysmenorrhea due to endometriosis. Gonadotropin downregulation by use of oral contraception or medroxy-progesterone acetate may likewise suppress ovarian cyst formation and has been suggested to relief pelvic vein engorgement in patients with pelvic congestion syndrome.[18] GnRH-agonist therapy has been proven useful in endometriosis and infertility associated with endometriosis. In a recent review by Anderson[22] the use of GnRH agonists in the treatment of uterine fibroids showed multiple benefits. Besides curbing the symptoms of fibroid disease prior to surgery, GnRH agonists decreased uterine volume by reducing blood supply to the myometrium, allowing for easier myomectomy with limited blood loss. In patients undergoing hysterectomy, the decrease in uterine size allows for the vaginal approach to hysterectomy and reduced preoperative anemia as well as postoperative requirements for blood transfusion. A concise review of the treatment intervals and dosing regimens is found in Anderson's article.[22]

Analgesics used as adjuncts to hormonal manipulation have been effective in treating pelvic pain. The use of nonsteriodal anti-inflammatory drugs (NSAIDs) has especially been beneficial because of the antiprostaglandin effect at these agents. By inhibiting production of prostaglandins, the major local pain mediator in primary dysmenorrhea, endometriosis, and uterine fibroids, NSAIDs have been successful in reducing pelvic pain and uterine bleeding with menses. Although NSAIDs alone (naprosyn, ibuprofen) may effectively control pelvic pain, they may be associated with significant gastritis. Narcotic analgesics alone or in combination with NSAIDs may be required in the patient with persistent pain; narcotics have their own well-known adverse effects (habitation, addiction). Treating patients with dysmenorrhea prior to the onset of menses with analgesics (1 to 2 days prior to the anticipated onset of menses) may ameliorate the pain cycle prior to the full-blown attack. Some advocate a trial of chronic narcotic therapy in the patient who fails to respond to the above measures, assuming that the patient has no treatable organic disease.

Surgical Treatment

Due to the growing knowledge of innovative hormonal agents (such as GnRH) and analgesics such as NSAIDs, the need for surgery as primary therapy for pelvic pain has diminished in recent years. Furthermore, because of the understanding that chronic pelvic pain is often out of proportion in severity to actual organic disease, pursuing an expectant course of medical management prior to surgery is warranted. Nevertheless, surgery should never be put off in the patient who demonstrates disease amenable to operative intervention.

The mainstay in primary operations for the relief of central pelvic pain is presacral neurectomy. Current indications include dysmenorrhea, deep dysparunia, and sacral back pain.[23] Presacral neurectomy is usually performed in concert with other procedures, mainly surgical management of endometriosis.[24,25] Several studies have shown that significantly more patients (75%) obtain relief when sacral neurectomy is combined with surgical treatment of endometriosis then when it is not (25%).[26] Both laparoscopy and laparotomy have been used for transsection of the presacral nerve. Pain recurrence, hypermenorrhea, and possibly uterine descent may occur postoperatively with pelvic denervation procedures. The availability of prostaglandin-synthetase inhibitors has further led to the decreased popularity of these procedures for the management of midline pelvic pain.

Operative laparoscopy, which has been used for ablation of adhesive disease and endometriosis with laser techniques or electrocautery, has demonstrated improvement in most patients,[27] albeit with some risk of pain recurrence. Laparoscopy is less likely than laparotomy to lead to postoperative adhesive disease and defeat the initial intent of surgical therapy.

Extirpative surgery for obvious pathology such as ovarian tumors or large uterine leiomyomata can clearly relieve the attendant pelvic pain. Available data on hysterectomy performed for pelvic pain are, however, somewhat puzzling. In up to 20% of hysterectomies performed in the United States where pelvic pain is the indication, one-third of hysterectomy specimens show no pathology.[28] Furthermore, more than one in five patients show no improvement in their pain.[29] This further underscores the difficulty in treating patients with chronic pelvic pain. The patient should undergo thorough evaluation with complete attention to ruling out even the most enigmatic conditions (For example, pelvic congestion, interstitial cystitis, and emotional factors) prior to undergoing surgical treatment for chronic pelvic pain.

Summary

The approach to the patient with chronic pelvic pain involves multiple factors that include individual patterns of pain perception and tolerance, emotional response, and psychological and social coping abilities. The varied causes of organic pelvic pain and the potential involvement of adjunct abdominal and urinary structures as well as the varied mechanisms of pain transmission in the pelvis should all be reviewed in each patient. Diagnosis should be methodical and individualized. To arrive at the appropriate diagnosis, repeat visits and reassessment of symptoms and physical findings may be necessary. All the while the patient should be allowed to involve herself freely in decisions regarding her care. As the usually prolonged diagnostic process unfolds feelings of inadequacy are common in these patients, and these feelings should be addressed with compassion. Treatment is aimed at simultaneously managing the varied components of pelvic pain while avoiding overtreatment. The symptoms of pelvic pain, if approached systematically and comprehensively, will usually yield to appropriate management and will usually result in relief. In the patient where accurate and timely diagnosis and treatment are elusive, the symptom may become the illness onto itself.

References

1. Dawood MY. Nonsteroidal anti-inflammatory drugs and changing attitudes toward dysmenorrhea. *Am J Med*. 1988;84(suppl 5A):23–29.
2. Pasquale SA, Rothhauser R, Dolese HM. A double-blind, placebo controlled study comparing three single-dose regimens of piroxicam with ibuprofen in patients with primary dysmenorrhea. *Am J Med*. 1988;84(suppl 5A):30–34.
3. Kresch AJ, Seifer DB, Sach LB, et al. Laparoscopy in 100 women with chronic pelvic pain. *Obstet Gynecol*. 1984;64:672.
4. Fedele L, Parazzini F, Bianchi S, et al. Stage and localization of pelvic endometriosis and pain. *Fertil Steril*. 1990;53:155.
5. Melzack R. Neurophysiologic foundations of pain. In: Sternbach RA, ed. *The Psychology of Pain*. New York: Raven Press; 1986:1–24.
6. Walker E, Katon W, Harrop-Griffiths H, Holm L, Russ J, Hickok LR. Relationship of chronic pelvic pain to psychiatric diagnoses and childhood sexual abuse. *Am J Psychiatry*. 1988;145:75–80.
7. Reiter RC, Gambone JC. Demographic and historical variables in women with idiopathic chronic pelvic pain. *Obstet Gynecol*. 1990;70:428–432.
8. Reiter RC, Shakerin RL, Gambone JC, Milburn AK. Correlation between sexual abuse and somatization in women with somatic and non-somatic chronic pelvic pain. *Am J Obstet Gynecol*. 1991;165:104–109.
9. Vargyas JM. Chronic pelvic pain. In: Mishell DR, Brenner PF, eds. *Management of Common Problems in Obstetrics and Gynecology*.

NJ: Medical Economics Company Inc; 1988: 322.
10. Steege JF. Assessment and treatment of chronic pelvic pain. In: Thompson JD, Rock JA, eds. *TeLindes Operative Gynecology* (Update Vol 1, No. 2). Philadelphia, Pa: JB Lippincott Company; 1992:3.
11. Williams TJ, Pratt JH. Endometriosis in 1,000 consecutive celiotomies: incidence and management. *Am J Obstet Gynecol*. 1977;116:245.
12. McArthur JW, Ulfelder H. The effect of pregnancy on endometriosis. *Obstet Gynecol Survey*. 1965;20:709.
13. Nunley WC, Kitchen JD. Endometriosis. In: Sciarra JJ, ed. *Obstetrics and Gyencology*. Vol 1, no. 20. Philadelphia, Pa: JB Lippincott Company; 1992:10.
14. Lundorff P, Hahlin M, Kallfelt B, et al. Adhesion formation after laparoscopic surgery in tubal pregnancy: a randomized trial verses laparotomy. *Fertil Steril*. 1991;55:911.
15. Peters AAE, Trimbos-Kemper GLM, Admiral C, Trimbos JB, Hermans J. A randomized clinical trial on the benefit of adhesiolysis in patients with intraperitoneal adhesions and chronic pelvic pain. *Br J Obstet Gynecol*. 1991;99:59.
16. Andreyko JL, Marshall LA, Kumesic DA, Jaffe RB. Therapeutic uses of gonadotropin releasing hormone analygos. *Obstet Gynecol Survey*. 1987;40:11–12.
17. Beard RW, Highman JH, Pearce S, et al. Diagnosis of pelvic varicosities in women with chronic pelvic pain. *Lancet*. 1984;2:946.
18. Farquhar CM, Rogers V, Franks S, et al. A randomized controlled trial of medroxyprogesterone acetate and psychotherapy for the treatment of pelvic congestion. *Br J Obstet Gynaecol*. 1989;96:1153.
19. Julian TM. Chronic pelvic pain. 1. Workup and diagnosis. *The Female Patient*. 1989;14:37.
20. Edelin KC. Evaluation of female pelvic pain. (pictorial) *Hosp Med*. 1983 March;19:201–203, 207, 210–212.
21. *Precis IV*. Washington, DC: American College of Obstetricians and Gynecologists; 1990:15.
22. Anderson F. How GnRH agonists facilitate fibroid surgery. *Contemp Ob Gyn*. 1992:37:55–65.
23. Tjaden B, Schlaff WD, Kimball A, Rock JA. The efficacy of presacral neurectomy for the relief of dysmenorrhea. *Obstet Gynecol*. 1990; 76:89.
24. Molinak LR. Endometriosis. In: Conn HF, ed. *Current Therapy*. Philadelphia, Pa: WB Saunders; 1980.
25. Garcia C-R, David SS. Pelvic endometriosis: infertility and pelvic pain. *Am J Obstet Gynecol*. 1977;129:740.
26. Lee RB, Stone K, Magelssen D. Presacral neurectomy for chronic pelvic pain. *Obstet Gynecol*. 1986; 618:517.
27. Key WR, Hanse LW, Astin M. Argon laser therapy for endometriosis: a review of 92 consecutive patients. *Fertil Steril*. 1987;47:208–212.
28. Steege JF. Assessment and treatment of chronic pelvic pain. In: Thompson JD, Rock JA, eds. *TeLindes Operative Gynecology* (Update Vol 1, no. 2). Philadelphia, Pa: JB Lippincott Company; 1992:7.
29. Stoval TG, Ling FW, Crawford DA. Hysterectomy for chronic pelvic pain of presumed uterine etiology. *Obstet Gynecol*. 1990;75:676.

12
Advances in Oral Contraception

Michael J. Hickey and Jay H. Levin

Introduction

Recent birth control surveys demonstrate a revival of interest in oral contraceptives (OCs), revealing that these medications have been the leading birth control method employed by women aged 15 to 44 during the past 5 years. In 1991 16.3 million healthy women across America and more than 60 million women worldwide took an oral medication to prevent unwanted pregnancy. The foremost three methods of birth control in 1991 were oral contraception (28%), sterilization of one of the partners (26%), and the condom (15%). In 1991 new users of contraception preferred the pill (42%), followed by the condom (38%), followed by sterilization (17%).[1] Oral contraceptives remain the most effective reversible method of contraception[2,3]; and of current users, 92% report satisfaction with OCs as a method of birth control.[1]

Health Advantages

Over half of the women surveyed in 1991 believe that OC use is safer today than it was 25 years ago; but unfortunately, only 18% indicated that they were aware of the significant noncontraceptive health benefits of OC use.[1] Epidemiologic studies have established conclusively that the use of OCs prevents ovarian and endometrial cancers. The use of OCs for 1 year reduces the risk of developing ovarian cancer by 40% and the risk of developing endometrial cancer by 50%.[4,5] Women who use OCs for 10 years or more have an 80% reduction in the risk of developing endometrial cancer, and, as with ovarian cancer, longer duration and earlier onset of OC use reduces the risk of endometrial cancer for up to 15 years after discontinuation of OC use.[4]

Noncontraceptive health benefits of combination OCs also include a reduction in the incidence of functional ovarian cyst formation, ectopic pregnancy, acute pelvic inflammatory disease, benign breast disease, and a variety of menstrual disorders, including associated iron deficiency anemia.[6-8] OC use appears to decrease the risk of developing uterine fibroids[9] and may confer protection against developing endometriosis.[10] There is evidence that OC use decreases the likelihood of developing rheumatoid arthritis[11] and may be protective against osteoporosis.[12]

Historical Perspective

The history of oral contraception is one of evolution. The present formulations bear little resemblance to the "pill" of the 1960s, and on balance the health advantages of using oral contraception significantly outweigh the risks in all but a minority of patients. The history of the development of this technology makes evident the biologic, scientific, and commercial reasons for the availability of those products prescribed.

The first OC formulations were released for

commercial use in the early 1960s. Because of their ease of administration and extraordinary contraceptive efficacy, they soon became the most widely used reversible method of contraception and were used by women of all ages.

International clinical trials had concentrated mainly on efficacy and cycle control. The diverse metabolic actions of orally active steroid hormones had not been anticipated, and the initial OC formulations contained supraphysiologic doses of these synthetic compounds. Common contraceptive-induced side effects such as breast tenderness, nausea, and weight gain, although occasionally severe enough to cause discontinuation, were largely tolerated. Gradually, reports of serious complications began to appear, and eventually epidemiologic studies confirmed significant health risks associated with OC use.

Scientific Foundation

The scientific foundation for the use of steroids in contraception was laid at the beginning of the 20th century in Germany, where researchers demonstrated that transplanted tissue fragments of corpora lutea produced infertility in rabbits and mice.[13] By 1923 Allen and Doisy had established that the ovary produces two different substances: one responsible for the growth and development of sexual organs and the other responsible for the development of secretory endometrium and the maintenance of pregnancy.[14]

By 1935 MacCorquodale et al had developed the ability to crystalize and extract "ovarian follicular hormone," isolating 17-β estradiol, the most important estrogen in humans.[15] That same year, Butenandt and Westphal received the Nobel Prize in chemistry for having isolated and synthesized progesterone.[16] In 1937 Makepeace and co-workers determined that progesterone could inhibit coitally induced ovulation in the rabbit.[17] In 1940 Sturgis and Albright, using estradiol benzoate for the treatment of dysmenorrhea, observed that this hormone could effectively inhibit ovulation.[18]

Researchers in steroid biochemistry responsible for the development of oral contraceptives worked on converging paths, developing orally active compounds and a process to mass-produce steroid hormones economically. In 1935 the ovaries from 2,500 pregnant pigs were required to produce 1 mg of progesterone. Russell Marker, a steroid biochemist, set out to find a plant steroid that could be used as the starting point for steroid hormone biosynthesis. In the jungles of Veracruz he discovered a species of Mexican yam that produced diosgenin, a plant steroid. Marker and colleagues soon elucidated the synthetic pathway to convert diosgenin to progesterone[19] that enabled them to make kilogram amounts economically. For the next quarter century, diosgenin was used as the starting material for virtually all commercial steroid production.

In 1938 it was discovered that the addition of an ethinyl group at the 17 position of the steroid molecule made estradiol orally active.[20] Ethinyl estradiol (EE), a very potent oral estrogen, is the form of estrogen used in all commercially available low-dose oral contraceptives today. Mestranol, the 3-methyl ether of ethinyl estradiol (MEE), another orally active estrogen, is used extensively in early contraceptive formulations. Because mestranol must be demethylated before it will be recognized by the estradiol receptor, its relative potency is less than that of EE.

The discovery that ethinyl substitution leads to oral potency was applied next to testosterone and led to the preparation of 17-ethinyltestosterone (ethisterone). Surprisingly, while this substitution conferred oral activity to androgenic compound, it also conferred a weakly progestational biological character to this hormone.[21]

In 1952, while working at Syntex, Carl Djerassi demonstrated that splitting off the C_{19} methyl group from ethisterone dramatically increased the progestational activity of this compound while significantly reducing its androgenic activity. This steroid, 17-α-ethinyl, 19-nortestosterone (norethindrone, norethisterone), proved to be an extremely active oral progestogen, with potency approximately

10 times that of naturally occurring progesterone.[22] In 1953 chemists at Searle synthesized another 19-nor (nor = no radical) synthetic progestogen, norethynodrel.

Although it was known that norethynodrel was metabolized in the intestine to norethindrone, Gregory Pincus chose norethynodrel as the preparation to be used in the first oral contraceptive trials. During early studies to determine an effective dose of norethynodrel, it was discovered that an estrogen contaminant, mestranol, the 3-methyl ether of ethinyl estradiol, was present in the norethynodrel preparation. Elimination of the estrogenic contaminant caused unpredictable vaginal bleeding and loss of cycle control, which had previously been satisfactory. The estrogen was then reintroduced at a dose that was theoretically determined to be 1.5% of the progestin dose. Thus the first oral contraceptive formulation was a "combination pill" containing an ovulation-suppressing dose of a 19-norprogestogen and an empirically determined dose of an orally active synthetic estrogen added primarily to prevent intermenstrual bleeding.

Enovid, the first commercially available combination pill, was marketed by Searle in 1960 and contained 9.85 mg of norethynodrel and 150 μg of mestranol. Ortho-Novum, the Syntex formulation containing 10 mg of norethindrone and 60 μg of mestranol, appeared in 1962. A short time later, reports of undesirable side effects and complications began to appear. As early as 1963 it was evident that these high doses of contraceptive steroids led to the deterioration of carbohydrate metabolism.[23] By the early 1970s the use of oral contraceptives was associated with venous thrombosis and pulmonary embolism, hypertension, stroke, and myocardial infarction.

When reports of these complications first surfaced, an era of medical research began that resulted in tremendous advances in the understanding of reproductive physiology as well as the specific metabolic impact of the various orally active contraceptive steroids. Finding that many of these complications were dose related, the pharmaceutical companies responded by (1) initiating trials to find the lowest effective doses of existing compounds, (2) altering dose scheduling to diminish total steroid exposure, and (3) continuing research for novel progestational agents exhibiting the smallest impact on human metabolism.

As a result, a gradual, but dramatic, lowering of daily steroid doses has been experienced. The dose of the estrogen component has fallen from 150 μg present in the first commercially available combination OC to 20 μg in the lowest-dose monophasic preparations available today. Reductions in progestogen content in a comparable monophasic formulation have been decreased by 60%. While dramatically improving the safety profile of these medications, no significant difference in clinical effectiveness has been demonstrated. As long as no pills are omitted, the unintended pregnancy rate at the end of 1 year of use is less than 0.2% with all combination formulations.[24]

Total monthly steroid exposure has been reduced by the development of multiphasic (biphasic and triphasic) formulations, with variations in the ratio of progestogen to estrogen being more similar to the pattern of circulating hormones in the natural cycle. As discussed in greater detail below, reduction in the total monthly exposure of specific progestogens has been a successful modification to diminish certain undesirable side effects.

Single-agent OCs, referred to as the "minipill," containing progestogen only have also been developed for specific indications. Although they do not inhibit ovulation as reliably as combination OCs, they provide a substantial barrier to conception. Proposed mechanisms of action include inducing changes in cervical mucus, altering endometrial receptivity, and possibly altering tubal motility. If taken daily, the pregnancy rate resulting from contraceptive failure rate is approximately 3%.[25]

In the search for novel progestogens, minor structural alterations of the synthetic progestogen norgestrel have produced a group of highly potent progestational agents that exhibit significantly less androgenic activity than the parent compound. While only recently available in the United States, desogestrel (DSG) has been commercially available in Europe since 1981 as Marvelon (DSG 150 μg of DSG plus 30 μg of EE); Mercilon (150 μg of DSG plus 20 μg of EE) was introduced in 1988. Gestodene (GSD) in combination with 30 μg of EE has been marketed outside the United States since 1987 under a variety of trade names.

The number of different OC products is truly remarkable. All low-dose combination OCs contain the same orally active estrogen, ethinyl estradiol, and the amount present within a given formulation varies in a limited range (50 to 20 μg). Until very recently, there were only essentially four distinct OC progestogens available in the United States. Yet, there are no fewer than 36 different OC products—monophasic, triphasic, progestogen only—each individually marketed. With the adoption of the products presently available only in Europe, there are 5 additional low-dose formulations, marketed under potentially 16 different product names, containing the newer progestogens.

The present variety of OC choices can, at times, seem bewildering. All combination OCs employ synthetic progestogens derived from the testosterone molecule belonging to one of two closely related classes of progestogens, the estranes and the gonanes. Despite their structural similarity, each of these progestogens may exhibit dramatic differences in biological impact within a given metabolic parameter. Because the action of the progestogen is frequently opposite that of the estrogen, the net metabolic effect of these combination OCs is the result of the balance between these two components. By understanding the biochemical character of the various progestogens the clinician can utilize the subtleties now present with these formulations to select the combination best suited to an individual patient's requirements.

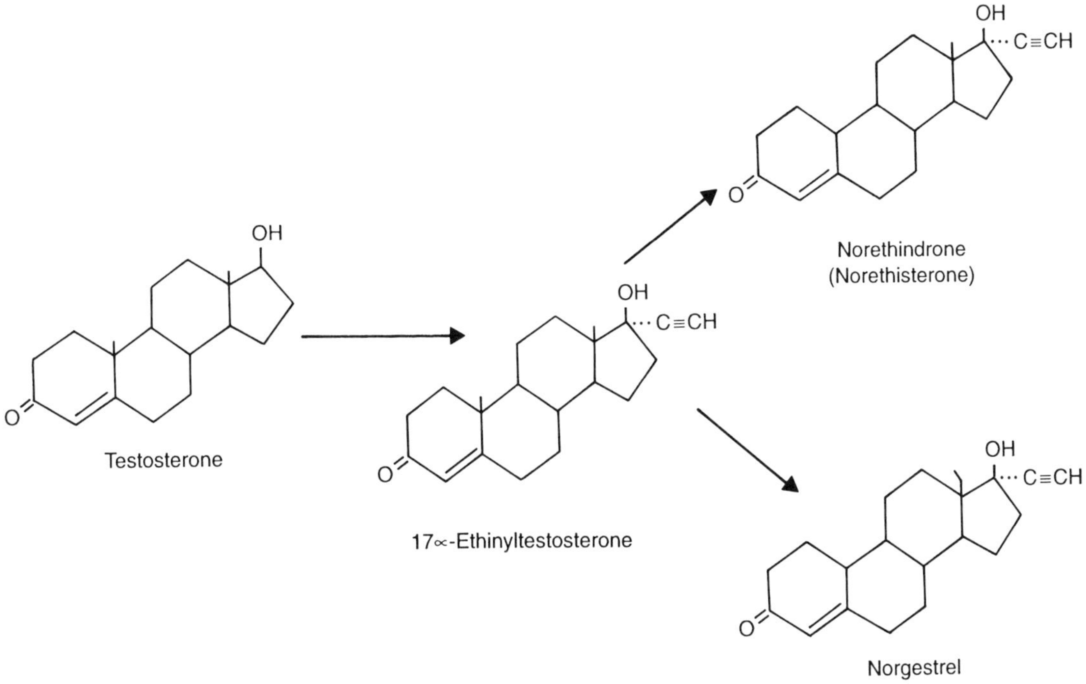

FIGURE 12.1. 19-Nortestosterone related progestogens. Chemical structures of the parent compounds of the estrane and gonane progestogens, norethindrone and norgestrel respectively. (Stanczyk FZ. Pharmacokinetics of progestogens. *International Proceedings Journal.* 1989;1:11.)

Pharmacology

Progestogens

All synthetic progestogens are derivatives of either 17α-hydroxyprogesterone or 19-nortestosterone, and they are classified into three chemical classes: pregnanes, estranes, and gonanes.

Pregnanes

Acetylation of the 17-hydroxy group of 17-hydroxyprogesterone gives oral potency, and substitution of the 6 position inhibits degrative metabolism, producing a group of progestational steroids with potent oral as well as parenteral activity (Fig. 12.2). The pregnanes include medroxyprogesterone acetate (Provera), chlormadinone acetate, and megestrol acetate. Cyproterone acetate, another compound in this class, is not only a potent progestogen but is also an extremely active antiandrogen. It is available for treatment of hyperandrogenism in Europe.

Estranes

The parent compound of the remaining two classes of progestogens is 19-nortestosterone (Fig. 12.1). Estranes are chemically characterized by the absence of a methyl group between ring A and B (19-nor) and by the presence of an ethinyl group in the 17-α position conferring oral activity (Fig. 12.3). The first orally active 19-nor steroid synthesized was norethindrone (17-α-ethinyl, 19-nortestosterone), which is still one of the most com-

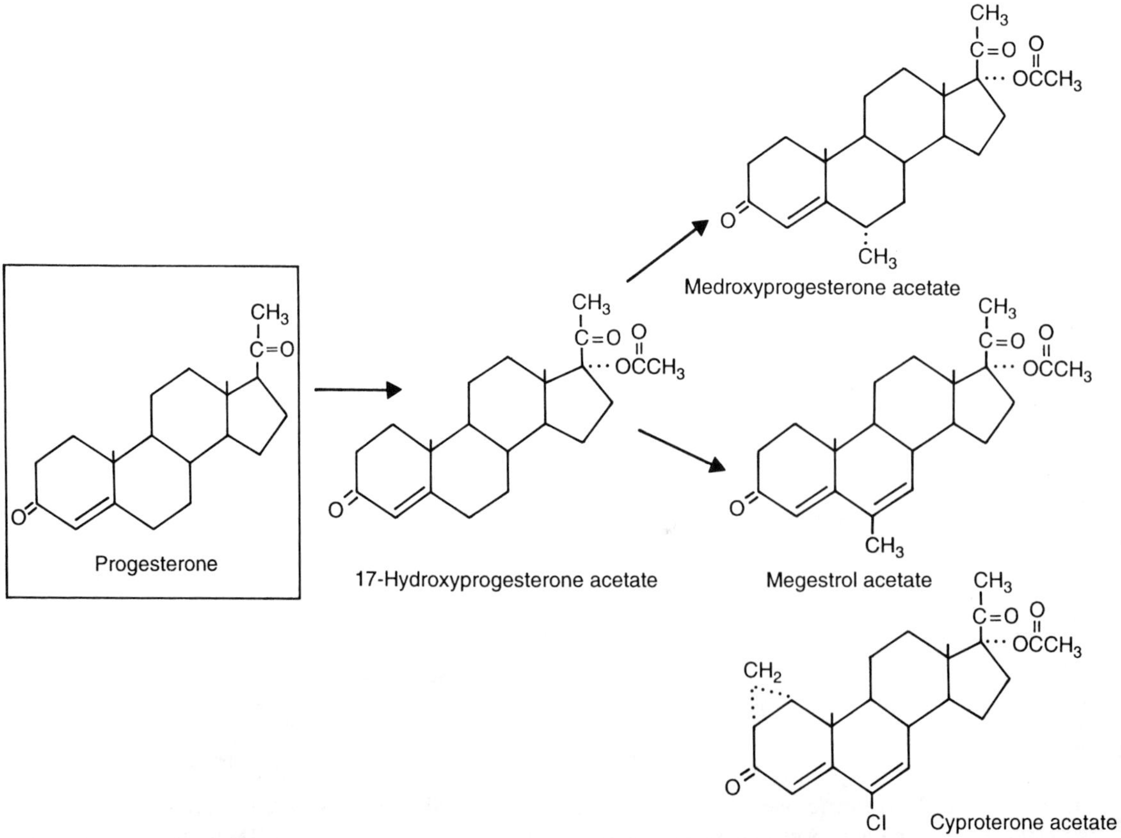

FIGURE 12.2. Pregnane progestogens. Chemical structures of several progestogens related to progesterone. (Stanczyk FZ. Pharmacokinetics of progestogens. *International Proceedings Journal.* 1989;1:11.)

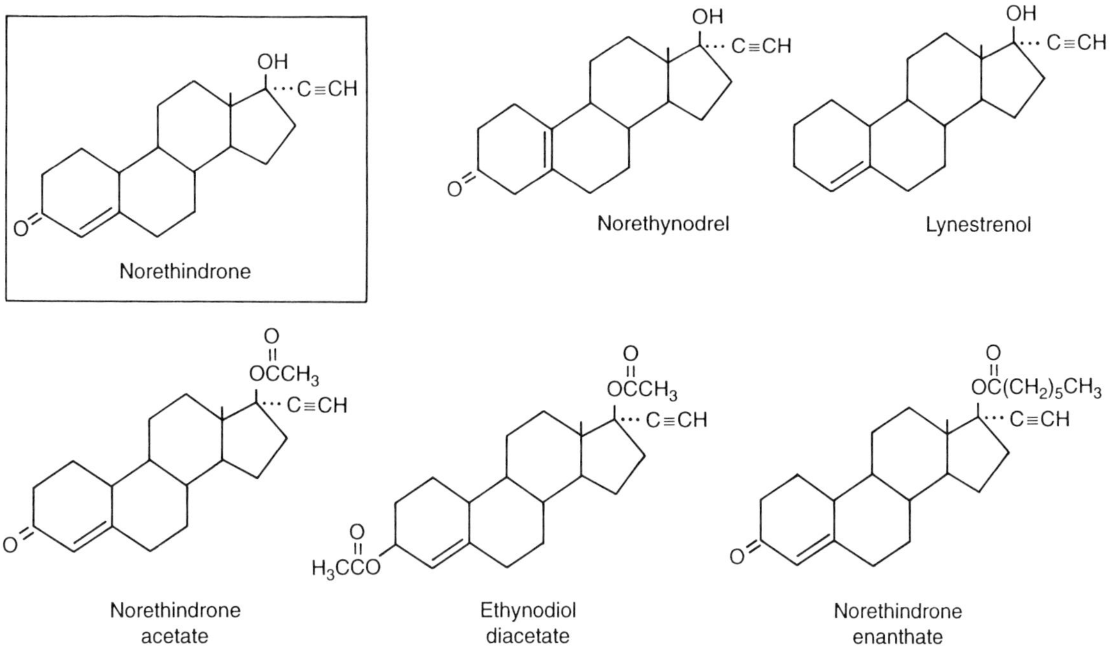

FIGURE 12.3. Estrane progestogens. Chemical structures of progestogens related to norethindrone. (Stanczyk FZ. Pharmacokinetics of progestogens. *International Proceedings Journal*. 1989;1:11.)

monly employed progestogens today. Acetylation of the 17-hydroxyl group of norethindrone gives rise to norethindrone acetate.

Norethynodrel differs from norethindrone by having a double bond between carbons 5 and 10 instead of between carbons 4 and 5. Etherification of the oxygen at position 3 produces ethynodiol diacetate, and elimination of the oxygen at position 3 gives rise to lynestranol. All estranes are metabolized to norethindrone to produce their biologic effect.

Gonanes

The principal difference between the estrane and gonane structure is simply the addition of an ethyl group in position 13. Attaching an ethyl group to C-13 of norethindrone produces levonorgestrel. This modification, however, has significant biological impact. Levonorgestrel is not only one of the most potent oral progestins; it is also the most androgenic of any progestogen employed in OCs.

The "novel progestogens" were derived from the levonorgestrel structure (Fig. 12.4). Addition of a methylene group at carbon 11 and elimination of the oxygen at carbon 3 produces desogestrel. Norgestimate results from the formation of an oxime at position 3 and an acetate at position 17. Gestodene differs from levonorgestrel only by the formation of a double bond between carbons 14 and 15.

In general, relative to levonorgestrel, these new progestogens bind more strongly at the progesterone receptor and less strongly at the androgen receptor. This allows OC formulations with less overall progestin exposure and diminished undesirable effects on carbohydrate and lipid metabolism while maintaining contraceptive efficacy.

Spectrum of Biologic Effects

The use of pregnane progestogens in oral contraceptives was abandoned when it was discovered that chlormadinone acetate was associated with the development of breast tumors in beagles.[26] Progestogens currently employed in OCs are the estranes and gonanes. Derived

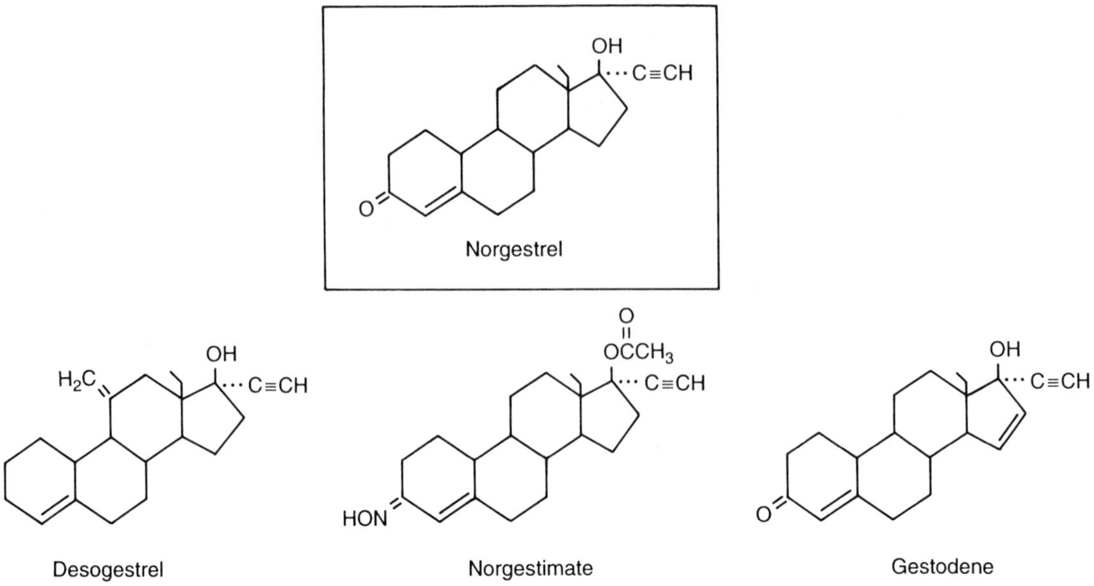

FIGURE 12.4. Gonane progestogens. Chemical structures of progestogens related o norgestrel. (Stanczyk FZ. Pharmacokinetics of progestogens. *International Proceedings Journal*. 1989;1:11.)

from 19-nortestosterone, they possess a broad range of biologic activity. Synthetic progestogens have, in varying degree, estrogenic, antiestrogenic, androgenic, and antiandrogenic effects.

Relative Potency

Relative progestogenic activity is frequently measured with respect to ovulation inhibition, binding affinity to the progesterone receptor, induction of secretory endometrium, and pregnancy maintenance. However, there are at least 27 different measurable effects of synthetic progestogens about which a bioassay can be constructed[27]; depending on which biologic parameter is tested, the relative potency of the progestins varies greatly.

Bioavailability

Given their great similarity in chemical structure, it is surprising how markedly the bioavailability of the various orally active progestins varies. These compounds are absorbed with varying efficiency in the gastrointestinal tract, undergo intermediate metabolism in the liver to various degrees, and circulate with varying affinity for their binding proteins SHBG and albumin. For example, 27% to 53% of an oral dose of norethindrone is metabolized on first pass through the hepatic circulation, while the first-pass degradation of levonorgestrel is virtually zero.[27] Certain progestins such as desogestrel are actually activated in the liver. The binding affinity of desogestrel to the progesterone receptor is very weak, but the active hepatic metabolite 3-ketodesogestrel has one of the highest affinities for the progesterone receptor of any substance known.[28]

The serum concentration following oral administration of these related compounds also differs significantly. Following administration of OCs containing widely varying amounts of progestogen—norethindrone (1 mg), levonorgestrel or desogestrel (150 μg), gestodene (75 μg), all combined with 30 to 35 μg of ethinyl estradiol—Fotherby found the relative steady-state serum progestogen concentrations to be gestodene > norethindrone > desogestrel > levonorgestrel.[29]

Androgenicity

Binding to intracellular steroid receptors is one index of the progestogens' biologic action. Of particular interest is their relative binding activity to the nuclear androgen receptor, which is felt to be related to the side effects occurring with oral contraceptives.[28] The relative potency of the progestogens, whether measured by steroid displacement or relative binding affinity, shows that the relative activities of the estrane progestins are roughly equivalent and 5 to 10 times less potent than levonorgestrel (1-NG). Desogestrel is 10 times more active than l-NG in binding affinity to the progesterone receptor but less than half as active at the androgen receptor, binding the androgen receptor with similar activity as the estrane progestin, norethindrone.[28]

Estrogenicity

The ability of synthetic steroids to bind to more than one receptor also has important biological consequences. The estrane progestogens norethindrone, norethynodrel, and ethynodiol diacetate bind weakly to the estrogen receptor and show slight estrogen activity.[30] In contrast, levonorgestrel does not appear to have any estrogenic metabolites and consequently shows no estrogenic action.[31] As mentioned above, the active metabolite of desogestrel, 3-ketodesogestrel, is strongly progestogenic and shows androgenic activity similar to the estrane progestins. An intermediate metabolite, 3β-hydroxydesogestrel, binds to the estrogen receptor,[32] accounting for the considerable estrogenic activity of this compound.

Induction of sex hormone binding globulin (SHBG) production is considered an index of estrogenic activity. The potential benefit of increased serum levels of SHBG, binding a greater amount of circulating testosterone and diminishing the amount that is unbound and biologically active, is recognized clinically in the treatment of hyperandrogenic states. As with all biologic parameters, a delicate balance exists between the metabolic impact of the ethinyl estradiol (EE) and the specific progestogen contained within a given OC formulation. For example, the combination of EE and desogestrel results in almost twice the amount of circulating SHBG than combinations of EE with levonorgestrel.[33]

Metabolic Effects

Carbohydrate Metabolism

Both estrogens and progestogens modify carbohydrate metabolism; as with other metabolic parameters, the net effect on glucose utilization is the result of the balance of these two components in a given formulation. Small amounts of estrogen improve carbohydrate metabolism[34] and may even improve hyperglycemia in noninsulin-dependent diabetes.[35] The mechanism underlying this improvement may be alteration of insulin receptor binding[36] or possibly the inhibition of insulin degradation.[37] In contrast, progesterone in high doses and synthetic progestins in moderate to low doses cause impairment of glucose utilization and insulin resistance.[38] An example of the effect of high levels of circulating progesterone on carbohydrate metabolism is seen in the emergence of insulin resistance in the third trimester of pregnancy, in which there is an exaggerated insulin response to a glucose load with subsequent blunted response to hypoglycemia. A similar impairment of glucose utilization was noted during therapy with the initial high (10 mg of NET) progestogen OC formulations.

Some deterioration of glucose utilization is evident with even moderate doses of progestins such as those found in the monophasic low- (estrogen) dose pills.[39–41] Certain authors feel that the degree of deterioration of glucose homeostasis varies according to the compound's intrinsic androgenicity and increases in series (pregnanes, estranes, and gonanes)[42]; others feel that in terms of inducing adverse carbohydrate effect, there is no inherent difference between biochemical classes of progestins but rather it is a question of using equipotent doses.[43]

In a series of experiments Spellacy and

colleagues[39-41] demonstrated that progestogens are primarily responsible for the deterioration of insulin response, showing that the progestin-only "mini-pill" produced insulin elevations. These experiments also demonstrated the importance of the balance of the steroid effects on the net glucose utilization, showing that the corresponding progestins in combination with low-dose estrogen induced fewer changes in carbohydrate metabolism than when employed alone.

Alterations in dose schedule that reduce total monthly progestogen exposure have also effectively minimized the impact of OC use on carbohydrate metabolism. Skouby[44] compared the effect of two levonorgestrel preparations on glucose utilization, a monophasic OC formulation and a triphasic formulation that reduced the content of levonorgestrel by approximately 30%. In contrast to the hyperinsulinemic response to a glucose challenge seen in the patients taking the monophasic formulation, no changes in glucose utilization were noted with the triphasic.

This is consistent with several other reports showing minimal or no deterioration of glucose homeostasis with the triphasic OCs,[44-47] even in women with a history of previous gestational diabetes.[48] Recent studies of carbohydrate metabolism also include assessment of glycosylated proteins, which reflect long-term serum-glucose levels. Examples of perfectly balanced monophasic formulations (at least in this metabolic parameter) of the progestogens deogestrel (150 μg of DSG + 20 μg of EE) and gestodene (75 μg of GSD + 30 μg of EE), as well as a triphasic levonorgestrel preparation, showed no increase in long-term glucose levels over controls.[49,50]

Lipid Metabolism

Contraceptive steroids have a diverse impact on lipid and lipoprotein metabolism. Individuals with certain genetic- and dietary-related lipoprotein profiles have been shown to be at significantly increased risk of morbidity from atherosclerosis and cardiovascular disease.[51-53] Although there is no direct experimental evidence that OC-induced alterations in lipid parameters lead to an increase in atherosclerosis, it has been hypothesized that any adverse alteration in lipid metabolism may accelerate the development of this disease, and therefore the search for "lipid-neutral" contraceptive formulations has been undertaken.

Lipid and Lipoprotein Metabolism

To understand the effects of OCs on lipid metabolism a fundamental understanding of lipid physiology is helpful. What follows is a simplified description, intended as a context for discussion.

Lipids, which are insoluble in plasma, combine with proteins to form soluble complexes called lipoproteins. Cholesterol is carried both in a free form on the lipoprotein's surface and within the core of the lipoprotein complex as cholesterol ester. Triglyceride (TG) is a nonpolar fat, also carried within the core of the lipoprotein complex. Electrically charged particles reside on the surface of the molecule and provide the polarity that allows for its transport in plasma.

In normal lipoprotein metabolism, very low-density lipoprotein (VLDL) is produced in the liver. The surface of the VLDL molecule contains free cholesterol. The core of the VLDL particle is triglyceride rich and also carries cholesterol esters. The synthesis of VLDL is regulated by diet and hormonal factors. Once released from the liver, VLDL is broken up by the action of lipoprotein lipase, which removes triglyceride, leaving a cholesterol ester–rich core. The remaining particle is then either taken up directly by liver receptors or further catabolized to low-density lipoprotein (LDL) through the action of hepatic triglyceride lipase. Low-density lipoprotein is the major carrier of cholesterol in the plasma. In peripheral tissues, receptors bind and transport the LDL particle into the cell where the cholesterol is either utilized or stored.

Cholesterol is returned to the liver by a process of reverse transport involving high-density lipoproteins (HDL). Two major subfractions of HDL are HDL2 and HDLc.

Cholesterol is taken up from cell membranes and is progressively converted to a more lipid-rich form called HDL2. HDLc is formed by combination of several HDL2 molecules.

Cholesterol transported by the HLD2 or HDLc subfractions can be utilized peripherally, reintroduced into the VLDL-LDL cascade, or returned to the liver. Hepatic receptors recognize the various cell surface proteins contained on these molecules. In the liver, HDL2 particles release their cholesterol and are degraded by hepatic endothelial lipase or are released back into the cycle.

Steriod Effects on Lipids

Exogenous estrogens increase plasma HDL concentrations, increasing the HDL2 subfraction, in part, by decreasing hepatic endothelial lipase activity[54,55] and diminishing HDL2 catabolism. Estrogens also decrease plasma LDL concentrations, presumably by up-regulation of hepatic LDL receptors.[56] Unfortunately, estrogens also induce an increase in the synthesis of VLDLs, leading to elevations in serum triglyceride levels.[57]

The 19-nortestosterone-derived progestogens oppose the effects of estrogen on lipoprotein physiology. The magnitude of this antagonism appears to be in proportion to their androgenic activity.[58] These progestins increase HDL catabolism by increasing hepatic endothelial lipase activity[59], decrease concentrations of VLDL/triglyceride, and increase LDL concentrations in the plasma. A unifying concept for these effects was proposed by Khokha and colleagues[60]: By increasing hepatic lipase activity, progestins enhance VLDL/triglyceride degradation, resulting in increased hepatic supply of cholesterol, down regulation of LDL receptors in the liver, and increased LDL release into the circulation.

Because the effects of synthetic progestins on lipoprotein metabolism are opposite those of estrogen, both the ratio of estrogen to progestogen and the degree of estrogenicity and androgenicity of the progestin determine extent and direction of OC-induced alterations in lipid physiology. In an attempt to produce a lipid-neutral formulation, researchers have sought to identify the proper dose and chemical structure of a progestin to offset the estrogenic increase in triglyceride without adversely affecting the other lipid parameters.

An example of a favorable balance between estrogen and progestogen is found in the lowest-dose combination estrane formulation, containing 35 µg of ethinyl estradiol and 400 µg of norethindrone, which shows no significant increase in the levels of total cholesterol or triglycerides, with a trend toward increased levels of HDLc.[61] Similarly, in studies of the triphasic combination containing mean daily doses of 32 µg of ethinyl estradiol and 92 µg of levonorgestrel, no significant changes in the lipid concentrations are noted.[62,63]

Desogestrel, the most estrogenic gonane progestogen, has an estrogen-dominant effect on lipoprotein metabolism. When compared with triphasic levonorgestrel combinations, an even higher mean daily progestin exposure of 150 µg of monophasic desogestrel produces less antagonism of the estrogen-stimulated decreases in LDL and increases in HDL but is associated with moderate increases in triglycerides.[64]

Lipid Hypothesis

It should be noted that the changes in lipid values induced by OCs are typically within the normal range for women and are of uncertain predictive value as to serious clinical consequence. If the lipid hypothesis is correct, atheromatous changes in vessel structure occurring during OC use should eventually place those women at greater risk of coronary heart disease than women who have never used oral contraception. Early epidemiologic data demonstrated significantly increased cardiovascular morbidity and mortality in women with high-risk profiles when exposed to the initial high-dose formulations. Of the high-risk factors, cigarette smoking over the age of 35 placed women at unacceptable levels of risk.[65] As a result, both the formulations and prescribing practices changed.

More recent epidemiologic studies indicate that there is no increased risk of myocardial infarction among former users of oral contraceptives.[66,67] Stampfer et al[68] performed a meta-analysis of an 8-year follow-up (1976–1984) of over 100,000 women who previously used OCs. They noted that the relative risk of developing coronary artery disease was reduced 20% and that the relative risk of experiencing a fatal cardiovascular event was reduced 10%. Further, the incidence of cardiovascular disease is not correlated with duration of use.[69]

Experimental evidence that combination OC-induced alterations in lipid physiology lead to atherosclerosis is also lacking. In animal studies macaque monkeys, while being fed an atherogenic diet, were maintained for 2 years on an oral contraceptive formulation containing ethinyl estradiol and levonorgestrel that reduced HDL cholesterol concentrations to levels that would presumably accelerate the development of atherosclerosis. At necropsy, despite their lower HDL values, OC-treated female macaques had less extensive atherosclerosis than did untreated controls.[70] Another group of monkeys, treated with levonorgestrel without estrogen, also had lowered HDL cholesterol levels. In this group, the extent of coronary atherosclerosis was significantly greater than that of the controls. The results of this study, confirmed in a larger study that also examined the effects of the progestin ethynodiol diacetate,[71] suggest that, despite changes in lipid profiles that would otherwise appear to be contributory to cardiovascular disease, the estrogen component of oral contraceptives may actually protect the user against coronary atherosclerosis.

The exact mechanism of this protective effect of estrogen is currently under investigation. As well as modifying the relative amount of circulating LDL, the principal carrier of cholesterol in the bloodstream, estrogen has been shown to alter its composition—increasing the content of triglyceride and phospholipid[72]—resulting in a less cholesterol-rich molecule that may reduce the uptake of cholesterol in vessel walls. Estrogen receptors have been shown to be present in vessel walls in several mammalian species.[71] Various direct effects of estrogen at the level of the arterial lumen have been demonstrated,[71,73] the most significant of which is a beneficial modification of prostacyclin and thromboxane synthesis.[71,74] Smoking is able to countermand this benefit by reducing prostacyclin production while increasing thromboxane production, leading to vasospasm of susceptible vessels.[75]

An increase in the incidence of thromboembolic disease was one of the first complications of estrogen-containing oral contraceptives to be noted. Ethinyl estradiol causes dose-dependent alterations in the hepatic production of several globulins involved in the coagulation process. The vitamin K-dependent clotting factors II, VII, IX, and X are elevated,[76] and antithrombin III production is decreased[77], but unlike the marked aberrations noted with the first oral contraceptive formulations, doses of less than 50 μg are associated with little or no impact on the various clotting factors.[78] It is thought that slight increases in thrombin formation are offset by increased fibrinolytic activity, and therefore platelet activation and clotting are not increased.[79]

The effect of the estrogen dose is underscored by a recent case-control study by Vessey[80] that compared the risk of fatal myocardial infarction in women taking low-dose versus high-dose (50 μg) formulations. The relative risk of myocardial infarction associated with current use of low-dose OCs as compared with nonuse was 1.1 (95% CL, 0.6–2.4) while current use of high-dose formulations was approximately fourfold greater. Risk factors significantly increasing the incidence of a fatal heart attack, such as overt diabetes, hyperlipidemia, hypertension, or previous pregnancy-induced hypertension, were found to place an individual at even greater risk (sevenfold). Current cigarette smoking increased the risk nineteenfold.

Hypertension

Approximately 5% of women using early, high-dose formulation oral contraceptives developed hypertension,[81] and a significant elevation in blood pressure is still observed in some women using formulations with the 30 to 35 μg dose.[82] Estrogen is primarily responsible for these blood pressure elevations, although there is evidence that the progestins may play a contributory role.

Estrogens induce the hepatic production of renin substrate-angiotensinogen with a secondary increase in angiotensin.[83] The elevation of aldosterone that results causes sodium and water retention. These are dose-related effects. With current OC formulations, most women experience only minor elevations in blood pressure and remain normotensive. If severe hypertension develops, it typically resolves several weeks or months after discontinuation of treatment.

Progesterone is known to increase sodium excretion. It has high affinity for the aldosterone receptor and competes with aldosterone at the level of the renal tubule. The synthetic progestins do not share this biologic activity.[84] Unlike natural progesterone, the synthetic progestogens do not inhibit the aldosterone-induced retention of sodium and have the potential for a gradual positive sodium balance and an increase in blood pressure.[85]

Although no significant elevations in blood pressure have been noted in women on triphasic formulations, patients with a history of hypertension or previous renal disease should be monitored carefully if placed on these preparations.

Hyperprolactinemia

Coincident with the advent of widespread OC use, laboratory techniques for the measurement of prolactin (PRL) and radiographic technologies for the detection of pituitary lesions improved dramatically, and an association between the use of contraceptive steroids and the development of PRL-secreting pituitary adenomas arose. Several observations fueled speculation that the estrogen contained in OCs caused pituitary adenomas in women. There is in vitro evidence of a specific receptor for estrogen on the lactotroph that regulates the transcription of the prolactin gene.[86] Endogenous estrogen stimulates PRL secretion, causing hyperprolactinemia and lactotroph hyperplasia throughout human gestation.[87] In addition, in animal studies pharmacologic doses of exogenous estrogen were shown to induce pituitary adenomas in rodents.[88]

However, the Pituitary Adenoma Study Group,[89] reporting the data from a large multicenter case-control study, found no causal relationship between OC use and the risk of developing a prolactin-secreting adenoma; this was consistent with the findings of two large prospective studies conducted by the Royal College of General Practitioners and the Oxford Family Planning Association, reported earlier.[90] As the authors discuss, the data collected were not designed to address the issue of whether estrogen may exacerbate a preexisting adenoma. A Swedish group studied 70 women with prolactinomas and compared those who had (61%) to those who had not previously used OCs.[91] Patients who had used OCs had a shorter duration of symptoms, lower serum prolactin levels, and less pronounced enlargement of the sella turcica, suggesting that the use of oral contraceptives either did not promote the growth of the pituitary adenoma or may have led to earlier detection of the lesion.

It has been reported in a number of studies that basal levels of prolactin are slightly elevated in normal women taking oral contraceptives.[92] Mishell and colleagues demonstrated that both initial and maximal prolactin elevations were exaggerated following thyrotropin-releasing hormone (TRH) stimulation. Estrogen potentiates this response, but the degree of elevation is unrelated to estrogen dose.[93]

Luciano and colleagues[94] reported that the incidence of hyperprolactinemia was significantly greater in a group of OC users than in a group of women who employed barrier methods of contraception (12% vs 5%), but that the hyperprolactinemia was often transient and resolved spontaneously in about 50% of both OC users and controls. He also identified a subset of patients who were more sensitive to the lactogenic effects of exogenous estrogens and who therefore might be at greater risk of developing hyperprolactinemia. Consistent with other reports, neither the duration of pill use nor the dose of estrogen in the pill had any effect on the development of hyperprolactinemia.

Menstrual Dysfunction

Although the Pituitary Adenoma Study Group[89] reported that the relative risk of developing a prolactinoma was not higher in OC users with a history of menstrual disorders, Luciano et al[94] noted that the patients in his study who developed hyperprolactinemia on OCs had a significantly higher prevalence of menstrual dysfunction in their histories. Women who have irregular menstrual periods are more likely to have hyperprolactinemia,[95] regardless of whether or not they take oral contraceptives. Before placing a woman on OCs for cycle control, her underlying etiology should be determined and counsel should be given as to what to expect on discontinuation.

Return of Fertility

In a review by MacLeod it was found that the incidence of post-pill amenorrhea and that of secondary amenorrhea unrelated to oral contraceptive use were similar (0.2% to 2.2% and 0.1% to 0.8%, respectively).[96] The rate of return of fertility after the discontinuation of oral contraceptives is lower than for women who have used barrier methods for 2 to 3 years, but eventually the percentage of women who conceive after ceasing either form of contraception becomes the same.[97] Neither the rate of spontaneous abortion[98] nor the incidence of chromosomal abnormalities[99] in abortuses is increased in women who conceive in the first or subsequent months after stopping oral contraceptives. Importantly, in contrast to earlier published reports, a recent large cohort study reported that ingestion of oral contraceptive steroids in the first few months of pregnancy does not significantly increase the risk of congenital malformations among the offspring of users overall or among those of nonsmoking users.[100]

Relationship to Cancers of the Female Reproductive Tract

As discussed in the context of health advantages to oral contraception, the incidence of both ovarian and endometrial cancers are conclusively and significantly reduced by even short-term OC use. Discussed in greater detail below, the incidence of breast carcinoma appears to be unaffected by OC use, and the possible contribution of OCs to the risk of developing cervical neoplasia is unclear.

Breast Carcinoma

The most substantive data on this subject are provided by the Cancer and Steroid Hormone (CASH) study, which reviewed case studies of nearly 10,000 women of reproductive age and found no overall adverse effect on risk of the development of breast cancer among OC users through the age of 54. In fact, the likelihood of a diagnosing breast cancer in women after the age of 45 was decreased with OC use.[101] Schlesselman reviewed 17 different epidemiologic studies in which the risk of breast cancer developing in women under 60 years of age was compared with dose and duration of OC use; no overall dose response was noted, and the length of duration of OC use was not associated with an increased risk.[102]

A large British case-control study did find a statistically significant increase in risk of breast cancer developing before the age of 36, after more than 5 years of OC use, but noted that the risk was diminished with OC formulations that contained less than 50 μg of estrogen.[103] In a subset of women who were nulliparous and had used OC for more than 8 years. Stadel et al[104] reported a significantly increased relative risk of diagnosing breast cancer before the age of 45. A trend of increasing risk of detecting breast cancer in women with more than 8 years of OC use prior to first term pregnancy had also been in noted Schlesselman's analysis.[102] This increase in earlier detection does not affect aggregate lifetime risk, which is not increased by OC use, but suggests long-term use of high-dose OCs in nulliparous women may promote breast cancer in susceptible women.

Since 1983 OCs containing fewer than 50 μg of estrogen have been used by the majority of women.[24] It will be several years before epidemiologic studies to determine if the current lower-dose formulations promote earlier development of breast cancer in this susceptible subpopulation are completed. A recent review article written by the chief of the Women's Health and Fertility Branch of the Centers for Disease Control, having analyzed the 40 epidemiologic studies published to date examining the relationship of OC use to the incidence of breast cancer, concludes that the "extensive" and "broadly reassuring" data reveal no aggregate change in the risk of developing breast cancer with OC use.[105] This is in concurrence with the findings of other health agencies including the World Health Organization and the U.S. Food and Drug Administration.

While there is speculation that the newer formulations could possibly decrease the risk of breast cancer[105] the benefits of regular self-examination, annual breast exams by trained health care personnel, adherence to the established guidelines for periodic mammography, and follow-up of clinically suspicious lesions must be stressed to all patients, regardless of their contraceptive histories.

Cervical Neoplasia

Certain epidemiologic studies have indicated that long-term use of OCs is associated with an increased risk of both preinvasive cervical neoplasia and carcinoma in situ when compared to matched control groups.[106] However, the development of cervical cancer is known to be multifactorial, and numerous potential confounding variables may well have influenced the results. Age at first sexual intercourse, number of sexual partners, cigarette smoking (an independent variable), more frequent cytologic screening of OC users, use of barrier contraceptives and spermicides by non–OC users, and different degrees of exposure to human papilloma virus are known to be significant factors.

Attempts to correct for these variables by statistical methods have resulted in conflicting conclusions. Certain authors report that no statistically significant increase in risk is apparent when the data are corrected for confounding variables.[107,108] Similarly, in an excellent, CDC-sponsored study Irwin et al[109] credited enhanced detection of disease as a result of more regular cytologic exams but found no true increase in incidence of cervical cancer in OC users versus nonusers. Conversely, Schlesselman's extensive review[102] of OC use and reproductive cancers reveals an approximate doubling of risk of developing preinvasive cervical neoplasia after more than 1 year of OC use, with the risk of invasive cervical cancer increasing with increasing duration of use, gradually becoming significant after 5 years of use, and resulting in a twofold increase of relative risk with 10 years of use.

With the exception of those women who consistently use barrier methods of contraception,[110] all sexually active women are at increased risk of cervical neoplasia. While it is uncertain whether OCs may be oncogenic themselves, act as cocarcinogens, or have no influence on the incidence of cervical cancer, the necessity of annual Pap smears, with appropriate clinical follow-up, must be stressed in this population.

Changing Trends

Over the past three decades advances in the knowledge of how the orally active steroid medications impact human metabolism have resulted in the manufacture of safer formulations and delivery schedules that provide a more physiologic dosing of OCs. Women's perceptions of the safety of using the pill have also undergone significant change, although slowly.

Reflecting the OC preference of new users of contraception, the age group most likely to use oral contraception is that of women 18 to 24 years of age, in which over 50% of those employing birth control are OC users.[1] However, over the past several years there has been an increasing tendency for women in their 30s and 40s who are healthy nonsmokers to use low-dose oral contraception.[1,111] This is the result of several factors, principally that the current formulations present fewer risks and diminished side effects than the earlier formulations. Another factor is demographic, reflecting the aging of the 'baby-boom' generation who have had access to oral contraception throughout their reproductive lives. Many of these women recognize the efficacy of these medications and prefer a nonsurgical and reversible method of contraception. Bridging these factors is the considerable evidence concerning the noncontraceptive health benefits of oral contraception, as is being increasingly conveyed in the medical literature and the press. Finally, physicians have become more adept at screening those individuals with significant risk factors and have begun to select the best candidates for continued OC use.

Because the age at which a woman will become naturally infertile is individual, a sexually active woman in her later reproductive years cannot safely assume that she is not exposed to the risk of pregnancy until menopause. The risk of obstetric complications increases with advancing age, as does the risk of genetic abnormalities such as Down syndrome.[112] Compared to women who give birth in their 20s, maternal mortality rates of women aged 40 to 44 and 45 to 49 are estimated to be 8 and 50 times greater, respectively.[111]

In March 1988 the American College of Obstetricians and Gynecologists revised the criteria for OC use in women over age 40 and set forth the following guidelines: An appropriate candidate is a woman without direct contraindication, is a nonsmoker, has a normal mammogram, has a normal lipid profile and glucose screening test, is not greater than 30% over ideal body weight, and has a negative family history for early heart disease.[113]

In January 1990 the Fertility and Maternal Health Drugs Advisory Committee of the FDA issued a statement that concluded that "the benefits of oral contraceptive use by healthy, nonsmoking women over 40 years of age may outweigh the possible risks" and recommended that the age limit on OC use be lifted for patients in low-risk categories. In the 2 years since that statement OC use among women aged 40 to 44 doubled from 4% in 1990 to 8% in 1991; additionally, 6% of the women surveyed between the ages of 45 to 50 were using OCs.[1] As discussed, the newer formulations have been shown to have minimal net metabolic impact on the populations studied, but until more data about OC use in this older age group is available, most physicians are prescribing the lowest effective dose and monitoring their patients closely.

Prescribing Oral Contraceptives

Strict contraindications to OC use include (1) a history of vascular disease, including thromboembolism, thrombophlebitis, atherosclerosis, and stroke, and systemic disease that may promote vasculitis and/or thrombosis such as lupus erythematosus, sickle cell disease, diabetes mellitus with vascular involvement, hyperlipidemia, and hypertension; (2) cigarette smoking after the age of 35; (3) functional heart disease; (4) active liver disease; and (5) undiagnosed abnormal vaginal bleeding. The FDA also lists breast or endometrial cancer as contraindications, although there are no data to confirm that women with these conditions are harmed by these medications.[24] Pregnancy is clearly a contraindication, although as discussed above there is no evidence of increased incidence of fetal malformations with accidental ingestion of OCs during early pregnancy.[100]

Relative contraindications to OC use include heavy cigarette smoking at any age, migraine headaches, depression, a history of pregnancy-induced hypertension (PIH) or gestational diabetes, gall bladder disease, and mild diabetes mellitus. Migraine headaches may worsen on OCs and have been associated with thrombosis and stroke. Patients with histories of gestational diabetes or mild insulin-dependent diabetes should be placed on a formulation containing the lowest effective dose of the least androgenic progestin available and should be watched closely for exacerbation; deterioration of glucose metabolism or any increase in insulin requirements should prompt discontinuation.

Before being placed on contraceptive steroids amenorrheic patients must be evaluated to determine the cause. The potential for missing a pituitary adenoma is the most significant threat to their health. Patients who develop galactorrhea on oral contraceptives should discontinue the medication and have a prolactin level checked 2 weeks after stopping. Estrogen deficiency in patients with exercise-induced amenorrhea may be adequately replaced by combination OCs, but conclusive data are not yet available; formulations that are more estrogenic in balance are preferable.

Follow-Up

After an initial history and physical to establish what, if any, risk factors are present for a given individual, baseline laboratory work-up in the uncomplicated patient involves a urinalysis, CBC, and Pap smear. The history should direct whatever additional baseline labs are indicated, such as cardiovascular lipid profiles, liver function tests, fasting blood sugar, and so forth. Patients with identifiable risk factors are told which side effects warrant a return visit, and all patient are encouraged to read the patient information accompanying the product prescribed.

In patients with positive histories these initial laboratory parameters should be rechecked after 3 months. The younger patient without risk factors need only have a blood pressure check and brief interview to determine her degree of satisfaction with the medication she is taking. After this, patients should be seen annually for a complete physical exam including cervical cytology. Care of patients with relative contraindications is individualized with respect to the severity of the underlying disorder and any untoward side effects that may develop.

Cycle Control

Breakthrough bleeding (BBT) is common with the low OC formulations and should be discussed with the patient before initiating therapy. In the first 3 months of therapy it is not uncommon for the patient to experience bleeding irregularities. BBT is variably reported to occur between 6.5% and 29% of first cycles, the great variance largely a result of definition of the term and the probability of reporting bias.[114] Alternatively, approximately 15% report little or no menstrual bleeding during the medication-free (or placebo) week, causing concern with the patient that she may be pregnant. Although this a normal and expected pattern in a certain number of OC patients, inadequate cycle control is a leading cause of discontinuation of OCs within the initial months of therapy, depriving the patient of the benefit of a highly reliable contraceptive choice, one with significant noncontraceptive health advantages.

Proliferation of the endometrial lining is under the control of estrogen. When taken in combination with a progestogen, estrogen is antagonized at the receptor level, resulting in a decrease in endometrial proliferation. Greater estrogenic stimulation of the endometrium will stabilize this tissue. Combination OCs with extremely low (20 μg of EE) estrogen doses or other low-dose (30 to 35 μg of EE) formulations appear to result in menstrual irregularity to a similar degree.[114] Prescribing additional estrogen, 1 to 2 mg of 17-β estradiol or 20 μg of EE, for 7 days during an episode of BBT regardless of where the patient is in her cycle pack is a very effective treatment. Having the patient take additional combination pills, two or three per day, will also deliver additional progestogen that will further antagonize the effect of additional estrogen and is therefore self-defeating. If the patient requires more than one additional course of therapy (the following month) or does not experience distinct withdrawal bleeding when anticipated, switch to a formulation with a more estrogenic balance (desogestrel, if available, or ethynodiol diacetate). There is no need to increase empirically to a high EE content formulation. Appreciating the distinctions in the biologic character of the various progestogens, clinical situations that require the use of even 50 μg of EE-containing formulations are exceedingly rare.

References

1. *1991 Annual Birth Control Study*. Raritan, NJ: Ortho Pharmaceutical Corporation; 1991.
2. Jones EF, Forrest JD. Contraceptive failure in the United States: revised estimates from the 1982 National Survey of Family Growth. *Fam Plann Perspect*. 1989;21:103.
3. American Health Consultants. 1987 oral contraception survey. *Contraceptive Technology Update*. 1987;8:109.
4. Cancer and Steroid Hormone Study of the Centers for Disease Control and the National Institute of Child and Human Development. Combination oral contraceptive use and the risk of endometrial cancer. *JAMA*. 1987; 257:796.
5. Cancer and Steroid Hormone Study of the Centers for Disease Control and the National Institute of Child and Human Development. The reduction in risk of ovarian cancer associated with oral contraceptive use. *New Engl J Med*. 1987;316:650.
6. Rubin GL, Ory HW, Layde PM. Oral contraceptives and pelvic inflammatory disease. *Am J Obstet Gynecol*. 1982;144:630.
7. Ory HW. The noncontraceptive health benefits from oral contraceptive use. *Fam Plann Perspect*. 1982;14:182.
8. Royal College of General Practitioners. *Oral Contraceptives and Health: An Interim Report*. New York, NY: Pitman; 1974.
9. Ross RK, Pike MC, Vessey MP, et al. Risk factors for uterine fibroids: reduced risk associated with oral contraceptives. *Br Med J*. 1986;293:359.
10. Burkman RT. Modern trends in contraception. *Obstet Gynecol Clin N Am*. 1990;17:759.
11. Vandenbroucke JP, Valkenburg HA, Boersma JW, et al. Oral contraceptives and rheumatiod arthritis: further evidence for a preventative effect. *Lancet*. 1982;2:839.
12. Lindsay R, Tohme J, Kanders B. The effect of oral contracetive use on vertebral bone mass in pre and post-menopausal women. *Contraception*. 1986;34;333.
13. Halban J. Über den Einfluss der Ovarien auf die Entwicklung des Genitales. *Mschr Geburtsh Gynak*. 1900;60:322.
14. Allen E, Doisy EA. An ovarian hormone: preliminary report on its localization, extraction, and partial purification, and action in test animals. *JAMA*. 1923;81:819.
15. MacCorquodale DW, Thayer SA, Doisey EA. Crystallin ovarian follicular hormone. *Proc Soc Exper Biol Med*. 1935;32:1182.
16. Butenandt A, Westphal V. Zur Isolierung und Charakterisierung des Corpus-Luteum-Hormones. *Berl Dtsch Chem Ges*. 1934;67: 1440.
17. Makepeace AW, Weinstein GL, Friedman MH. Effect of progestin and progesterone on ovulation in the rabbit. *Am J Physiol*. 1937;119:512.
18. Sturgis SH, Albright R. Mechanism of estrin therapy in the relief of dysmenorrhea. *Endocrinology*. 1940;26:68.
19. Marker RE, Rhormann E. Sterols LXXI, Conversion of sarsasapogenin to pregnandiol-3 (α), 20 (α). *J Am Chem Soc*. 1939;61:3592.
20. Inhoffen HH, Logemann W, Hohlweg W, Serini S. *Berl Dtsch Chem Ges*. 1938;71:1024.
21. Inhoffen HH, Hohlweg W. Neue per os wirksame weibliche Keimdrusenhormon-Derivate. *Naturwissenschaften*. 1938;26:96.
22. Djerassi C, Miramontes L, Rosenkranz G. 17α-ethynyl-19-nortestosterone. American Chemical Society Meeting, 1952, Abstract 18J.
23. Waine H, Frieden EW, Caplan HI, Cole T. Metabolic effects of Enovid in rheumatiod patients. [Abstract]. *Arthritis Rheum*. 1963; 6:796.
24. Mishell DR Jr. Oral steroid contraceptives. In: *Infertility, Contraception, & Reproductive Endocrinology*. 3rd ed. Cambridge, Mass: Blackwell Scientific Publications; 1991:839–871.
25. *Physician's Desk Reference*. 46th ed. Montvale, NJ: Medical Economics Data; 1992: 1673.
26. Daniel GR. Chlormadinone contraceptive withdrawn. *Br Med J*. 1970;1:303.
27. Goldzieher JW. Progestogen pharmokinetics and the unknown. In: Lobo RA, Whitehead MI, eds. Consensus Development Conference

on Progestogens. *International Proceedings Journal.* 1989;1:21.
28. Kloosterboer HJ, Deckers GHJ. Desogestrel: a selective progestogen. In: Lobo RA, Whitehead MI, eds. Consensus Development Conference on Progestogens. *International Proceedings Journal.* 1989;1:26.
29. Fotherby K. *Contraception.* 1990;41:533–550.
30. Edgren RA. Progestogens. In: Givens J, ed. *Clinical Uses of Steroids.* Chicago: Yearbook Medical Publishers; 1980;1–29.
31. Fotherby K. A new look at progestogens. *Clin Obstet Gynecol.* 1984;11:701.
32. Ojasoo T, Raynaud JP. Unique steroid cogeners for receptor studies. *Cancer Res.* 1978;38:4186.
33. Van Der Vange N, Blankensetein MA, Kloosterboer HJ, et al. *Contraception.* 1990; 41:345.
34. Spellacy WN. Menopause, estrogen treatment, and carbohydrate metabolism. In: Mishell DR, ed. *Menopause: Physiology and Pharmacology.* Chicago: Yearbook Medical Publishers Inc.; 1987:253.
35. Kalkhoff RK. Effects of oral contraceptive agents on carbohydrate metabolism. *J Steroid Biochem.* 1975;6:949.
36. Ballejo G, Saleem TH, Khan-Dawood FS, et al. The effect of sex steroids on insulin binding by target tissues in the rat. *Contraception.* 1983;28:413.
37. Tsbris JCM, Ballejo G, Kramer NC, et al. Estrogens inhibit the degradation of insulin by human placental preparations in the presence of glutathione. *Biochem Biophys Res Commun.* 1988.
38. Wynn V. Effects of progesterone and progestins on carbohydrate metabolism. In: Bardin CW, Milgrom E, Mauvis-Jarvis P, eds. *Progesterone and Progestins.* New York: Raven Press; 1983:395–410.
39. Spellacy WN, Buhi WC, Birk SA. Effects of norethindrone on carbohydrate and lipid metabolism. *Obstet Gynecol.* 1975;46:560.
40. Spellacy WN, Buhi WC, Birk SA. Carbohydrate and lipid metabolic studies before and after one year of treatment with ethynodiol diacetate in "normal" women. *Fertil Steril.* 1976;27:900.
41. Spellacy WN, Buhi WC, Birk SA. Prospective studies of carbohydrate metabolism in "normal" women using norgestrel for 18 months. *Fertil Steril.* 1981;35:167.
42. Wynn V, Godsland BA. Effects of oral contraceptives on carbohydrate metabolism. *J Reprod Med.* 1986;31:892.
43. Skouby S. Progestogens and carbohydrate metabolism: effects at the cellular level. In: *Consensus Development Conference on Progestins. International Proceedings Journal.* 1989;1:81.
44. Skouby SO. Oral contraceptives: effects on glucose and lipid metabolism in insulin-dependent diabetic women and women with previous gestational diabetes. *Dan Med Bull.* 1988;35:157.
45. Rabe T, Runnebaum B, Kohlmeyer M, et al. Lipid, carbohydrate and androgen metabolism in women using a triphasic oral contraceptive containing norethindrone for one year. *Int J Fertil.* 1986;31:46.
46. Knopp RH, Walden CE, Wahl PW, et al. Oral contraceptive and postmenopausal estrogen effects on lipoprotein triglyceride and cholesterol in an adult female population: relationships to estrogen and progestin potency. *J Clin Endocrinol Metab.* 1981;53:1123.
47. Spellacy WN, Ellingson AB, Tsibris JC. The effects of two triphasic oral contraceptives on carbohydrate metabolism in women during 1 year of use. *Fertil Steril.* 1989;51:71.
48. Skouby SD, Kuhl C, Molsted-Pedersen L, Petersen K, Christensen MS. Triphasic oral contraception: metabolic effects in normal women and those with previous gestational diabetes. *Am J Obstet Gynecol.* 1985;153:495.
49. Luyckx AS, Gaspard UJ, Romus MA, Grigorescu F, De Meyts P, Lefebve PJ. Carbohydrate metabolism in women who use oral contraceptives containing levonorgestrel or desogestrel: a 6-month prospective study. *Fertil Steril.* 1986;45:635.
50. Fotherby K. Desogestrel and gestodene in oral contraception: a review of European experience. *J Drug Dev.* 1991;4:101–111.
51. Castelli WP. Epidemiology of coronary heart disease: the Farmingham study. *Am J Med* 1984;2A (suppl):4.
52. Criqui MH. Epidemiology of atherosclerosis: an updated overview. *Am J Cardiol.* 1986; 153:68.
53. Miller NE, Hammett F, Saltissi S, et al. Relation of angiographically defined coronary artery disease to plasma lipoprotein subfractions and apolipoproteins. *Br Med J.* 1981;282:1741.
54. Schaefer EJ, Foster MD, Zech LA, et al. The

effects of estrogen administration on plasma lipoprotein metabolism in premenopausal females. *J Clin Endocrinol Metab.* 1983; 57:262.
55. Applebaum DM, Goldberg AP, Pykalisto OJ, et al. Effect of estrogen on post-heparin lypolytic activity: selective decline in hepatic triglyceride lipase. *J Clin Invest.* 1977;50:601.
56. Kovanen PT, Brown MS, Goldstein JL. Increased binding of low-density lipoprotein to liver membranes from rats treated with 17-alpha-ethinyl estradiol. *J Biol Chem.* 1979; 254:11367.
57. Kissebah AH, Harrigan P, Wynn V. Mechanism of hypertriglyceridemia associated with contraceptive steroids. *Horm Metab Res.* 1974;5:184.
58. Wahl P, Walden C, Knopp R, et al. Effect of estrogen/progestin potency on lipid/lipoprotein cholesterol. *N Engl J Med.* 1983; 308:862.
59. Tikkanen MJ, Nikkila EA. Regulation of hepatic lipase and serum lipoproteins by sex steroids. *Am Heart J.* 1987;113:562.
60. Khokha R, Huff MW, Wolfe BM. Divergent effects of d-norgestrel on the metabolism of rat very low-density and low-density apoprotein B. *J Lipid Res.* 1986;27:699.
61. Krauss RM, Roy S, Mishell DM Jr, Casagrande J, Pike MC. Effect of two low-dose oral contraceptives on serum lipids and lipoproteins. *Am J Obstet Gynecol.* 1983;145:446.
62. Gaspard UJ, Buret J, Gillian D, Romus MA, Lambotte R. Serum lipid and lipoprotein changes induced by new oral contraceptives containing ethinylestradiol plus levonorgestrel or desogestrel. *Contraception.* 1985,31:395.
63. Percival-Smith RK, Morrison BJ, Sizto R, Abercrombie B. The effect of triphasic and biphasic oral contraceptive preparations on HDL-cholesterol and LDL-cholesterol in young women. *Contraception.* 1987;35:179.
64. Capitiano GL, Bertolini S, Croce S, et al. Lipidemic changes induced by two different oral contraceptive formulations. *Adv Contraception.* 1985;1:238.
65. Further analyses of mortality in oral contraceptive users: Royal College of General Practioners' oral contraceptive study. *Lancet.* 1981;1:541.
66. Layde PM, Ory HW, Schlesselman JJ. The risk of myocardial infarction in former users of oral contraceptives. *Fam Plan Perspect.* 1982;14:78.
67. Stampfer MJ, Willet WC, Colditz GA, et al. A prospective study of past use of oral contraceptive agents and risk of cardiovascular diseases. *N Engl J Med.* 1988; 319:1313.
68. Stamfer MJ, Willet WC, Colditz GA, et al. Past use of oral contraceptives and cardiovascular disease: a meta-analysis in the context of the Nurse's Health Study. *Am J Obstet Gynecol.* 1990;163:285.
69. Ory HW, Forrest JD, Lincoln R. *Making Choices: Evaluating the Health Risks and Benefits of Birth Control Methods.* New York: The Alan Guttmacher Institute; 1983.
70. Adams MR, Clarkson TB, Koritnik DR, Nash HA. Contraceptive steroids and coronary artery atherosclerosis in cynomologus macaques. *Fertil Steril.* 1987;47:1010.
71. Clarkson TB, Shively CA, Morgan TM, Koritnik DR, Adams MR, Kaplan JR. Oral contraceptives and coronary athersclerosis of cynomolgus monkeys. *Obstet Gynecol.* 1990;75:217.
72. Rossner S, Larsson-Cohn O, Carlson LA, Boberg J. Effects of an oral contraceptive agent on plasma lipids, plasma lipoproteins, the intravenous fat tolerance and the post-heparin lipoprotein lipase activity. *Acta Med Scand.* 1971;190:301.
73. Fischer-Dzoga K, Wissler RW, Vesselinovitch D. The effect of estradiol on the proliferation of rabbit aortic medial tissue culture cells induced by hyperlipemic serum. *Exp Mol Pathol.* 1983;39:355.
74. Husmann F. Long-term metabolic effects of estrogen therapy. In: Greenblatt RB, Heithecker R, de Gruyter W, eds. *A Modern Approach to the Perimenopausal Years: New Developments in Biosciences, 2.* New York: DeGrayter 1986:161.
75. Milekowsky GN, Nadler J, Huey F, et al. Evidence that smoking alters prostaglandin formation and platelet aggregation in women who use oral contraceptives. *Am J Obstet Gynecol.* 1988;159:1547.
76. Meade TW. Oral contraceptives, clotting factors, and thrombosis. *Am J Obstet Gynecol.* 1982;142:758.
77. Ambrus JL, Ambrus CM, Lillie MA, et al. Effect of various estrogen treatment schedules on antithrombin III levels. *Res Commun Chem Pathol Pharmacol.* 1976;14:543.
78. Beller FK, Ebert C. Effects of oral contraceptives on blood coagulation: a review. *Obstet Gynecol Survey.* 1985;40:425.

79. Bonnar J. Coagulation effects of oral contraception. *Am J Obstet Gynecol.* 1987;157:1042.
80. Thorogood M, Mann J, Murphy M, et al. Is oral contraceptive use still associated with an increased risk of fatal myocardial infarction? Report of a case-control study. *Br J Obstet Gynecol.* 1991;98:1245.
81. Royal College of General Practitioner. *Oral Contraceptives and Health.* New York: Pitman Publishing; 1974.
82. Kaw K-T, Pearl WS. Blood pressure and contraceptive use. *Brit Med J.* 1982;285:403.
83. Beck WJ Jr. Complications and contraindications for oral contraceptives. *Clin Obstet Gynecol.* 1981;24:893.
84. Wambach G. How do gestational hormones modify sodium balance? *Fortscher Med.* 1986;104:561.
85. Carr BR. Progestogens: effect on water/salt metabolism and blood pressure. In: Lobo RA, Whitehead MI, eds. Consensus Development Conference on Progestogens. *International Proceedings Journal.* 1989;1:87.
86. Maurer RA. Estradiol regulates the transcription of the prolactin gene. *J Biol Chem.* 1982;157:2133.
87. Rigg LA, Lein A, Yen SCC. Pattern of increase in circulating prolactin levels during human gestation. *Am J Obstet Gynecol.* 1977;129:454.
88. Jacobi J, Lloyd HM, Meares JD. Induction of pituitary tumors in male rats by a single dose estrogen. *Horm Metab Res.* 1975;7:228.
89. J Pituitary Adenoma Study Group. Pituitary adenomas and oral contraceptives: a multicenter case-control study. *Fertil Steril.* 1983;39:753.
90. Wingrave SJ, Kay CR, Vessey MP. Oral contraceptives and pituitary adenomas. *Br Med.* 1980;280:685.
91. Hulting AL, Werner S, Hagenfeldt K. Oral contraceptive steroids do not promote the development or growth of prolactinomas. *Contraception.* 1982;27:69.
92. Reyniak JV, Wenof M, Aubert JM, Jose M, Stangel JJ. Incidence of hyperprolactniemia during oral contraceptive therapy. *Obstet Gynecol.* 1980;55:8.
93. Mishell DR, Kletsky OA, Brenner PF, et al. The effect of contraceptive steroids on hypothalamic-pituitary function. *Am J Obstet Gynecol.* 1977;128:60.
94. Luciano AA, Sherman BM, Chapler FK, Hauser KS, Wallace RB. Hyperprolactinemia and contraception: a prospective study. *Obstet Gynecol.* 1985;65:506.
95. Cowden EA, Ratcliffe WA, Beastall GH, et al. Laboratory assessment of prolactin stress. *Ann Clin Biochem.* 1979;16:113.
96. MacLeod SC. Endocrine effects of oral contraception. *Int J Gynaecol Obstet.* 1979;16:518.
97. Vessey MP, Lawless M, McPherson K, Yeats D. Fertility after stopping use of intrauterine contraceptive device. *Br Med J.* 1983;286:106.
98. Vessey MP, Meiser L, Flavel R, Yeates D. Outcome of pregnancy in women using different methods of contraception. *Br Med J.* 1979;86:548.
99. Jacobson C. Cytogenic study of immediate postcontraceptive abortion. Report of a study under Food and Drug Administration contract; 1974.
100. Harlap S, Shinono PH, Ramcharan S. Congenital abnormalities in the offspring of women who used oral and other contraceptives around the time of conception. *Int J Fertil.* 1985;30:39.
101. The Cancer and Steroid Hormone Study of the Centers for Disease Control and the National Institute of Child Health and Human Development. Oral contraceptive use and the risk of breast cancer. *N Engl J Med.* 1986;257:796.
102. Schlesselman JJ. Cancer of the breast and reproductive tract in relation to use of oral contraceptives. *Contraception.* 1989;40:1.
103. UK National Case-Control Study Group. Oral contraceptive use and breast cancer risk in young women. *Lancet.* 1989;1:973.
104. Stadel BV, Lai S, Schlesselman JJ, et al. Oral contraceptives and premenopausal breast cancer in nulliparous women. *Contraception.* 1988;38:287.
105. Peterson HB, Wingo PA. Oral contraceptives and breast cancer: any relationship? *Contemp Ob/Gyn.* 1992;37:31–40.
106. Beral V, Hannaford P, Kay C. Oral contraceptive use and malignancies of the reproductive tract. *Lancet.* 1988;2:1331.
107. Clarke EA, Hatcher J, McKeown-Eyssen GE, Lickrish GM. Cervical dysplasia: association with sexual behavior, smoking, and oral contraceptive use. *Am J Obstet Gynecol.* 1985;151:612.
108. Hellberg D, Valentin J, Nilsson S. Long-term

use of oral contraceptives and cervical neoplasia: an association confounded by other risk factors. *Contraception*. 1985;32:337.
109. Irwin KL, Rosero-Bixby L, Oberle MW, et al. Oral contraceptives and cervical cancer risk in Costa Rica: detection bias or causal association? *JAMA*. 1988;259:59.
110. Celtentano DD, Klassen AC, Weisman CS, et al. The role of contraceptive use in cervical cancer: the Maryland cervical cancer case-control study. *Am J Epidemiol*. 1987; 126:592.
111. Mosher WD. Contraceptive practice in the United States 1982–1988. *Fam Plann Perspect*. 1990;22:198.
112. Working Group on the Health Consequences of Contraceptive use and Controlled Fertility. *Contraception and Reproduction: Health Consequences for Women and Children*. Washington, DC: National Academy Press; 1989.
113. ACOG Committee on Gynecologic Practice. *Contraception for Women in Their Later Reproductive Years*. ACOG Committee opinion no. 41; December 1985 (revised March 1988).
114. The Association of Reproductive Health Professionals Clinical Proceedings. *Maximizing Oral Contraceptive Effectiveness: Changing Perceptions*. Califon, NJ: Gardiner-Caldwell SynerMed; 1992.

13
Management of Menopause, Including Hormonal Replacement Therapy

Erica G. Sinsheimer, Pauline J. Shipley, and Jay J. Gold

In 1900, only 3.1 million people in the United States were 65 or older. In 1980, 25 million were 65 or older, some 10 million of whom were 75 or older. By 2025, it is estimated that almost 60 million individuals will be over age 65 and about 30 million will be over age 75. In the past, women did not live long enough to reach the age of normal menopause. Only recently has the life span of the human encompassed the postmenopausal years; therefore, the complications associated with menopause rarely occurred.

Menopause is a deficiency state. The normal hormones produced by a functioning ovary disappear with ovarian senescence. The most prominent deficiency, estrogen, leads to a variety of symptoms. These include vasomotor symptoms in the early stages and in the later stages, osteoporosis, cardiovascular disease, dyspareunia, bladder dysfunction and atrophy, and collapse of the pelvic floor (Table 13.1). In addition, certain nonspecific symptoms such as depression, irritability, insomnia, and headache may also occur.

Hormonally, there is an absolute decrease in total estrogen production, an elevation of follicle-stimulating hormone (FSH) and luteinizing hormone (LH), a decrease in the androgen/estrogen ratio, a decrease in sex hormone binding globulin, and a loss of cyclic progesterone production. These deficiencies, in turn, lead to the changes previously mentioned. There may be a perimenopausal period during which the woman is still menstruating but changes in hormone production

TABLE 13.1. Effects of estrogen deficiency: hormonal or related effects.

Vasomotor symptoms
Atrophy
   Vaginal mucosa
   Pelvic floor
   Cystocele-rectocele
Incontinence
Dyspareunia
Loss of libido
Osteoporosis
Cardiovascular disease

occur. At this time the woman may complain of vasomotor symptoms, including hot flashes, sweating, and the like, even though the menses are normal and even regular. During this stage hormonal therapy may be indicated to alleviate some of these symptoms. Further, these symptoms indicate that the woman will soon enter menopause and that therapy will be indicated in the near future.

On average the final menstrual period occurs at about age 50. The normal range for the onset of menopause is from 42 to 54, and loss of menses prior to that time is considered premature menopause. Women who develop premature menopause, either spontaneous or surgical, require estrogen replacement therapy, since complications appear at a young age. Further, they may require a larger replacement dose of estrogen.

After 1 to 2 years of estrogen deficiency, women develop atrophy of the vaginal mucosa, which may result in vaginal discomfort, dyspareunia, and bleeding. Soon thereafter, atrophy of the pelvic floor with cyst-rectocele may occur, resulting in such bladder symptoms as increased frequency of infection and urinary incontinence.

It left untreated, the postmenopausal state is not a benign condition. By age 60, i.e. 7 to 10 years after menopause, osteoporosis appears. A major postmenopausal problem is the onset of clinical heart disease. This results from the loss of the protective effect of estrogen on lipids and on the induction of atherosclerosis. Studies have shown that 105 women per 1,000 die per year of ischemic heart disease and 18.8 women per 1,000 die per year of breast cancer. Although clearly more postmenopausal women die of ischemic heart disease than breast cancer, generally women are more concerned about developing breast cancer than ischemic heart disease. This concern makes them fearful of taking estrogen. Women must understand that not taking estrogen leads to an increased incidence of ischemic heart disease that is six times that of breast cancer. In addition, in women not receiving estrogen therapy, hip fractures occur in 9 women per 1,000 per year and endometrial carcinoma is seen in 4 women per 1,000 per year.

Long-Term Complications of Menopause

Osteoporosis

Osteoporosis is a major disabling disease in postmenopausal women,[1] although it does not cause death as often as heart disease or breast cancer. The condition can be prevented with prophylactic treatment. This complication is caused by a reduction in bone mass from increased bone resorption. The resulting porous bone has an increased risk of fracture even in the absence of trauma. It can affect cortical or trabecular bone, but after age 70 cortical bone is particularly at risk. Osteoporosis is a serious problem, which has not been fully appreciated until recent years. At least 20 million women in the United States are affected by it. There are almost 1.5 million fractures related to osteoporosis per year in the over-45 age group. There are approximately 250,000 hip fractures annually in the United States. The medical costs and the loss of ability to work make osteoporosis a costly disease, perhaps as costly as $10 billion per year.

Osteoporosis has been divided into two categories.[2,3] Type I occurs in women up to age 70. In this type of osteoporosis, trabecular bone is primarily affected. More women than men are affected. The most frequent fractures in Type I are crush fractures of the vertebrae and Colles fractures of the radius. Estrogen, acting through a specific receptor in bone, inhibits bone resorption. After menopause, absence of estrogen leads to increased bone resorption and increased release of calcium into the bloodstream. This, in turn, inhibits parathyroid hormone (PTH) and decreases Vitamin D production, causing decreased calcium absorption from the gut. This sequence of events aggravates the osteoporotic changes.

Type II osteoporosis occurs in individuals over age 70 and mainly affects cortical bone. These individuals are more prone to fractures of the hip and pelvis. Vertebral fractures are usually wedge-type and, unlike crush fractures, are not associated with pain. Ultimately, vertebral fractures may lead to the so-called dowager's hump, and crush fractures lead to a marked decrease in height.

Risk factors for osteoporosis include race (Caucasian or Oriental), premature menopause, inactivity, smoking, alcohol use, chronic steroid use, or amenorrhea. Black and Latino women are less likely to have osteoporosis (Table 13.2).

Although calcium intake does not play a direct role in osteoporosis, a lack of adequate dietary calcium in childhood or up to the time of maximal bone mass at age 35 results in lower maximal bone mass, which leads to an increased risk of osteoporosis in later years. At age 35 a 2% to 3% loss of calcium each year begins, resulting in thinning of the bones and predisposing to osteoporosis. It is well known that prolonged inactivity, such as prolonged bedrest, predisposes to calcium loss. Prolonged steroid therapy induces loss of calcium from bone and interferes with calcium absorption

TABLE 13.2. Risk factors for osteoporosis.

1. Caucasian or Oriental race
2. Prolonged estrogen deficiency
3. Premature menopause
4. Prolonged inactivity or recluse
5. Thin body build
6. Prolonged steroid therapy
7. Cigarette smoking
8. Excess alcohol use

from the gastrointestinal tract. Women with anorexia nervosa or women who exercise so rigorously that they develop amenorrhea tend to be estrogen deficient, which predisposes to osteoporosis. Cigarette smoking and excess alcohol use interfere with the efficacy of estrogen, which may predispose to osteoporosis.

Routine X rays are not very effective in diagnosing osteoporosis or identifying a patient at risk because they are not sensitive. When X rays do show osteoporosis, it is usually far advanced.

There are new procedures to evaluate a woman for osteoporosis.[4] These include single photon absorptiometry, which measures bone density in the wrist; quantitative computed tomography; dual photon absorptiometry, which measures bone density in the lumber vertebrae and in the hip; and the newest procedure, dual energy X ray absorptiometry (DEXA), which also measures bone density in the spine and the hip but uses a radiologic source rather than a nuclear source.[4] DEXA is less expensive than dual photon absorptiometry and requires less time to perform.

Screening bone density should not be done on all women. Indications for evaluation include confirming the diagnosis of osteoporosis in an individual in whom it is strongly suspected, determining the extent of osteoporosis when its presence is well known, determining whether to initiate preventive therapeutic modalities in a patient at risk, and monitoring therapeutic responses.

Cardiovascular Disease

Estrogen has a positive effect on lipids by raising high-density lipoprotein (HDL) and lowering low-density lipoprotein (LDL).[5,6] This occurs while producing a minimal decrease or no change in total cholesterol. Estrogen is also thought to have a beneficial effect on cardiovascular disease in the female, unrelated to its effects on cholesterol. Premenopausally, cardiovascular disease is not as common in the female as it is in the male of comparable age, unless there is congenital heart disease, diabetes, or hypertension. After menopause, with the loss of estrogen, the incidence of cardiovascular disease approaches that of the male.

Since estrogens are most commonly administered with progestational agents, concern exists that these agents may counteract the beneficial effects of estrogen. Specifically, 19 nortestosterone progestational agents that have an androgenic derivation have been shown to cause a reduction in HDL cholesterol. In contrast, medroxyprogesterone acetate has been found to be neither beneficial nor detrimental.[7]

In summary, administration of estrogen is a preventive factor in the development of cardiovascular disease and thus should strongly be considered for use in the postmenopausal female.

Therapy of Menopause: Estrogen Therapy

Estrogen therapy of menopause is the mainstay of treatment.[8] Estrogen replacement has been used to treat vasomotor symptoms but recently has been given to treat or to prevent osteoporosis and to prevent ischemic heart disease. Estrogen replacement should be strongly considered, except in those patients who have a strong contraindication to the use of estrogen.

Such contraindications include a history of breast cancer in either the patient or a first-degree relative, a strong history of vascular thrombosis, migraine headaches, hypertension, and gallbladder disease. Abnormal uterine bleeding in the patient with an intact uterus should be evaluated for endometrial carcinoma before estrogen therapy is started.

As previously stated, estrogen acts directly on estrogen receptors on bone.[9] Activation of these receptors rapidly and effectively inhibits bone resorption. Studies on metacarpal mineral content have shown that the initiation of estrogen replacement at the earliest stage of bone loss will negate this bone loss and will continue to negate it as long as the estrogen is continued.[10] Individuals have also been treated 2 to 4 years after the onset of bone loss, and bone loss stops almost immediately. This inhibition persists for as long as the estrogen is continued. This beneficial effect of estrogen can be seen if estrogen is instituted as late as 8 years after menopause.

A study measuring bone density at menopause and 3 years later in individual age groups of 51 to 55, 56 to 60, 60 to 65, and 66 to 70 showed acute losses in bone density in this period of time. In the 71 to 75 age group, the rate of bone loss was much less. When similar groups were studied, patients taking estrogen had a reduced level of bone loss.[10] Estrogen has also been shown to decrease the rate of bone loss in patients over 70, although the difference was not as great as in younger individuals. In addition, studies comparing the incidence of vertebral fractures in individuals receiving estrogen with individuals not receiving estrogen showed the incidence of fractures to be markedly reduced in the group treated with estrogen.[11]

The unopposed use of estrogen in a woman with an intact uterus has been clearly shown to increase the incidence of adenocarcinoma of the endometrium. This known and accepted fact is one of the major concerns of estrogen replacement therapy. Any woman with an intact uterus treated with estrogen in the postmenopausal state must be treated with progesterone as well. The incidence of endometrial carcinoma in women not on estrogen is about 200 cases per 100,000 women. In women treated with unopposed estrogen, the incidence may double. Thus, it is essential that such women receive progesterone therapy along with their estrogen therapy.[12,13]

Estrogen may be given in a variety of regimes. The most popular is to give estrogen for 25 days and concomitant progesterone the last 10 to 14 days. Estrogens that have been used include Premarin (conjugated estrogens), 0.625 mg daily; Estrace (estradiol), 1 mg daily; Estinyl (ethinyl estradiol), 0.02 mg daily; or Estraderm Patch (estradiol), 0.05 mg twice a week for the first 25 days of the month. Progesterones include Provera (medroxyprogesterone acetate), 10 mg daily; Norlutin (norethindrone) or Norlutate or Aygestin (norethindrone), 5 to 10 mg daily; and/or crystalline micronized progesterone, 200 to 400 mg daily for 10 to 14 days (Table 13.3). An alternative is to give estrogen continuously, adding progesterone the first 10 to 14 days of each month.

All of these regimes will usually result in withdrawal periods. Newer regimes designed to try to prevent withdrawal bleeding involve giving estrogen in similar doses continuously, accompanied by continuous progesterone in smaller doses of 2.5 to 5 mg daily. In many women no withdrawal bleeding occurs, but in some women there will be intermittent spotting, which will usually disappear within the first year of treatment. Some women on the continuous regimen will not stop bleeding intermittently, and they should be switched to the cyclic forms of treatment mentioned above. If bleeding occurs frequently or at inappropriate times on any one of these regimes, an endometrial biopsy should be performed to rule out endometrial pathology. A study by Whitehead, et al[14], showed that if bleeding occurred before the 11th day of therapy, an endometrial

TABLE 13.3. Estrogen regimes.

A. INTERMITTENT
1. Estrogen, day 1 to day 25
   Conjugated estrogen, 0.625 mg per day
   Estradiol, 1 mg per day
   Ethinyl estradiol, 0.02 mg per day
   Estradiol patch, 0.5 mg two times per week
2. Progesterone, day 16 to day 25
   medroxyprogesterone acetate, 10 mg per day
   norethindrone, 10 mg per day
   norethindrone acetate, 10 mg per day

B. CONTINUOUS
1. Estrogen: one daily or one patch two times per week
2. Progesterone: 2.5 to 5.0 mg daily
   If intact uterus, may also be given first 10 days of month with continuous estrogen

biopsy should be done. If bleeding occurred after the 11th day of therapy, the biopsy almost always showed endometrial suppression rather than hyperplasia, making a biopsy unnecessary. If a patient has no uterus, continuous estrogen without progesterone can be given, since there is no concern for the development of endometrial carcinoma.

When estrogen is contraindicated, the administration of continuous progesterone will sometimes alleviate some vasomotor symptoms. Whether this has any beneficial effect on the development of osteoporosis is not known.

In women, androgen is produced in both the ovary and the adrenal gland. At the time of menopause, women lose 50% of their androgen production. This may contribute to a decrease in libido and a loss of a sense of well-being. Small doses of androgen replacement, such as Halotestin (fluoxymesterone), may alleviate these symptoms.

In summary, women who take estrogen have one-fifth the incidence of osteoporosis, one-half the incidence of fractures, and about one-third the incidence of cardiovascular disease compared with women who do not take estrogen. Thus, it is clear that estrogen therapy is essential in the treatment of menopausal women for the prevention of osteoporosis and heart disease, unless there is an absolute contraindication.

Breast Cancer

Breast cancer is a potential complication of estrogen replacement therapy and thus a major deterrent for many women (Table 13.4). Many years ago the literature suggested there was an association between estrogen and breast cancer. Most recent studies have not indicated such a relationship.[15,16] When evaluating a patient for estrogen replacement, the patient's family history must be considered. For a long time fibrocystic disease of the breasts has been a contraindication to the use of estrogen. However, fibrocystic disease is so common in women that it is no longer considered a risk factor, although any woman with fibrocystic disease who is on estrogen should be evaluated

TABLE 13.4. Risk factors for breast cancer.

1. Heredity
2. Increases with advancing age
3. Family or personal history of breast cancer
4. Regional diets (low in Japan)
5. Early menarche, late menopause
6. Late age at first pregnancy
7. Fibrocystic disease of breasts

with mammograms and breast exams at regular intervals. A strong family history of breast cancer is generally considered a contraindication. Other contraindications include hypertension, gallbladder disease, and smoking.

The incidence of breast cancer increases with advancing age. Regional diets may also be a factor in the development of breast cancer. Japanese women living in Japan have a low incidence of breast cancer. When living in the United States and eating an American diet, their risk of breast cancer is that of American women.

Women who have an early menarche, late menopause, or late age of first pregnancy may also have an increased incidence of breast cancer. Early pregnancies help produce the desired differentiation of normal breast tissue. In women with a first pregnancy at a late age, the breast has not undergone the same cell differentiation, which may make those individuals more predisposed to malignant changes.[17]

Similarly, if there is damage to breast tissue, the breast cells proliferate rather than differentiate, making the cells more susceptible to malignant change.

Therapy of Postmenopausal Osteoporosis

The hormonal treatment of osteoporosis is basically the same as the treatment of menopause (Table 13.5). However, in addition to estrogen, there should be adequate calcium taken as a supplement to serve as a substrate for the bony skeleton. Although calcium alone does not improve trabecular density, calcium and estrogen given together improve bone density. Calcium must be given in

TABLE 13.5. Medical therapy of osteoporosis.

1. Calcium: 1,500 mg of elemental per day with food
2. Vitamin D:
    400 units per day or 50,000 units once per week
    0.25 to 0.50 mg per day
    1.25 Dihydroxyvitamin D
3. Estrogen and progesterone
4. Etidronate (diphosphonate): 400 mg per day on empty stomach, ×15 days
5. Calcitonin: 100 to 200 units three times per week
6. Fluoride
7. Nonaerobic exercise

TABLE 13.6. Elemental calcium content.

TUMS (SmithKline Beecham) Calcium Carbonate	40%
OS-CAL (Marion) Calcium Carbonate	40%
CITRICAL (Mission) Calcium Citrate	20%
NEO-CALGLUCON (Sandoz) Calcium Gluconate	65%

adequate quantity: 1,000 mg of elemental (not total) calcium in the premenopausal woman and 1,500 mg in the postmenopausal woman. There are many calcium products on the market: calcium carbonate, calcium citrate, oyster shell calcium, and the like. Each contains varying amounts of elemental calcium (Table 13.6).

Calcium should be given with vitamin D in doses of at least 400 units a day. When giving calcium and vitamin D, serum calcium must be monitored three to four times per year, and a 24-hour urine for calcium excretion should be done at least once a year. The calcium dose should be titrated so the individual is not excreting abnormally high amounts of calcium in the urine, since this may predispose to renal problems and kidney stones.

There are alternative forms of therapy that can be added to cyclic estrogen therapy or used alone if estrogen is contraindicated. Medications can be given to the patient with osteoporosis that interfere with the bone remodeling cycle. Bone remodeling is an ongoing event that begins with osteoclast activation and leads to resorption of a protion of bone. After resorption, osteoblasts cause bone formation, followed by a resting phase and then repetition of the cycle. Exercise, activity, or mechanical stress will initiate activation, resorption, and then deposition of bone. After menopause this cycle alters in favor of resorption.

One group of medications useful in altering the bone remodeling cycle are the diphosphonates known in the United States as Etidronate or Didronel.[18-20] These medications have not yet been officially approved by the FDA for use in the treatment of osteoporosis, although they have been approved for use in other bone diseases. Didronel is administered once daily for 15 days, cycled every 90 days. This cycling activates the remodeling cycle but shortens the resorption cycle, thereby increasing bone density. The increased density can vary from about 4% to 6% yearly and can be followed by bone density studies.

The side effects of diphosphonates are minimal if not used properly. The fracture rate in individuals treated with diphosphonates may be reduced by about 50% in the early stages and over a period of more than 1 year may be reduced as much as 90%. In women who already have osteoporosis, Didronel can be given alone or with estrogen. Calcium should also be given with vitamin D, either 400 units a day or a 50,000-unit vitamin D once a week.

The thyroid parafollicular cells normally produce calcitonin[21], a substance that inhibits bone resorption. Calcitonin has been obtained from both salmon and human sources; the salmon source is the most frequent. Presently, it is available only by subcutaneous injection, and is used 3 to 7 days a week. An intranasal preparation is under study and is likely to improve efficacy by improving compliance and reducing side effects. Calcitonin is expensive and so is generally used only in certain circumstances. It has been shown to improve bone density but has not definitely been shown to reduce fracture frequency. It is extremely useful in the Type I postmenopausal osteoporotic patient who has severe pain since it has the ability to reduce that pain remarkably. The exact reason for this is not understood at this time. It also may be used in women who cannot receive estrogen, and it can be given along with Didronel occasionally. It can also be used in men with osteoporosis. Calcitionin dosage is titrated to the patient response with doses of 50 to 100 IU, three to seven times weekly. Side effects may include flushing, nausea, diarrhea and, rarely, anaphylaxis.

Fluorides[22] were popular several years ago. However, they are no longer as popular because they can have serious side effects, including gastrointestinal irritation, stress fractures, and articular pain. They promote osteoblastic activity and can increase trabecular bone mass in the spine. However, some of the new bone that forms is soft, so fluorides are not generally used at this time. Studies are ongoing to develop a safer and more effective form of fluoride.

Vitamin D was recently shown to have a modest beneficial effect in the treatment of osteoporosis when given by itself.[11] It is frequently given as part of the overall treatment program.

Summary

The menopausal state does not require therapy in all women, but in those women who do require it, therapy can relieve the undesirable symptoms of estrogen deficiency and the complications of osteoporosis and cardiovascular disease. The relationship between estrogen therapy for menopause and breast cancer is not clear-cut, but the risk of breast cancer is six to eight times less than that of heart disease. Thus, the benefits of estrogen therapy often outweigh the risks.

References

1. Christiansen C, ed. Consensus development conference on osteoporosis *Am J Med.* 1991; 91(5B):68.
2. Albright F. Osteoporosis. *Ann Int Med.* 1947;27:861–882.
3. Riggs BL, Melton LJ III. Evidence for two distinct syndromes of ihvolutional osteoporosis *Am J Med.* 1983;75:899–901.
4. Genant HK, Faulkner KG, Gluer CC. Measurement of bone mineral density: current status. *Am J Med.* 1991;91(5B):49–53.
5. Hazzard WR. Estrogen replacement and cardiovascular Disease: serum lipids and blood pressure effects *Am J Obstet Gynecol.* 1989;161:1847.
6. Wolf PH, Madans JH, Finucane FF, Higgins M, Kleinman JC. Reduction of cardiovascular disease–related mortality among postmeopausal women who use hormones: evidence from a national cohort. *Am J Obstet Gynecol.* 1991; 164(2):489–494.
7. Hirvonen E, Malkonen M, Manninen V. Effects of different progestagens on lipoproteins during postmenopausal replacement therapy. *N Engl J Med.* 1981;304:560–563.
8. Notelovitz M. Estrogen replacement therapy: indications, contraindications and agent selection. *Am J Obstet Gynecol.* 1989;161:1832–1841.
9. Erikssen ER, Colvard DS, Berg NJ, et al. Evidence of estrogen receptors in normal human osteoblast-like cells. *Science.* 1988; 241:84–86.
10. Lindsay R. Sex steroids in the pathogenesis and prevention of osteoporosis. In: Riggs BL, Melton LJ III, eds. *Osteoporosis: Etiology, Diagnosis and Management.* New York: Raven Press; 1988:335–358.
11. Consensus development conference: prophylaxis and treatment of osteoporosis. *Am J Med.* 1991;90:107–110.
12. Cummings SR. Evaluating the benefits and risks and postmenopausal hormone therapy. *Am J Med.* 1991;91(5B):14–18.
13. Padwick MB Pryse-Davies J, Whitehead MI. A simple method for determining the optimal dosage of progestin in postmenopausal women receiving estrogen. *New Engl J Med.* 1986; 315:930.
14. Whitehead MI, Hillard TC, Crook D. The role and use of progestogens. *Obstet Gynecol.* 1990;75:59S–76S.
15. Henderson BE. The cancer question: An overview of recent epidemiologic and retrospective data. *Am J Obstet Gynecol.* 1989;161:1859.
16. Dupont W, Page, D. Menopausal estrogen replacement therapy and breast cancer. *Arch Intern Med.* 1991;151:57–72.
17. Nachtigall LE. Estrogen and breast cancer. Current Perspectives on Managing the Menopausal and Post-Menopausal Patient: A Symposium. Ciba Pharmaceutical Company. Irving, Texas, December 8, 1990.
18. Storm T, Thamsborg G, Steiniche T, Genant HK, Sorensen OH. Effect of intermittent cyclical etidronate therapy on bone mass and fracture rate in women with postmenopausal osteoporosis. *New Engl J Med.* 1990;332:1265–1271.
19. Watts N, Harris ST, Genant HK, et al. Inter-

mittent cyclical etidronate treatment of postmenopausal osteoporosis. *New Engl J Med.* 1990;323:73–79.
20. Parfitt AM. Use of biphosphonates in the prevention of bone loss and fractures. *Am J Med.* 1991;91(5B):49–53.
21. Reginster JY. Effect of calcitonin on bone mass and fracture rates. *Am J Med.* 1991;91(5B):23–28.
22. Riggs BL. Treatment of osteoporosis with sodium fluoride or parathyroid hormone. *Am J Med.* 1991;91(5B):37–41.

14
The Dominant Breast Mass

John H. Isaacs

The number of breast carcinomas discovered by the physician largely depends on his or her interest, diligence, and skill during the examination. The most common physical sign of mammary cancer is a mass in the breast. Fortunately, most complaints concerning the breast are not cancer related; benign conditions are far more frequent.[1]

Most women in the United States consider their obstetrician/gynecologist their primary physician. If the gynecologist does not accept this mantel of responsibility, the results will be a failure to diagnose a breast cancer, possibly a tragic mortality, and an inevitable lawsuit. The physician may do a hasty, and most likely incomplete, breast examination and assure the patient that the mass is "fibrocystic disease" or that it is "nothing" and not to worry. Bearing this in mind, a dominant mass found in the breast by either the patient or the physician must be diagnosed. This does not mean that all breast lumps must undergo biopsy, but they do need to be followed until the nature of the lump has been determined.

Most breast lumps are discovered by the patient.[2] This is probably due to the quasi-proprioceptive sense that enables the patient to detect subtle changes in her breast that may not be detected by the physician. Even if the physician cannot feel any mass, a careful follow-up after a meticulous breast examination is essential. Until the cause of the breast lump is determined, the physician has not fulfilled his or her responsibility. Breast cancer can mimic almost all benign conditions; thus, it must always be a suspected cause.

Steps in the Diagnosis of a Dominant Breast Mass

When a discrete lump is noted, a differential diagnosis should run through the physician's mind. In the breast clinic at Loyola University, Chicago, the conditions listed in Table 14.1 have been encountered and should be considered by the examining physician.

After a definite mass has been discovered, the physician must consider all breast cancer risk factors. Although there is only one method for making a definitive diagnosis of breast can-

TABLE 14.1. Conditions associated with the dominant breast mass.

Fibrocystic change
Fibroadenoma
Mammary duct ectasia
Gross cysts
Tietze's syndrome
Lipoma
Fat necrosis
Fibrous disease
Intraductal papiloma
Superficial thrombophlebitis (Mondor's disease)
Sebceous cysts
Sclerosings adenosis
Metastatic carcinoma

cer (biopsy and histologic examination of the tissue), many indications may decrease or increase suspicion, including medical history, physical examination, mammography, ultrasonography, and needle biopsy. Except for mammography and ultrasonography, which are discussed elsewhere in this text, these methods will be discussed in this chapter. Nipple discharges and nonpalpable masses detected on mammography will be discussed in other chapters.

Medical History

Age

There is a definite relationship between age and breast carcinoma. Only 0.2% of women under age 25 will develop breast cancer.[3] The incidence rises gradually between the ages of 25 and 30. After 30, the incidence increases sharply and continues to rise with age. Thus if a lump is detected at an early age, careful observation and reexamination following a menstrual period may establish a diagnosis of a benign condition and a biopsy may be avoided. It is of little help to order a mammogram in young patients since the breasts are often quite dense, making the mammogram difficult to interpret. In addition, ordering a mammogram for a young woman is probably unwise because low-dose radiation is possibly carcinogenic.

Family History

A positive family history of breast cancer seems to increase the patient's risk. This becomes particularly important if the family relative developed the breast cancer before menopause. In such incidences, the physician should not procrastinate, and a biopsy should be done soon.

Reproductive Factors

Although still controversial, some studies suggest that patients with breast cancer had earlier menarche than control patients.[4] In addition, patients who develop breast carcinoma tend to have a later menopause.

Oophorectomy protects women to a considerable degree against subsequent breast carcinoma. The protective effect seems to be greatest when oophorectomy is performed on patients under age 35.

In most studies, nulliparity emerged as a limited but definite risk factor in breast carcinoma. Wynder et al found that late age (25 or older) at first birth significantly increased the risk of breast cancer for premenopausal and perimenopausal patients.[5]

Benign Breast Diseases

Certain benign breast diseases predispose the patient to breast cancer. These diseases are specific pathologic entities and should not be lumped under the catchall term *fibrocystic disease*.

Gross cystic disease of the breast increases the risk of developing breast cancer two to three times, even though the malignant changes do not usually develop in the gross cysts themselves.[6] These cysts usually occur in women between the ages of 35 and 50; they are rare before age 30. Simple aspiration is the treatment of choice. Large cysts are much easier to aspirate than small cysts. If aspiration fails either to yield fluid or to cause total collapse of the mass, the cyst should either undergo biopsy or at least be reexamined in 2 to 4 weeks to determine whether the cyst wall has diminished. If clinical examination suggests a cyst but aspiration fails to obtain fluid, aspiration via needle ultrasound guidance should be attempted before proceeding directly to an excisional biopsy. This is often possible, and an excisional biopsy can be avoided.

Patients with a history of a biopsy that was reported as multiple intraductal papilomas have a greater risk of developing breast cancer.[3] These lesions are usually located on the periphery of the breast, forming an ill-defined mass. Nipple discharge may also be present. If such a patient presents with a mass on the periphery of the breast, the physician should obtain and review the pathology of the previous biopsy. If the previous diagnosis was multiple intraductal papilomas, biopsy of the new area is definitely indicated.

The third and last benign breast disease that predisposes the patient to breast carcinoma is lobular neoplasia. This condition is usually an incidental finding on a breast biopsy specimen done for some other reason. Lobular neoplasia puts a women at a five times greater risk of developing breast cancer than patients without this finding. If such a patient develops a palpable mass at a later date, biopsy is mandatory.

Physical Examination

After taking a careful medical history, a breast examination must be done. The breasts are examined with the patient in both the sitting and supine positions. Often, the physical examination will establish a diagnosis.

Start the examination with the patient in the sitting position. The supraclavicular and axillary areas should be checked for palpable masses. If nodes are felt, note their number, size, consistency, and mobility. Bear in mind that palpable nodes in the axilla may be inflammatory and not neoplastic. It is not unusual to find small, freely movable nodes under either axilla.

Next, inspect the breasts with the patient's hands at her side. Look for signs of early skin retraction. Have the patient press her hands into her waist; this will cause contraction of the pectoral muscles and may reveal skin retraction that would go undetected otherwise. Ask the patient to raise her arms above her head and bend forward: This will also help uncover early evidence of skin retraction. Retraction is a fairly good sign of the presence of a carcinoma. However, benign lesions—including mammary duct ectasis, fat necrosis, and thrombosis of the superficial veins of the breast (Mondor's disease)—also produce retraction.

Look at the skin of the breasts for signs of redness, which usually results from acute or chronic inflammation but could be caused by inflammatory carcinoma. Also be alert for skin edema. Although edema may be caused by infection, malignant cells may be plugging the lymphatics in the corium. Furthermore, malignant lesions deep in the breast often show signs of edema in the areola or just inferior to it.

Finally, look for changes in the nipples: flattening, broadening, retraction, or deviation in direction. All these can be subtle signs of carcinoma. However, these changes should not be confused with nipple inversion, which is a normal condition present in many women.

After completing the sitting portion of the exam, have the patient lie down. Check the skin of the nipples and areola for thickening, reddening, or erosion, which may signal Paget's disease. The changes caused by Paget's disease always begin on the nipple epithelium and then extend to the areola. A skin biopsy will show intact epithelium infiltrated by Paget's cells.

With the visual inspection completed, palpate the breasts. Early diagnosis of breast cancer by physical examination requires a meticulous but gentle palpation of the entire breast. Use the pads of the fingers, not the tips, to gently compress the breast tissue against the chest wall. Palpation can be augmented by molding the breast around a suspicious area, which sometimes reveals skin retraction. If the physician feels a sharply defined dominant mass, a biopsy of the area may be indicated. Note the degree of mobility of the lump. Fibroadenomas and macro cysts have a greater degree of mobility than carcinomas, adenosis, and fibrocystic changes, all of which are relatively fixed. Gross cysts of the breast usually feel fluctuant and may be slightly tender.

Following this thorough examination, carefully record the information on the chart, noting any areas of nodularity or irregularity. If possible, attach a drawing to the chart indicating location and size and a tentative diagnosis (Fig. 14.1).

Mammary duct ectasia is one of the most difficult lesions to distinguish from carcinoma. Its central location and chronic localized inflammatory reaction may help to clinically differentiate it from breast cancer. Painful, costochondral swelling (Tietze's syndrome) may also be confused with a carcinoma. Tender swelling of the cartilage at the costochondral junction of one of the ribs usually points to

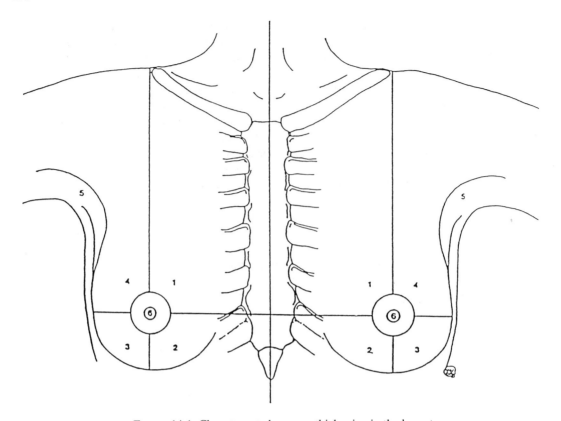

FIGURE 14.1. Chart to note lumps or thickening in the breast.

Tietze's syndrome. Finally, always keep in mind that a malignant lesion of the breast may be metastatic in origin. When an unusual malignant tumor is discovered in the breast, search for other possible sources, such as the ovary, endometrium, kidneys, adrenal gland, or a melanoma.

Breast Biopsies

Although a medical history and a thorough physical examination will greatly influence how soon a breast biopsy should be done, the basic reasoning remains the same: If a mass is present and a diagnosis cannot be established with a reasonable degree of certainty, then a biopsy must be performed.

A dominant mass or thickening in the breast or axilla still remains the most common indication for a breast biopsy. Biopsy techniques can be divided into closed and open methods. Closed techniques include aspiration for suspected breast cyst, aspiration for cytology with a fine needle, and core biopsy for histology. Open techniques include incisional biopsy, excisional biopsy, needle localization breast biopsy, nipple biopsy, and biopsy of the other breast. The only biopsies to be considered in this chapter are for a dominant mass.

Aspiration for Suspected Breast Cysts

If upon examination it seems obvious that the patient has a dominant cystic mass, a 20-gauge or 22-gauge needle attached to a 10 mL or 20 mL syringe is used for aspiration. The patient should be lying on her back. The skin is cleaned with alcohol. The cyst is fixed between the fingers of one hand while the needle is directed into the cyst with the other hand (Fig. 14.2). If the mass is indeed a cyst and aspiration is successful, the aspirated fluid will vary in color depending on the age of the cyst. The

14. The Dominant Breast Mass

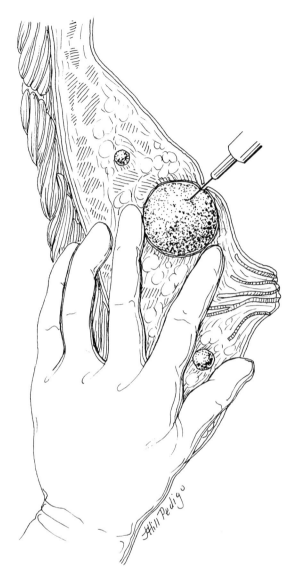

FIGURE 14.2. Breast cyst is stabilized with the fingers, and the fluid is aspirated.

longer the cyst has been present, the darker the fluid will be. If the fluid is bloody, a biopsy of the cystic mass should be planned. After aspiration, a small pressure dressing is applied. Once the cyst has been aspirated, the mass should no longer be palpable. If the area still feels thick and irregular and does not disappear in a week or two, a biopsy is required. If the cyst reforms in a short period of time, malignancy should be suspected and an excisional biopsy must be considered. The question always arises as to what should be done with the aspirated fluid. It may be sent for cytological examination, but a positive cytologic result for cyst fluid is rare.[7] Most often, the fluid is discarded.

Pneumocystogram is a technique used to determine whether a cyst wall has a smooth inner lining. Air is injected into the cystic cavity after the fluid has been aspirated. An X ray of the breast is then taken to outline the cyst wall. This is a time-consuming procedure that is not routinely indicated.

Aspiration for Cytology with a Fine Needle

Needle aspiration for the diagnosis of breast lesions was first described by Martin and Ellis.[8] The technique uses a short, beveled 22-gauge needle attached to a 20-mL syringe fitted into a pistol grip (cameco 20 mL Pat NR 3819091) syringe holder. The skin is cleaned with antiseptic solution, and the needle is passed through the skin into the center of the mass. When the needle enters the mass, suction is applied via the pistol grip holder. While maintaining suction in the syringe at all times, the physician points the needle in different directions in the mass (Fig. 14.3). The suction is released before withdrawing the needle from the mass. One word of caution concerning this seemingly simple technique: The physician must take care that the needle does not pierce the pleura and lung, resulting in a pneumothorax.

The contents of the needle are expelled onto a glass slide. This is best accomplished by removing the needle, attaching it to a second 20 mL syringe filled with air, and forcing the contents of the needle onto the slide. A second slide is drawn over the first slide to make the smear. Both slides are then placed in a fixative solution and sent to the laboratory for staining. Several slides (about six) can usually be prepared this way: The greater the number of slides, the greater the accuracy of this technique.

This technique is of value only if positive or suspicious cells are aspirated. Failure to aspirate positive or suspicious cells may indicate

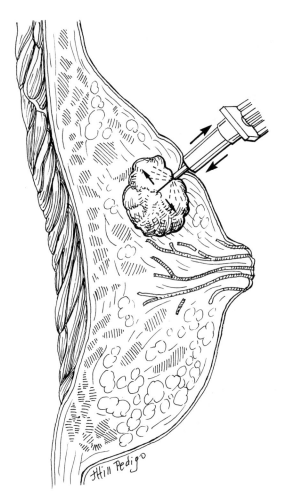

FIGURE 14.3. Fine needle aspiration of a solid nodule in the breast. The needle is directed in several areas to obtain an adequate aspirate.

eliminating time-consuming preoperative discussion. If the patient knows that the lump is almost certainly malignant, the histologic confirmation can be obtained at the same time definitive surgery is performed.

Despite enthusiastic support, fine needle aspiration has been slow to gain acceptance in the United States. The principal reason for this resistance is false-negative aspirates. In a study performed at the Mayo Clinic, however, the false-negative rate was only 6%, there were no false positive results, and the accuracy was 94%. These results were similar to other reported series in the literature.[9]

Core Biopsy for Histology

Core biopsy for histology may be used when a clinically obvious advanced carcinoma is present and histological confirmation is necessary. A cutting-edge core needle biopsy, such as the Tru Cut needle, is used to obtain tissue for histologic examination. The larger the tumor, the greater will be the accuracy of diagnosis. The sensitivity of this technique ranges from 67% to 90%.

A 1-mm to 2-mm skin incision is made, and the needle point is advanced. The outer sheet is pulled back and then forward. The cutting edge entraps a core of the tumor (Fig. 14.4). The needle is removed. The specimen is placed in formula and sent to the laboratory for histologic diagnosis.

Incisional Biopsy

Although rarely used today, incisional biopsy has a place in establishing the diagnosis of an advanced cancer as well as obtaining sufficient tissue for hormone receptor determinations. An incisional biopsy may also be indicated to help confirm the benign nature of a broad area of what is thought to be fibrocystic breast changes.

The procedure is relatively simple, requiring only a local anesthetic and a scalpel to excise a small wedge of tissue. Bleeding points can be cauterized. No attempt should be made to close the breast tissue defects. Simply be

the mass was not actually aspirated, and further investigation is necessary. If the results are positive, the various treatment methods should be discussed before any actual surgical intervention. Even with positive cytologic results, definitive treatment should not be performed without histologic verification of cancer. These points should be carefully discussed with the patient before beginning therapy.

Fine needle aspiration can save considerable time. If aspiration indicates malignancy or is highly suspicious, the patient's care can be streamlined by completely avoiding open biopsy as a separate operation and often

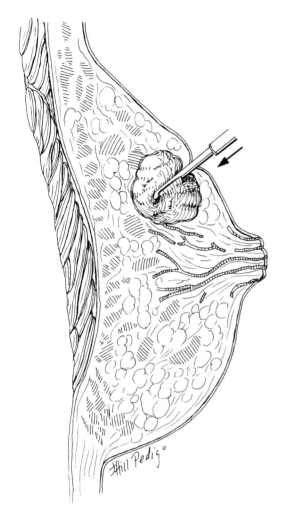

FIGURE 14.4. Tru Cut needle is inserted into the tumor mass, and a specimen is obtained.

Excisional Biopsy

An excisional biopsy is more desirable than an incisional biopsy because it provides the pathologist with the complete specimen. Such biopsies are usually done on outpatients under local anesthesia. The specimen can be sent for frozen section and, if indicated, hormone receptor determinations. Since there can be a 2% to 5% error rate in frozen section, it is usually wise to advise the patient that the final diagnosis will not be determined until the permanent sections have been reviewed by the pathologist.

The procedure involves preparing and draping the breast as for any surgical procedure. The skin incision should be a gentle curvilinear line over the suspected lesion, following the skin's Lines of Langer. The area of incision should be outlined with a marking pencil, and the incision line should be hashmarked directly over the breast mass (Fig. 14.5). This hash-

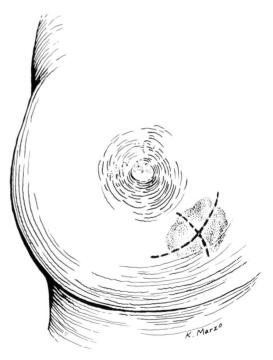

FIGURE 14.5. Incision is along Langer's line. The area of incision is outlined with a marking pencil and is the hashmarked over the mass.

certain of hemostasis and then close the skin with interrupted 5–0 nylon vertical mattress stitches.

To establish a diagnosis of inflammatory cancer, a representative eclipse of skin and subcutaneous tissue must be removed for pathologic determination of dermal lymphatic involvement. The lesion is often associated with pain, tenderness, warmth, erythema, and peau d'orange. The biopsy site must be carefully selected, or the characteristic lesion may be missed. Often an erroneous diagnosis of mastitis or cellulitis is made.

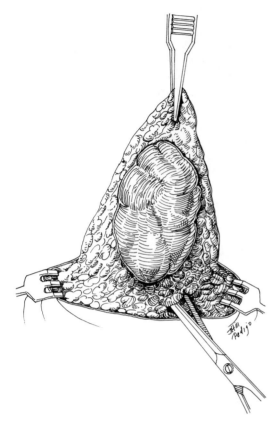

FIGURE 14.6. Breast lump being excised with a rim of normal tissue around the mass.

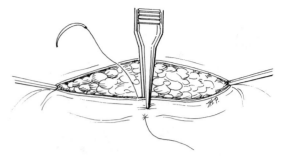

FIGURE 14.7. The skin is closed with 5-0 nylon interrupted stitches.

mark will be helpful after the local anesthetic has been infiltrated into the skin should the mass in question become "lost." The ideal skin incision is along the areolar margin, because the resulting scar is much less noticeable. However, if the lesion is peripherally located, this incision may be difficult to accomplish. The incision should also be directed in such a way that it will be contained within the planned eliptical mastectomy incision should a subsequent mastectomy be necessary.

The incision should extend down through the skin into the superficial subcutaneous tissue. The lesion in question should be completely excised, ideally having a rim of normal breast tissue around the mass (Fig. 14.6). All bleeding points should be cauterized, achieving total hemostasis. No attempt should be made to close the dead space. The small seroma that will form in the dead space will pull the breast tissue together, giving a much better cosmetic effect. The physician should close the skin with 5-0 nylon interrupted stitches, being careful not to touch the skin edges (Fig. 14.7).

Complications from an excisional biopsy should be few and include hematoma and stitch abscess. To minimize the chance of hematoma developing, the entire chest should be bound with an Ace bandage after the biopsy has been completed and the skin closed. The most serious complication is an inaccurate biopsy leaving a residual mass that could contain a malignancy. For this reason, there are certain indications for performing a biopsy under general anesthesia. Such conditions include a small mass located deep in the tissue of a large breast, a patient's sensitivity to local anesthetic, lesions that are not palpable and only detected by mammography, and those patients with a nipple discharge requiring meticulous duct dissection.

The clinician should be critical of his or her reaons for biopsy: A ratio of about 20% to 25% positive findings is generally considered acceptable.

Summary

A dominant mass detected by either the patient or the physician can be dismissed as inconsequential only after a firm diagnosis is made. All risk factors and physical findings must be taken into consideration. If further investigation is necessary, some type of closed biopsy technique or, if indicated, an open biopsy should be performed. If these guidelines are followed, no patient will be denied a definitive diagnosis of a dominant breast mass.

References

1. Donegan W. Diagnosis. In: Donegan W, Spratt JS, eds. *Cancer of the Breast*. Philadelphia, Pa: WB Saunders Co; 1988:126.
2. Donegan W, Donnegan W, Spratt JS, eds. *Carcinoma of the Breast*. Philadelphia, Pa: WB Saunders Co; 1988:126.
3. Isaacs JH. How to tell if that breast lump is malignant. *Modern Medicine*. 1984; 56–72.
4. Isaacs JH. History and physical examination and breast self-examination. In: Isaacs JH, ed. *Textbook of Breast Disease*. St. Louis, Mo: Moseby Year Book Inc; 1992:21–38.
5. Wynder EL, MacCornack FA, Stellman SD. The epideimology of breast cancer in 785 United States caucasian women. *Cancer*. 1978;41:2341–2354.
6. Haagensen CD, Bodian C, Haagesen DE. *Breast Carcinoma Risk and Detection*. Philadelphia, Pa: WB. Saunders Company; 1981:69.
7. Ciatto S, Cariaggi P, Bulgaresi P. The value of routine cytological examination of breast cyst fluid. *Acta Cytol*. 1987;3:301–304.
8. Martin HE, Ellis EB. Biopsy by needle puncture and aspiration. *Ann Surg*. 1930; 92:169–181.
9. Grant CS, Goellner JR, Welch JS, et al. Fine-needle aspiration of the breast. *Mayo Clinic Proceedings*. 1986;61:377–381.

15
Nipple Discharge

Antonio J. Bravo

The problem of nipple discharge and its clinical evaluation and management beginning in the office setting of the gynecologist, primary care physician, or surgeon are the focus of this chapter. Nipple discharge is fairly common, comprising 3% to 7% of chief complaints relating to the breast. This problem is very distressing to women because they feel that it may indicate the presence of a serious breast disease or cancer. To allay the patient's fear, gentleness, concern, and reassurance must be expressed in this and all breast disease management. The assessment of nipple discharge aims to distinguish physiologic from pathologic discharge.

TABLE 15.1. Frequency of types and association with cancer of 503 surgically significant nipple discharges.*

Type	Frequency (%)	Cancer-Associated (%)
Serous	207 (41.2)	13 (6.3)
Serosanguineous	160 (31.8)	19 (11.9)
Sanguineous	125 (24.9)	30 (24.0)
Watery	11 (2.2)	5 (45.5)

* From Leis et al[1] by permission.

The majority of bloody or blood-tinged discharges result from benign rather than malignant disease.[1,3]

Characteristics of Nipple Discharge

Nipple discharge has been reported in 10% to 15% of women with benign breast disease and in 2.5% to 3% of those with carcinoma.[1,2] There are several basic types of nipple discharge: milky, multicolored, purulent, watery (clear), serosanguineous (pink), serous (clear-yellow), or sanguineous (bloody). It is beneficial to establish the type, color, and consistency of the discharge. Discharges of significance are *unilateral, persistent, spontaneous, nonlactational,* and *uniductal.* Watery, serous, serosanguineous, and sanguineous types are the most relevant and indicative of possible cancer or precancerous lesions (Table 15.1).[1]

Physiologic Nipple Discharge

Nipple discharge occurs physiologically and is normal during pregnancy (colostrum) and during lactation. On rare occasions, during late pregnancy a bilateral bloody discharge may be produced as a result of extreme hyperplasia of the ductal epithelium. Oral contraceptive pills, estrogens, phenothiazines, methyldopa, tricyclic antidepressants, and other drugs may lead to nipple discharge or galactorrhea. The discharge produced is due to the systemic effect of the medication, which induces diffuse changes in breast tissue. Multiple ducts are affected in both breasts. Thus discharge can be elicited nonspontaneously or spontaneously and is seen bilaterally from multiple ducts. Women, particularly in the perimenopausal period, can commonly elicit discharge from their breasts; this is considered normal.[4] This should not be considered as a primary breast disorder, and expectancy and reassurance should be offered. Discontinuation or a dose reduction in a medication-induced galactorrhea may be effective.

Care must be exercised that the discharge originates truly from the nipple and not from surrounding areolar lesions. In patients with inverted nipples, eczema, traumatic erosions, duct fistulae, or areolar infections, secretions and exudates may appear that may mimic true discharges.[1]

Bilateral spontaneous milky secretion or galactorrhea is seen in patients with hyperprolactinemia secondary to a pituitary adenoma or other pituitary lesion. Amenorrhea, oligomenorrhea, or infertility are commonly associated. In these cases serum prolactin levels must be followed, and evaluation of the pituitary gland and sella turcica by computerized tomography (CT) scan or a magnetic resonance imagery (MRI) scan are recommended (Table 15.2).

TABLE 15.2. Causes of galactorrhea.

Physiological
  Mechanical stimulation
  Extremes of reproductive life (puberty, menopause)
  Postlactational
  Stress
  Chest wall trauma

Drugs associated with hyperprolactinemia
  Dopamine receptor-blocking agents
    Phenothiazines (e.g. chlopromazine)
    Haloperidol
    Metoclopramide
  Dopamine-depleting agents
    Reserpine
    Methyldopa
    Estrogens
    Opiates

Pathological
  Hypothalamic and pituitary stalk lesions
  Pituitary tumors
    Macroadenoma
    Microadenoma
    Craniopharyngioma

Miscellaneous
  Ectopic prolactin secretion (e.g. bronchogenic carcinoma)
  Hypothyroidism
  Chronic renal failure

Nonphysiologic (Pathologic) Nipple Discharge

Abnormal nipple discharge that requires investigation and possible surgical intervention is generally unilateral and uniductal. Despite the importance of these factors, bilateral and multiple-duct discharges should not be ignored if they are of the watery, serous, serosanguineous, or sanguineous types. In contrast to the physiologic secretions produced by diffuse involvement of many or all mammary ductal units, a discharge coming from one breast and one duct suggests a focal isolated lesion involving a specific duct or duct unit. Only 8% to 10% of patients with nipple discharge harbor serious pathology (Table 15.3).[5]

Intraductal Papilloma

An intraductal papilloma is the most common cause of unilateral, uniductal nipple discharge

TABLE 15.3. Cause, frequency, and median age of occurrence in 503 surgically significant nipple discharges.*

Cause	Frequency (%)	Median age
Intraductal papilloma	231 (45.9)	40
Fibrocystic disease	181 (36.0)	34
Cancer**	67 (13.3)	56
Advanced duct ectasia	24 (4.8)	43

* From Leis et al[1] by permission.
** Cancer over age 50 occurred in 53/104 patients (51%).

and is usually of the serosanguineous type. It is a benign lesion characterized by a solitary proliferation of duct epithelium. There may be a single pedunculated or sessile stalk arising from the duct wall. The architecture is villous and delicate. Papillomas rarely measure more than 5 mm and are often within 1 cm from the areola.

The median age of occurrence of intraductal papilloma is 40, although it can be found from adolescence to old age. Because papillomas are soft and small they rarely present as a mass. Benign solitary papillomas are not premalignant but must be differentiated histologically from multiple papillomatosis or from papillary carcinoma.[6]

Duct Ectasia

Duct ectasia may produce a multicolored (gray, green, black), sticky, spontaneous, multiple- or single-duct nipple discharge that is often bilateral; however, it may be unilateral. The condition begins with dilation of the terminal ducts. Lipid fluid, acellular material, and desquamated epithelium accumulate within the duct and produce inflammation and nipple discharge. Itching and burning are symptoms that improve or resolve after the discharge is expressed. This usually occurs in the peri- or postmenopausal period.

Purulent Discharges

Purulent discharges may be noted through one or more ducts and can occur with infections such as mastitis, central (subareolar) breast abscesses, or advanced duct ectasia.[1]

Cancer

The types of discharge most often associated with cancer are the watery, serous, serosanguineous, and sanguineous types.[1,7] Nipple discharge originating from underlying malignancy occurs at a rate of 2.5% to 13.3%.[1,2] Carcinoma may be more likely in the presence of an associated mass. Seltzer et al[8] and Leis et al[1] found an approximately 11.8% incidence of carcinoma in patients with discharge and *no* palpable mass. Seltzer also called attention to the finding that with increasing age the incidence of carcinoma rises. In his series, 32% of patients over age 60 with spontaneous bloody discharge and no associated mass had carcinoma.[7] Murad et al[9] found the average age for malignant discharge to be one decade older than that for benign lesions (55 to 60 versus 45 years of age). Chaudary et al[10] found that 5.9% (16 of 270) patients with nipple discharge and no associated lump or mass to have carcinoma. It can therefore, be concluded that in the absence of a mass, nipple discharge is a rare presenting symptom of cancer.[3,10]

Fibrocystic Changes

Fibrocystic changes are a common cause of nipple discharge, ranging from 36 to 42 percent of patients.[1,2,7,9] The average age is 45. Definitive diagnosis is made histologically.

Papillomatosis

Much less common are peripheral papillomas that are much smaller and more numerous than intraductal papilloma. Careful histological examination must distinguish this from papillary carcinoma.

Clinical Evaluation and Management of the Patient with Nipple Discharge

The evaluation of a patient with nipple discharge should begin with a complete history and physical examination. It is important to inquire as to when the discharge began and whether it is spontaneous or elicited only on compression of the nipple or on touching the breast, and to obtain a description of its color and consistency. A family history of breast disease should be noted. Determine if the patient is taking any medications that may be producing the discharge; if so, consideration should be given to discontinuing the medication or to reducing the dosage.[8]

The patient with bilateral, multiple-duct milky discharge (galactorrhea) may have a pituitary adenoma or other pituitary tumor. Irregular periods, amenorrhea, and/or infertility due to oligo- or anovulation are common features. The hyperprolactinemia due to the prolactin-secreting adenoma inhibits gonadotropin by inhibiting gonadotropin-releasing hormone (GnRH) release and, hence, ovulation. Not all women with galactorrhea have hyperprolactinemia; this is seen in about one-third of affected women.[11] Galactorrhea with normal serum prolactin levels is not a cause for concern. Treatment for galactorrhea in this circumstance is necessary only if the secretions are sexually, cosmetically, or socially bothersome.[11] The drug of choice for galactorrhea, as for hyperprolactinemia secondary to pituitary adenoma, is bromocriptine (Parlodel). The starting dose is 2.5 mg daily, increased to twice daily after 1 week if well tolerated. If galactorrhea is confirmed or discovered on physical examination, a serum prolactin level is indicated. If the serum prolactin level is elevated or if there are associated headaches or visual disturbance, a CT scan or an MRI scan is needed. Speroff et al[11] deem a coned-down view of the sella turcica as adequate in the patient whose serum prolactin level is less than 100 and who shows no symptoms of sellar enlargement. Response to treatment with bromocriptine is good, with complete cessation of galactorrhea in 50% to 60% of patients in an average time of 12.7 weeks and 75% resolution achieved in 6.4 weeks.[11]

The physical examination of the breasts aims to identify the source and quality of the nipple discharge. The discharge can usually be expressed from the nipple by palpation and mild pressure in the periareolar area. One should observe whether the discharge comes from one or several points on the nipple; each quadrant can be examined for more precise localization of the involved area. Routine cytologic examination of the discharge is not generally helpful in light of its low accuracy.[12,13] Examination for occult blood by Hemostix was advocated by Chaudary et al[10] but generally is nonspecific and will not have an impact on management. The discharge may be collected on white gauze or a glass slide and assessed for color and character.

The finding of a dominant breast mass warrants intensive investigation until a definitive histologic diagnosis is made. Fine-needle aspiration (FNA) is a safe and effective initial step in the evaluation of the breast mass. If a definitive diagnosis is not obtained by FNA, an open excision biopsy is necessary. If the breast mass is the cause of the discharge, its excision will usually resolve the problem.

Mammography is helpful in discerning the origin of the discharge. Prior to any surgical intervention, mammography is beneficial in evaluating the presence of occult lesions, particularly in older women. Occult lesions, including carcinoma, fibrocystic changes, fat necrosis, and calcification associated with duct ectasia, may be recognized. The location of any abnormality should be correlated with the quadrant from which the discharge originates.

An absence of mammographic abnormalities does not exclude the presence of cancer and should not defer a biopsy in patients with a surgically significant discharge.[1] In the study by Leis et al.[1] 7 (10.4%) of the 67 patients with a diagnosis of cancer and associated nipple discharge had no mammographic abnormality. Intraductal papilloma cannot usually be identified mammographically. Funderburk and Syphax[7] and others have found the injection of mammary ducts with contrast material (galactography) to be beneficial in patients with a nipple discharge but no palpable mass. Intraductal papillomas may be identified by this method. The duct responsible for the discharge is identified and cannulated with a fine catheter and contrast is injected. Its use has not gained widespread popularity, because it is time-consuming, not available in all centers, and often uncomfortable for the patient. The technique may be helpful in the patient with large breasts and negative mammograms in whom complete excision of the duct system is difficult.

Patients who have single-duct nipple discharge that is spontaneous and persistent and who have no associated abnormalities on physical examination or mammography should have an excision of the involved duct system. The excision should include the duct from its terminus of the nipple to its proximal branching within the breast parenchyma. The excision is performed through a radial or, more commonly, a circumareolar incision. The duct or ducts in question are identified by pressure on the nipple and areola and cannulated with a fine probe such as a lacrimal duct probe. The areola is elevated, and the entire major duct

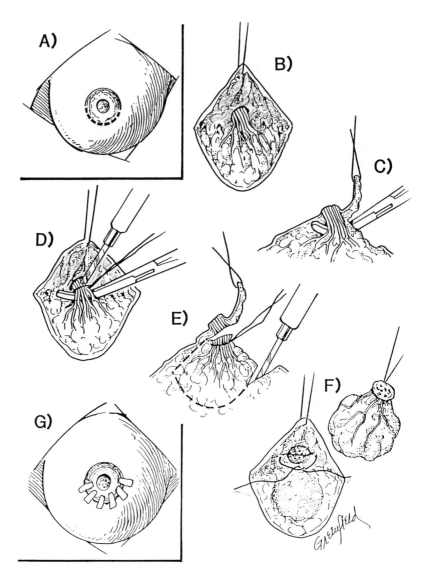

FIGURE 15.1. Excision of the major duct system. (*A*) Circumareolar incision. (*B*) Undermining areola. (*C*) Isolating lactiferous sinuses. (*D*) Transecting end of major duct system. (*E*) Removing core of surrounding breast tissue. (*F*) Specimen has been removed and cavity is being closed. (*G*) Incision closed. (From Love et al[14] by permission.)

system is isolated with sharp and blunt dissection. An inverted cone of tissue is excised, extending 3 to 5 cm into the breast (Figure 15.1).[2] The cosmetic result should be quite good. Excellent descriptions of the excision technique can be found in Urban and Egeli,[2] King and Goodson,[15] and Leis et al.[1] The surgical approach to patients with nipple discharge and an occult (mammographic) lesion depends on whether the location of the mammographic abnormality is in the same quadrant as the discharge. If the lesion is in the same quadrant, one may consider excision of the occult lesion alone; however, if the nipple discharge persists, a second surgery to excise the duct system would be required. If the lesion is not in the same quadrant and meets the criteria for excision, the lesion should be localized preoperatively and removed along with the duct system in the aforementioned technique. These procedures are performed under either local or general anesthesia, and permanent sections are awaited.

Duct ectasia does not require surgery unless there is a mass, unless the discharge changes to serous, serosanguineous, or sanguineous, or unless the discharge produces excessive discomfort or emotional trauma for the patient. Patients are advised to wash both nipples gently with cotton or gauze soaked in disinfecting solution (Phisohex) or surgical scrub (Betadine) for 2 to 5 minutes daily.[1]

Purulent discharges occasionally respond to antibiotics; however, abscess formation requires incision an drainage. Cultures and sensitivity testing may be performed and may help direct appropriate antibiotic choice. Initial antibiotic treatment should cover Staphylococcus aureus (for example, Dicloxacillin or Ciprofloxacin).

Summary

Nipple discharge is a significant clinical problem, comprising 3% to 7% of breast complaints. The majority of nipple discharge is benign. Physiologic causes of nipple discharge (excluding pregnancy and lactation) may be secondary to drug ingestion or high serum prolactin levels that lead to galactorrhea. Nonphysiologic (pathologic) nipple discharge of significance is generally spontaneous, persistent, unilateral, and uniductal. The most commonly found lesions are the intraductal papilloma, duct ectasia, fibrocystic disease, and cancer. Watery, serous, serosanguineous, and sanguineous discharges are of greater concern. Surgical exploration involves complete excision of the involved duct system.

References

1. Leis HP, Cammarata A, LaRaja RD. Nipple discharge: significance and treatment. *Breast*. 1985;11:6–12.
2. Urban JA, Egeli RA. Non-lactational nipple discharge. CA. 1978;131–140.
3. Devitt J. Maanagement of nipple discharge by clinical findings. *Am J Surg*. 1985;149:789–792.
4. Isaacs, JH. Diagnosis and treatment of benign lesions of the breast. In: Hindle WH, ed. *Breast Disease for Gynecologists*. New York: Appleton and Lange; 1990:193–201.
5. Wertheimer MD. Diagnosis of malignant disease. In: Hindle WH, ed. *Breast Disease for Gynecologists*. New York: Appleton and Lange; 1990:173–181.
6. Stehman FB. Infections and inflammations of the breast. In: Hindle WH, ed. *Breast Disease for Gynecologists*. New York: Appleton and Lange; 1990:151–165.
7. Funderburk WW, Syphax B. Evaluation of nipple discharge in benign and malignant diseases. *Cancer*, 1969;24:1290–1296.
8. Seltzer MH, Perloff LJ, Kelley RI, et al. The significance of age in patients with nipple discharge. Surg *Gynecol Obstet*. 1970;131:519–522.
9. Murad TM, Contesso G, Mouriesse H. Nipple discharge from the breast. *Ann Surg*. 1982;259–264.
10. Chaudary MA, Millis RR, Davies GC, et al. Nipple discharge: the diagnostic value of testing for occult blood. *Ann Surg*. 1982;196:651–655.
11. Speroff L, Glass RH, Kase NG. Amenorrhea. In: Speroff L, Glass RH, Kase NG, eds. *Clinical Gynecologic Endocrinology and Infertility*. 4th ed. Baltimorc, Md: Williams and Wilkins; 1989:165–211.
12. Knight DC, Lowell, DPT. Aspiration of the breast and nipple discharge cytology. *Surg Gynecol Obstet*. 1986;163:415–420.
13. Lippman ME, Lichter AS, Danforth DRN. Evaluation of the patient with nipple discharge. In: Lippman ME, Lichter AS, Danforth DRN, eds. *Diagnosis and Management of Breast Cancer*. Philadelphia, Pa: WB Saunders; 1988:75–80.
14. Love SM, Schnitt SJ, Connolly JL, et al. Benign breast disorder. In: Harns JR, Henderson IC, Hellman et al, eds. *Breast Diseases*. Philadelphia, Pa: JB Lippincott; 1987:15–53.
15. King EB, Goodson WH. Discharges and secretions of the nipple. In: Bland KI, Copeland EM, eds. *The Breast: Comprehensive Management of Benign and Malignant Diseases*. Philadelphia, Pa: WB Saunders; 1991:46–67.

16
Complications of Lactation

John M. Hobart and Cheryl L. Dorenbos-Hobart

Breastfeeding complications are primarily caused by improper feeding techniques and schedules and by inadequate patient education, support, and follow-up.[1-3] Consequently, a thorough review of breastfeeding techniques is imperative for preventive and therapeutic care.

Breast milk is the optimal nutritional source for term[4,5] and preterm[6,7,8] infants, but unfortunately breastfeeding in the United States continues to decline. In 1989, only 50% of mothers were breastfeeding at discharge from a hospital, a drop from 63% in 1984. Disappointingly, after 6 months, only 20% were still breastfeeding.[9] The reasons for this progressive decline are many, but the medical community's ambivalence toward breastfeeding is a contributing factor.[10-13]

According to Newton and others,[12,14] breastfeeding research has been flawed by weak design and biased conclusions. Many practitioners have consequently assumed that there is little difference between breastfeeding and bottle feeding where good hygiene is present.[11] In addition, many practitioners lack knowledge of breastfeeding management, exposure to successful lactating women, and time to manage lactation problems. They would prefer to direct questions and problems to a breastfeeding specialist if available.

"Quit nursing" or "Give the baby a bottle" is a frequent but unfortunate answer to breastfeeding problems.[15] Early supplementing promotes early weaning[15] and may not be in the best interest of the mother or baby. Moreover, the mother who has been forced to quit nursing may experience feelings of failure and great loss.[16] The purpose of this chapter is to give the health care provider comprehensive knowledge of normal lactation and management of lactation complications.

Antepartum Lactation Management

The key to successful breastfeeding begins in the first trimester of pregnancy with the initial obstetric visit. The advantages of breastfeeding should be discussed (Table 16.1), educational literature and resources should be provided (see resource guide), and the breasts should be examined. Expectant mothers may also be directed to a lactation consultant, breastfeeding classes in the local hospital, or a community support group such as La Leche League. Evidence has suggested that some neonatal feeding problems may be eliminated by patient education.[13] The mother's nutritional status should be evaluated for pregnancy and future lactation requirements. The initial breast examination should check for the presence of lumps, cysts, breast elasticity, and any breast or nipple malformations that may preclude or hinder successful breastfeeding or milk production. Breast masses may need further evaluation and treatment prior to delivery. The size and shape of the breasts does not correlate with the ability to produce milk. Severely

16. Complications of Lactation

TABLE 16.1. Advantages of breastfeeding.

1. Breast milk contains immune factors that protect and moderate infant disease.[17]
   Breastfed babies have fewer incidences of:
   a. respiratory illness.[18,19]
   b. gastrointestinal disease.[19,20]
   c. necrotizing enterocolitis.[6]
   d. otitis media.[21]
   e. influenza, bacteremia, and meningitis.[19]
   f. Hodgkin's disease.[22]
   g. lymphomas.[23]
   h. insulin-dependent diabetes mellitus.[24,25]
   i. urinary tract infections.[26]
2. Breast milk has a protective effect against allergies.[11,27,28]
3. Breast milk contains long chain fatty acids necessary for optimal brain development.[7,29,30]
4. Breast milk contains the correct proportions of nutrients and continues to change to keep pace with individual term and preterm infants' needs.[31,32]
5. Breast milk is easily digestible: iron is readily absorbed 49% as opposed to 10% in formula.[32,34] Nitrogenous compounds from proteins are readily absorbed as well.[35]
6. Breast milk is convenient; it need not be purchased, prepared, or warmed.
7. Breast milk cannot easily become contaminated.[35,36]
8. Breast milk enhances vaccine response.[37]
9. Breastfeeding may help prevent obesity.[38]
10. Extended breastfeeding protects babies against illness for the first 2 to 3 years of life.[39,40]
11. Breastfed babies may be advantaged in IQ development.[7,40]
12. Breastfeeding promotes strong jaw, gum, and tooth development, less malocclusion, and reduced dental caries.[41]
13. Breastfeeding promotes maternal infant bonding, eye-to-eye and skin-to-skin contact.[42]
14. Breastfeeding is less expensive than formula. Formula may cost $2 to $8 per day.
15. Breastfeeding stimulates uterine contractions and hastens involution.
16. Breastfeeding stimulates the production of "prolactin," the mothering hormone.[43]
17. Breastfeeding helps metabolize fat deposits laid down in pregnancy.[44]
18. Less incidence of ovarian cancer in the mother who has breastfed.[45]
19. Breastfeeding brings feelings of physical comfort to mother and babies.[12]

lactation unless glandular tissues, blood, or nerve supply have been interrupted.[10]

Lactation after reduction mammoplasty might be possible if the integrity of the nipple and ducts has been preserved.[46] Infant weight gain should be closely monitored after any history of breast surgery, including biopsy.[46–48] Breast texture should be assessed by palpation.[10] An inelastic breast appears fixed to underlying tissue and cannot be picked up. It is more prone to engorgement and may respond to periareolar massage[12] and early frequent nursing (see management for engorgement). The skin of an elastic breast is looser and the tissue more compliant (Fig. 16.1).

Nipple and areolar assessment identifies malformations and inversion of the nipple that may interfere with breastfeeding. Gross malformations are easily identified, but inverted asymmetric or hypoplastic breasts or breasts that do not enlarge during pregnancy may be indicative of inadequate glandular tissue that can preclude breastfeeding.[46] Augmentation mammoplasty does not usually interfere with

FIGURE 16.1. Elastic and inelastic breast.

FIGURE 16.2. Inverted nipple.

FIGURE 16.3. Nipple shell for inverted nipple.

nipples or congenital "tethering" may not be apparent. A nipple that is inverted or fixed to underlying fascia is easily identified by gently squeezing the outer edge of the areola (Fig. 16.2). A normal nipple will protrude. An inverted or "tethered" nipple will retract. Flat and inverted nipples require prenatal intervention. The most acceptable preparation is the use of a breast shell,[10,44] a hard plastic tent with holes in the center through which the nipple will be forced to protrude when a brassiere is worn (Fig. 16.3). The shell should have many holes to allow for adequate aeration to the nipple and can be worn daily for the last trimester. Initially, the shell can worn for a few hours at a time, eventually it can be used for most of the day.[44] It should also be worn for the first few days postpartum or until the baby routinely latches onto the mother's nipple. Shells may be purchased through Medela or Ameda Egnell (see resource guide).

To help the nipple protrude, postpartum nipple stretching or pulling or use of a breast pump may be done prior to feedings antepartum nipple stretching, pulling, or expression of colostrum may cause the production of oxytocin and stimulation of premature uterine contractions and has not proven beneficial.[12,49] Surgery using a purse-string suture and traction of holding sutures has been suggested by Hauben to correct inverted nipples.[50] A severely inverted nipple may be indicative of abnormal ducts.[10]

Nipple Preparation

1. Use nipple shells if applicable.
2. Avoid soap and other drying agents and abrasive towel-drying techniques.[51]
3. Avoid nipple creams. Pure vitamin E or lanolin is acceptable for dry nipples, although some women may be sensitive to these substances.[51,52]

Nutrition for Lactation

Most well-nourished mothers can successfully breastfeed without significant changes in their diets. The recommended caloric intake for breastfeeding mothers aged 15 to 23 is 2,600 kcal per day. Over 23 years of age, 2,500 kcal per day is preferred.[53] Recent studies, however, have suggested that this caloric intake may be too high for most women.[54,55] Conversely, less than 2,000 kcal per day in the first 2 months has been associated with unsuccessful breastfeeding.[55] Butte et al have suggested that after the first 2 months, 2,200 kcal per day is adequate for milk production; this level also facilitates gradual weight loss for the mother.[56] A weight loss of no more than 2 to 3 pounds per month is recommended.

All nursing mothers should continue taking prenatal vitamins. A mother nursing more than one infant should increase her caloric intake by an additional 800 cal per day.[53] For the vegetarian mother who does not use milk products, 4 μg of Vitamin B12, 12 mg of calcium, and 400 IU of vitamin D should be taken daily. She should be encouraged to select complete proteins and iodized salt.[10] Any diet containing high amounts of fiber and whole grains may impair absorption of iron, zinc, and calcium.[57] Strict vegetarian and zen-macrobiotic diets are nutritionally inadequate and contraindicated for nursing mothers.[10] Referral to a registered dietician familiar with the vegetarian diet is indicated.

Much has been said in folklore about foods to avoid while breastfeeding, but research in this area is minimal. Of the research that exists, cow's milk, volatile oils (including garlic and onions), teas, and caffeine have been shown to affect some infants.[10,58–60] Some herbal teas may contain coumarin or other potential psychogenic compounds.[59] Caffeine may cause hyperstimulation of the infant.[60] The dyes in some foods may color the appearance of breast milk.[10] Highly seasoned foods and gassy vegetables, which result in babies with gassy or colicky behavior 8 to 12 hours after maternal ingestion, have been cited in folklore as foods to avoid.[10,58] If food intolerance is considered, a trial diet, without the suspected food, should be adhered to for 1 week, followed by a rechallenge. If cow's milk is the offender, a calcium supplement is necessary for the mother.

Postpartum Lactation Management

Immediate Postpartum

Early stimulation of the breast is requisite for establishing milk supply and decreasing breast engorgement.

1. Nurse as soon after birth as possible. Immediately after delivery is optimal because the baby is in the quiet alert state and is most receptive to the new stimulus of breastfeeding.[2,61]
2. If the baby is sleepy, attempts should be made to encourage nursing for at least 5 minutes per breast.
3. If the baby is alert, at least 15 to 20 minutes of sucking at each breast is recommended every 2 to 3 hours.[3]

First 24 Hours

Many babies are too sleepy or too congested to nurse, but that should not reduce attempts to feed.

1. The average healthy newborn *needs* to feed at least 8 times in a 24-hour period, with 12 feedings still within the range of normal (hence feedings every 2 to 3 hours).
2. If the baby will not latch on within a 15-minute period, the mother should discontinue attempts and try again in $\frac{1}{2}$ to 1 hour.
3. The mother should not become overly anxious if the baby won't latch on; this is not uncommon.

Second 24 Hours

Most babies are alert enough to feed every 2 to 3 hours, sucking at least 15 to 20 minutes at each breast, with 8 to 12 feedings per day.

1. Healthy full-term infants have no need for formula or water supplements,[62] which reduce breast milk intake.[63]

TABLE 16.2. Difficult latch.

Symptoms	Causes	Treatment
Baby refuses to nurse, or nurses only briefly. Abnormal wake/sleep cycles. Abnormal suck: clicking noise, dimpling of cheeks, absent lip line, failure to remain on breast, tongue at roof of mouth. Absent or infrequent swallow Absence or decrease in urine or stool, <4–5 BMs and 6–8 diapers by day 5 of life, failure to gain weight. Sore nipples. Prolonged engorgement. Insufficient lactation.	Sleepy or congested baby first 24 hours. Normal disorganization of newborn.[68] Infant in pain; fx. clavicle, forceps bruise, vigorous suctioning, vacuum extraction (pulls on cervical vertebra).[69] Tongue-tied infants.[69] Torticollis.[69] Illness. Prematurity. Infants with neurological problems (CNS, immaturity, maldevelopment, or injury).[70] Infants with systemic problems (congenital heart disease, sepsis, hypothyroid).[70] Addicted or labor-drugged babies.[70] Abnormalities of muscles (c.p., muscular dystrophy, myasthenia, rumination).[71] Infant anatomical malformations (cleft lip/palate, tracheoesophageal anomalies).[8] Breast and nipple abnormalities. Engorgement. Inadequate milk supply.	Sterile saline drops into congested baby's nostrils aspirated with a bulb syringe may help baby breathe, but overuse should be avoided since it may increase mucous production. Utilize proper positions and feeding techniques and schedules. Switch nursing. A baby with a normal physical exam and gag reflexes may be stimulated to suck with the use of a supplemental nutrition system (see Fig. 8).[70] Continue to wear shells with flat/inverted nipples. Pump breast prior to feeding to pull nipple out. Use heat prior to feeding to soften breast. Use hand to reshape nipple. Support infant's chin during feeding. Refer impaired infants (neurologist, lactation consultant, NDT). Suck training may be necessary.[72]

2. If the baby still refuses the breast after two 15-minute trials, the mother should pump with a piston action electric breast pump (available through Medela or Ameda). The baby should then be supplemented with pumped colostrum/breast milk or infant formula.
3. The mother should be receiving help for difficult latch (see Table 16.2).

Breastfeeding Technique

Waking of Baby

The newborn must be awake enough to feed.

1. Unwrap the baby and strip to diaper.
2. Raise and lower the baby from prone to vertical position.
3. Walk fingers up spine to base of skull.

Calming of Baby

Most crying infants will calm at the breast. Some infants will not nurse if they are overly irritable, crying, and arching.

1. Wrap the infant's upper body, securing arms across chest. With the baby in sitting position, rock the baby in the direction of the ears (Fig. 16.4).
2. If the infant is arching, bring the head and neck forward to remove the tongue from the roof of the mouth.
3. Slide the wrapped baby into a football hold for feeding.

Positioning at Breast

The most common feeding positions are the football (Fig. 16.5), cradle (Fig. 16.6), or side-lying (Fig. 16.7) positions. The mother should

FIGURE 16.4. Calming position.

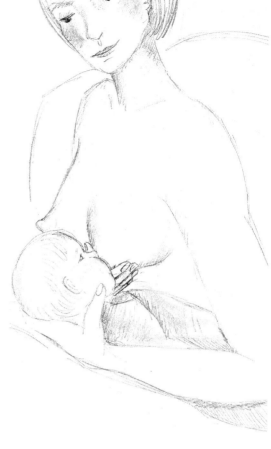

FIGURE 16.5. Football hold.

practice *All* positions before hospital discharge (Fig. 16.5). Rotation of nursing positions will reduce suction pressure on portions of the nipple and ensure complete emptying of all breast lobes.[12] In Figures 16.5 to 16.7, the appropriate positioning for successful breastfeeding is illustrated. This positioning includes:

1. Supporting the breast with fingers below and thumb above, well behind the areola.
2. Exerting no thumb pressure on top of the breast. Pressure is applied under the breast to aim the nipple down in to the baby's mouth.
3. Supporting the breast throughout the feeding to prevent the nipple and areola from slipping out of the baby's mouth.
4. Placing an index finger under the chin to support chin and stimulate sucking.
5. Aligning the baby's head and stomach tummy-to-tummy with the mother.
6. Holding the baby closely, with the baby's nose touching the breast.

Latch on Breast

A proper latch can be assured by:

1. Removing the bra to facilitate breast access.
2. Lifting the breast with the thumb and index finger well behind the areola.

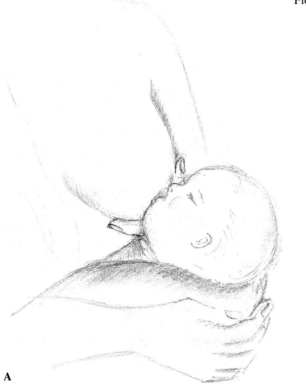

FIGURE 16.6. Cradle holds showing hand positions.

A   B

3. Tickling the baby's lips with the nipple or a finger until the baby's mouth opens wide.
4. Rapidly pulling the baby on breast.
5. Taking pressure off the top of the breast and aiming the nipple down.

If the baby comes off, repeat the above procedure. At the end of feeding, break the suction with a finger at the corner of the baby's mouth. *If the latch hurts, it is wrong!* Take the baby off by breaking the suction, and repeat the above steps. After feeding, express colostrum/breast milk and massage it gently into the areola and nipple. Air dry for 10 minutes.

Breastfeeding Follow-Up

Rule of Thumb

The baby is probably getting adequate nutrition if by day 5 of life the baby is having at

FIGURE 16.7. Side-lying nursing.

least 4 to 5 bowel movements and 6 to 8 wet diapers in 24 hours.

Early follow-up, before the traditional 2-week pediatric visit, has been recommended by Niefert and others to assess infant breastfeeding success and well-being.[1,64,65] It cannot be assumed that *all* nursing mothers are supplying their infants with adequate nutrition. Infant dehydration and malnutrition have been reported as early as 1 to 2 weeks of age.[65] Further, many mothers change feeding methods without any consultation prior to the 2-week visit.[66] Early supplementation seems to be a factor.[63,64] Niefert recommends a follow-up visit on the third to fifth day.[1]

Lactation Complications

The correct diagnosis of a breastfeeding problem should begin with an observation of the mother and baby throughout one or more feedings. The mother's positioning and the baby's latch and sucking should be observed. Frequency and quality of feedings should be acquired through a careful neonatal history of the mother and baby. Depending on observation and history, a physical assessment of the mother and baby may also be necessary. Certain lactation problems are best managed by a multidisciplinary approach, possibly utilizing an obstetrician, pediatrician, surgeon, neurologist, lactation consultant, or a neurodevelopmental therapist (NDT) for suck evaluation and treatment. An NDT is an occupational, speech, or physical therapist with oral-motor training in lactation. It is appropriate to refer to an NDT for the following cases:[67]

1. High-risk infants
2. Persistent breastfeeding problems
3. Abnormal muscle tone
4. Newborns with absent or weak reflexes
5. Absent or weak righting reflexes
6. Persistent primitive reflexes
7. Excessive or abnormal extensor tone
8. Infant who likes to "stand a lot," quality is still
9. Abnormal quality of movement
10. Persistent ankle clonus
11. Delay in developmental milestones

The most common lactation problems are outlined in Tables 16.2 through 16.8. These tables can be used as a quick reference guide for diagnosis and treatment.

Table 16.2 Difficult Latch
Table 16.3 Let-Down Failure
Table 16.4 Nipple Problems
Table 16.5 Breast Problems
  a. Engorgement
  b. Plugged Ducts
  c. Mastitis
  d. Recurrent Mastitis
  e. Thrush Mastitis
  f. Breast Abscess
  g. Galactocele
  h. Breast Rejection
Table 16.6 Milk Supply Problems
Talbe 16.7 Jaundice

Nursing under Special Circumstances

Most mothers can successfully breastfeed their infants, but some may need special guidance and monitoring. Close adherence to appropriate feeding techniques and nutritional adjustments and assessment of maternal disease and drug therapy on infant health are crucial to the management of these lactating mothers.[12] Recommended texts include *Drugs During Pregnancy and Lactation* by Briggs, Freeman, and Yaffe[80]; *Breastfeeding: A Guide for the Medical Profession* by Lawrence[10]; and *The Nursing Mother's Companion* by Huggins.[44]

Diabetic Mothers

Diabetic mothers should be encouraged to nurse; however, close monitoring of serum glucose levels to avoid ketonuria and ketonemia is required. Immediately after delivery, insulin needs may drop dramatically.[44] Actively breastfeeding diabetic women may experience a further reduction in insulin requirements.[10]

A nurse diabetic educator (RN CDE) can assist the mother in adjusting fluctuating insulin needs. Due to glucose-rich vaginal secre-

TABLE 16.3. Let-down failure.

Symptoms	Causes	Treatment

Early signs of let-down include:
  Sound of baby swallowing every couple of sucks.
  Uterine cramping or increased vaginal flow during nursing.
  Softening of breasts after feeding, leaking, dripping, or spraying of milk.
 	
Signs of let-down failure:
  No swallowing or only swallowing initially.
  Baby pulls away from breast crying.
  No cramping.
  No leaking.
	As the infant sucks, oxytocin is released contracting the mammary myoepithelial, and the milk is let down into the ductal system. During the first several weeks of breastfeeding, the let-down reflex is being developed. The nursing mother will eventually feel a "tingling" sensation as the milk lets down (a few weeks).
	A true let-down failure is extremely rare, but may follow a mom experiencing pain, or following inadequate feeding schedules.
		Find a quiet, comfortable place to nurse.
		Use relaxation techniques and focus on let-down.
		Nurse every 2–3 hours, for at least 20 min each breast.
		Apply warm compresses shortly before feeding to get milk to let down.
		Massage breast prior to nursing.
		Take pain reliever one-half hour prior to feedng if needed.
		Drink water or juice before and during feeding.
		Use correct position and latch.
		Switch breasts every 5 minutes: "switch nursing."
		Massage breast gently during feeding.
		Avoid smoking and ingestion of alcohol.
		If swallowing is not heard after a few feedings, contact physician for an oxytocic nasal spray.
		Pharmacologic treatment: Syntocinon (40 USP) nasal spray, one spray in each nostril 3 min before. Let-down will usually function normally in 1–2 days.

TABLE 16.4. Nipple problems.

Symptoms	Causes	Treatment
Sore nipples: painful, red, blistered, cracked, bleeding, bruised nipples and areola may cause the nursing mother to quit nursing or dangerously decrease the number of feedings.[1]	Not caused by failure to toughen nipples prenatally, failure to restrict nursing times postnatally, and a vigorous infant suck. Breast creams and nipple cleanings are of no actual benefit. Research has shown that sore nipples are caused from trauma from improper positioning and latch, and irritation primarily from allergic sensitivities to nipple creams.[51,52,73]	Never use shield or rubber nipple over nipple (significantly obstructs milk flow).[74] Ice to nipple prior to feeding may have numbing effect. If overly engorged, pump breast first to get nipple to become erect. Pump affected breasts (at least eight times per day) with correctly centered flange, if too sore to nurse. Express colostrum/breast milk after each feeding and smooth over nipple and areola. Air dry 5–10 min with hair dryer on low. A mild hydrocortisone cream may be ordered for nipple application after air drying such as Des Owen 0.05%.

16. Complications of Lactation

TABLE 16.4. (continued)

Symptoms	Causes	Treatment
Nipple injuries: site of the nipple injury is a clue to the incorrect latch and position being used.		
Injured nipple tip.	Damage results when nipple enters baby's mouth at an upward angle.	Aim nipple down in mouth throughout feeding, avoid pressing on top of breast, lean in to baby, not back into chair.
Injured nipple base.	Damage results from closure of baby's gums on nipple instead of back on areola.	Hold breast with thumb on top and fingers below, not covering the areola. Tickle baby's mouth with nipple until it opens wide, then rapidly pull baby fully on breast. (Do not let baby chew onto breast.) Use a pillow under arm and baby to reduce tiring and dropping of arm, as baby may slip off areola. Maintain tummy to tummy position. Break suction with index finger to end feeding. (Never pull baby off or breast away from baby.)
Sore underside of nipple.	Damage results when baby nurses with lower lip in, or from tongue-tied baby.	Pull baby's lower lip out while nursing. Apply petroleum jelly to underside of nipple prior to nursing to reduce friction. If "tongue-tied," consider having frenulum clipped to release tongue.
Irritated nipple: Red, slightly swollen, cracked, "burning nipples."	Nipple dermatitis is most commonly caused from an allergic response to nipple creams or oils. Many contain potential allergens such as lanolin (wool allergy/sensitivity) and coca butter (chocolate); some contain drying agents such as alcohol. A few women may be allergic to Vitamin E preparations or laundry detergents used to wash nursing bras. Irritated nipples may also be caused from retained moisture and skin breakdown from the continuous use of shells, which promote leaking; or wearing wet breast pads and/or pads with plastic liners.	Discontinue use of the allergen and keep breast dry. Cool compress may be applied after feedings. Mild hydrocortisone cream may be ordered. Remove breast pads when wet. Discontinue shell.
Thrush nipples Mother: Itching, flaking, burning nipples. Infant: white patches in mouth, tongue, cheeks or gums, or diaper rash that does not respond to normal treatment.	Yeast infection (monilial) in baby's mouth that has spread. Infant may have contracted thrush from the vagina during delivery; may result after administration of antibiotics to mother or baby.	Pharmacological treatment: Mother/baby require Mystatin. Nysatin ointment to nipples after each feeding for 14 days. Drop 0.5 mg Nystatin suspension into each side of the baby's mouth 4x/day/14 days. General treatment guidelines: Rinse nipples after feedings with a vinegar and water solution (1 Tbls. vinegar in 1 cup of water) and pat dry. Do not use breast milk on nipples. Apply ointment after air drying. Change nursing pads after each feeding. Pacifiers, rubber nipples, breast shells, or pump parts should be boiled for 20 min/day during treatment.

TABLE 16.5. Breast problems.

Symptoms	Causes	Treatment
Engorgement: Painful, swollen, red, lumpy breasts, swelling may extend to collar bone and under arms. Slight fever 99°–100° typically occurs 24–48 hours after birth.	Increased flow of blood and lymph to mammary glands causing swelling, and tension in surrounding tissue.	Frequent early breast feedings and peri-areolar massage in women with nonelastic breasts may help prevent. Frequent feedings 2–3 hours around the clock. Length of feeding time beyond 15–20 minutes. Moist heat prior to feedings may help soften breast (showers, warm packs) (may also aggravate situation).[44] Ice packs to breast prior to or after feedings may help reduce swelling and discomfort. Supportive, but not tight, nursing bra. Brief breast pumping may help relieve pressure prior to or in between feedings, but too much pumping may stimulate the production of milk. Hand expressing should be managed with extreme gentleness. Lean into baby to encourage nipple to stand out. This condition will resolve in 2–3 days.
Plugged ducts: An area of the breast becomes reddened, lumpy, and painful, and does not soften with nursing. It usually occurs in only one breast.	During the course of breastfeeding, a collecting milk duct may become obstructed. Plugged ducts may be caused by dried secretions on the nipple tip, missed feedings, tight and underwire bras (esp. sports bras), and tight baby carrier straps.	Frequent nursing or pumping (baby is usually most effective). Nurse affected side first, unless too painful. Can begin on least sore side first until milk lets down, then switch (baby sucks best on first side). Increase fluid intake (to point when urine is pale yellow). Moist heat 20 minutes prior to nursing. Gentle massage above sore area while nursing. Position baby with chin close to plugged duct to promote drainage. Apply ice in between feedings to reduce swelling. Check nipple for dried secretions or clogged pore and soak off. Visible plug may be removed with sterile needle. If flu-like symptoms occur such as fever, chills and achiness, a mastitis could be developing and antibiotic therapy is indicated.
Mastitis: Most common first three months; usually in one area of *one breast**; swollen, red, tender area; fever, chills, achiness, headaches, nausea and vomiting. *Bilateral mastitis is rare, but may be a sign of a Strep B infection from the baby. Penicillin is the drug of choice.[10] The baby's doctor should be contacted for assessment and treatment.	A breast infection can follow a plugged duct or a cracked nipple, and usually unilateral. The common organisms involved include Staphylococcus aureus, and Escherichia coli. Recurrent mastitis results from inadequate or delayed treatment of initial infection. (Patient needs to be seen, milk cultured, and antibiotics continued or resumed).	Nursing or pumping every 2–3 hours. Pharmacologic treatment: Dicloxacillin 250–500 mg (or Erythromycin) every 6 hours for 10-day course is the recommended treatment. Continue measures for a plugged duct. Anti-pyretics as necessary. Expect 10% recurrence without culture, 4% with culture.[75] In cases of infectious mastitis, $>10^6$ leukocytes and $>10^3$ bacteria per mL of milk, culture is recommended.[75]

TABLE 16.5. (continued)

Symptoms	Causes	Treatment
A Thrush mastitis: The breast may not look unusual, but the mother complains of hot, burning sensation when infant nurses.	This is a complication of a prolonged antibiotic therapy; it is a fungal infection of the ducts.	Nystatin cream needs to be massaged into the nipples and areola after feedings, the baby is treated with oral Nystatin as with any thrush. If the mother is prone to vaginal yeast infections, post antibiotic therapy, prophylactic use of Nystatin should be considered.[10]
Breast abscess: rare. Symptoms of mastitis do not subside within to 2 days; lump persists that does not change with nursing. Elevated fever, 39°–40°C. Fluctuant area may be difficult to palpate.[12] Frequently in upper outer quadrant.	Lactation failure. Delayed or inadequate mastitis therapy.	Surgical treatment: Surgical incision and drainage is necessary. An incision avoiding the lactiferous ducts is made followed by drainage of the abscessed cavity. The cavity must be explored, loculations broken up, and a biopsy submitted. A biopsy is essential since 10% to 15% of breast carcinomas in women under 40 are diagnosed during pregnancy and lactation.[76] Gram stain and aerobic and anaerobic cultures should be done to determine antibiotic therapy. Additional treatment: Complete emptying of the breast is necessary every 2 hours by pumping or nursing. Authorities say some nursing can be maintained unless the abscess ruptures into the ductal system and the incision and drainage tube are far enough from the areola. Other support pumping the affected breast with a piston action pump. Equally as important are rest and adequate nutrition for the mother, with the infant monitored for infection.[10]
Galactocele: It is characterized by a smooth rounded lump. Compression may cause milk to exude from the nipple.	Caused by a blockage of a milk duct.	The duct may be aspirated, but it will fill up again. Surgical removal under local anesthetic should be considered. This does not require discontinuation of breastfeeding. A firm diagnosis can be made by ultrasound or mammography. A cyst and milk in the ducts will appear the same; a tumor will be distinguishable.
Breast rejection: Occasionally nursing babies under 6 months refuse to nurse. This can last from one feeding to several days and can involve one or both breasts.	Infant rarely ready for weaning, may be reacting to foods in the mother's diet, (8–12 hours after ingestion) return of the mother's menstrual cycle, change in let-down. Illness may affect baby's desire to nurse. Babies who have upper-respiratory infections may be too congested to nurse. Others just go on temporary nursing strikes; this may be as a result of their particular stage of development. Some infants prefer one breast over another.	Reject both sides: Supplement (breast milk/formula, pump breasts when baby feeds. Change positions. Reduce external stimuli (i.e. quiet room, under blanket). Reject one side: Express milk from least preferred side prior feed to soften breast and pull out nipple. Begin with least favorite side first, when baby drowsy. Begin with favorite side until initial hunger satisfied, then slide baby to other breast, without changing baby's position. Increase milk supply in least favored breast by pumping after feedings for a few days. Use SNS, on least favored side. Accept baby's preference, but feed more often. Pump other breast, gradually decreasing. Monitor baby's weight for adequate nutrition. Examine breast for presence of abnormalities.

TABLE 16.6. Milk supply problems.

Symptoms	Causes	Treatment
Insufficient milk supply: Sign of let-down failure, less than 4–5 bowel movements in 24 hours, fussy baby, poor weight gain (less than 4–5 ounces per week over discharge), baby resting or sleeping at breast, sore nipples, dysfunctional suck.	Limitation of frequency and duration of feedings (limiting to one breast per feeding). Overuse of pacification methods: pacifiers, swings, etc. Supplementing with water or formula. Maternal illness, medication, malnutrition, smoking. Early introduction of solids or solids given prior to breastfeeding. Normal distractibility of older infant. Possible failure to thrive. Improper feeding techniques or poor suck. Use of breast shield or bottle nipple over mother's nipple. Infant anomalies, illness or injury. Prematurity. Placid baby. Dysfunctional suck.	Feed every 2 hours during day and every 3 hours at night. Rest. Increase fluid and/or caloric intake. Pump after feedings and possibly supplement with breast milk or formula if baby's weight loss warrants intervention. Stimulate baby prior to feedings and while at breast. Discontinue pacifiers, water. Reevaluate weight gain in 3–4 days. Refer to lactation consultant or NDT therapist for further evaluation and treatment. Refer to baby's doctor for further physical assessment. Evaluate maternal health and medications (not drying up milk) and make necessary changes. Alternate massage may increase milk volume and fat content.[77] Alternate massage: When baby changes from long slow mouth movements to rapid shallow mouth movements or sleeping, start massage. Massage breast near armpit and massage to nipple. Baby may stop, but then responds with long slow strokes. Massage again when baby suck changes, changing areas when area softens.
Overabundant milk supply: Chronic engorgement, leaking, spraying, baby gasps and chokes as milk lets down.	Common first 2 months. Prolonged wearing of plastic shells and excessive pumping after and in between feedings promotes milk production.	Problem is usually temporary. Pump or express milk in between feedings just to relieve pressure. Discontinue wearing plastic shells. After 2 months can offer only one breast per feeding every 2 hours. May pump unused breast slightly, if uncomfortable. Don't reduce fluid intake.

tions the diabetic mother is more prone to infections (for example, mastitis) and the baby to thrush. Diabetic infants may be initially hypoglycemic or hypocalcemic and may require admission to a special care nursery. They may also have a weak suck initially. The diabetic mother should meet with a knowledgeable registered dietician (CDE) to ensure proper caloric and nutritional intake for both mother and baby. Abrupt weaning is always contraindicated.

Thyroid Disease

The mother with thyroid disease should be followed carefully throughout pregnancy and the postpartum period. During lactation, the hypothyroid mother should continue on full replacement therapy, equivalent to 3 grains of desiccated thyroid daily. A small amount of thyroid is present in the milk, but breastfeeding is not contraindicated. The infant should be screened for hypothyroidism. For the hy-

TABLE 16.7. Jaundice.

Symptoms	Causes	Treatment
Early jaundice: associated with starvation.[78] Jaundice, 2–5 days of age: Delayed, infrequent stools, Bilirubin level < 15 mg/dL.	Infrequent feeds, ≤ 8x/24 hours water and dextrose supplements.	Stimulate stool if none first 24 hours. Early frequent breastfeeding, at least 8x/24 hours. Discontinue water or formula supplements when bilirubin approaches 15 mg; stop breastfeeding and supplement with formula for 12 hours; mother should pump with piston action pump. Photo therapy may be considered, if bilirubin is above 15 mg, previous sibling with jaundice, or jaundice after 6 days.[79]
Breast milk jaundice: extremely rare, bilirubin > 15 mg for more than 24 hours; jaundice on 4–5th day after birth, peaking at 10–14 days; history of previous infant with jaundice.	Inhibition of hepatic enzyme, glucuronyltransferase, to conjugate bilirubin.	Bilirubin > 20 mg: medical emergency; stop breastfeeding and start phototherapy; feed baby formula and water; mother should pump with a piston action pump and save milk temporarily. Bilirubin < 20 mg: draw bilirubin 2 hours after breastfeeding; discontinue breast feeding for 12 hours; supplement baby with formula; mother should pump; draw bilirubin after 12 hours; if drops > 2mg/100, it is breastmilk jaundice; if bilirubin does not drop significantly, wait 18–24 hours with bilirubin levels every 4–6 hours. If bilirubin does not drop, it is *not* breast milk jaundice. Evaluate for other causes of infant jaundice (i.e., hepatitis, biliary atresia). When bilirubin drops < 15 mg, resume breastfeeding. Expect a slight rise in bilirubin, then a steady drop. Periodic bilirubin levels need to be drawn.[79]

perthyroid mother, the drug of choice is PTU (propylthiouracil) and perhaps tapazole (methimazole). Thiouracil is contraindicated. These drugs may cause infant goiter and hypothyroid.[7,9,10]

The infant should be monitored with thyroid function tests biweekly and may need 0.125 to 0.25 mg of thyroid daily.[81,82] During any radioactive thyroid tests, the mother must not breastfeed but should pump and discard.[60]

Gastrointestinal Diseases

A lactating mother with malabsorption from Crohn's disease, ulcerative colitis, or postintestinal bypass surgery is at risk for nutritional deficiencies. Potential losses of protein, fat, fat-soluble vitamins, and minerals such as calcium, magnesium, and zinc can be quantitated by 24-hour stool collection.[83] The mother's diet may then need to be supplemented with oral or parenteral nutrition. During lactation the mother and baby need to be closely evaluated.

Hypertensive Disease

Lactating hypertensive mothers may require pharmacologic agents for adequate control. Propanolol, a commonly used antihypertensive, is considered safe because it is present only in very low levels in breast milk. The American Academy of Pediatrics considers hydralazine to be compatible with breastfeeding. Other antihypertensives, such as Methyldopa, and high-dose diuretics may suppress lactation.[10]

Epilepsy

Infants of epileptic mothers need to have drug serum levels checked at birth. Some breastfeeding mothers will initially need to pump their breasts and discard until the infant serum levels taper off.[85] Any drug therapy should help the mother remain seizure free and, at the same time, not accumulate in the breast milk. The American Academy of Pediatrics considers phenytoin and carbamazepine compatible with breastfeeding; the use of phenobarbital requires special caution since infant serum levels may become excessive and result in induced sedation.[60] Epileptic mothers need to take special precautions for infant safety should they be temporarily incapacitated with a seizure. (See *The Nursing Mother's Companion*[44] for suggestions.)

Herpes

Mothers with active herpes infections can nurse if the lesion is not on the breast and if hands are washed properly before feeding. The lesion should be covered. If the lesion is on the breast, the breasts should be pumped and the milk discarded until the lesion is gone.[86]

Working and Breastfeeding

Many mothers return to work and continue breastfeeding. Options available include:

1. Taking the baby to work
2. Finding child care at work site
3. Bringing the baby to work for feedings during breaks and lunch hour
4. Going to the baby's location during work hours
5. Part-time work and job sharing
6. Working at home, part-or full-time
7. Expressing breast milk at work during usual feeding times and giving the milk to the baby the next day
8. Weaning daytime feedings and nursing mornings and evenings ("minimal nursing")[87]

"Minimal nursing" is an alternative to total weaning. Partial breastfeeding may also give protection against infection in the day-care setting. When feedings with formula are more numerous than feedings at the breast, milk supply is diminished and has limited nutritional benefit. A decrease in urinary output will indicate inadequate intake. The baby's doctor should be contacted about the amount of desired oral intake for the baby's age and weight. Mothers who choose to express their breast milk may do so manually or with the help of a breast pump. Manual expression can be taught by the hospital nurses postpartum, a La Leche League leader, or a lactation consultant. The most effective breast pumps are the piston-active electric breast pumps rented through a Medela or Ameda/Egnell rental station. "Double-pumping" (both breasts at the same time) reduces pumping time and helps to maintain adequate milk supply by increasing prolactin levels.[88] Manual or small AC/DC pumps are used successfully by some mothers and are less expensive, but many mothers find that pumping takes longer and that the pumps are less efficient than electric pumps.[89] Breast milk can be refrigerated for up to 48 hours and frozen for 3 months in a refrigerator freezer.[44] In a zero-degree freezer, it can be stored for 6 months.[10]

Whatever choices a breastfeeding mother makes, she may wish to discuss them with successful lactating working women. In recent years La Leche League has been supportive to working mothers. See the text *Of Cradles & Careers* by La Leche League.[90]

Contraindications to Breastfeeding

There are only a few situations where breastfeeding is either absolutely or temporarily contraindicated.

Absolute Contraindications

1. Infant galactosemia, inability to tolerate lactose
2. HIV-positive mother or high-risk mother

seronegative less than 4 months;[91] HIV isolated in breast milk
3. Breast cancer; increased prolactin may advance tumor and treatment should not be delayed
4. Drug therapy with the following drugs:[12,60]
   anticancer agents
   radioactive materials
   lithium
   chloromphenical
   phenylbutazone
   atroprine
   ergot alkaloids
   bromocriptine
   phenindione

Before any drug therapy is suggested, consult a recent text such as *Drugs in Pregnancy and Lactation* by Briggs, Freeman, and Yaffe[80] or refer to the *Committee on Drugs* by the American Academy of Pediatrics.

Potential Contraindication

1. Sputum positive tuberculosis; until mother is treated for 1 week and the baby is receiving Isoniazid.[10,92]
2. Active hepatitis B or carriers; nursing is safe only if infant HBIG immunized within 1 hour after birth, then at 1 month and 6 months of age.[10,92]
3. Oral contraceptives; lowest dose estrogen should be chosen and infant weight gain should be monitored.[60,91]
4. PCB- or PBB-contaminated breast milk; women who have eaten one or more Great Lakes fish per week or live in the PBB-contaminated cattle area in Michigan should have their milk tested and make a decision based on that information.[93]

Full lactation may be impossible in the following situations[1]

Prolactin deficiency
Hypoplastic breast
Insufficient glandular tissue
After breast surgery when neurohumeral pathways have been disrupted
"Sheehan's syndrome" or hypopituitarism after postpartum hemorrhage[10]
Prior breast radiation therapy

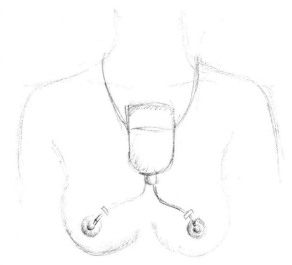

FIGURE 16.8. Supplemental Nutrition System (SNS) or Lactaid.

Nursing at breast is possible, however, using a supplemental nutrition system (SNS); see Fig. 16.8.

Relactation

Some mothers who have discontinued nursing or have begun bottle feeding choose to relactate. If the baby is under 1 week of age or if it has been less than 1 week since cessation of breast feeding, nursing every 2 hours can stimulate a full milk supply in a few days.[44] If the baby is older or if more time has passed, the degree of breast involution may preclude full milk production. An SNS should be used to achieve adequate infant nutrition. Pumping to increase milk supply and a syntocinal spray to stimulate "let down" may be used.

Nursing an Adopted Baby

Full milk production is possible only if the adoptive mother has carried a former pregnancy at least 4 months.[69] Pumping prior to the baby's arrival and after feedings at breast may be recommended. The use of an SNS is necessary to ensure adequate infant nutrition. A referral to a lactation consultant or La Leche League may be helpful.

Acknowledgment. The authors extend their appreciation to Catherine Hobart, RN, for her thoughtful illustrations, and Arlene Stach, FACCE, for her creative concepts and managerial skills. A special thank you is extended to Diane Bachrach and Veronica McKinney for their patience and assistance in manuscript preparation.

Resource Guide

Breastfeeding equipment, supplies, and electric (piston-action) breast pump rental.

Medela

Medela, Inc.
P.O. Box 660
McHenry, IL 60051
800-435-8316
815-363-1166
FAX: 815-363-1246
Breast Pump Station Info:
800-Tell You
Breastfeeding hotline,
electric pump rental,
nipple shells,
manual pumps,
AC/DC pumps,
breastfeeding video,
reprints,
S.N.S.

Ameda/Egnell

Ameda/Egnell, Inc.
765 Industrial Drive
Cary, IL 60013
800-323-5750
708-639-2900
FAX: 708-639-7895
Electric pump rental,
manual pumps,
battery pumps.

Lactaid

Lact-Aid Int.
P.O. Box 1066
Athens, TN 37303
800-228-1933
615-744-9090
Supplemental nursing device.

Lactation Consultants

ILCA

International Lactation
Consultant Association
201 Brown Avenue
Evanston, IL 60202
Professional organization of lactation consultants;
breastfeeding support and problem consultation.

UCLA

Lactation Training Program
UCLA Extension
Department of Health Sciences
10995 LeConte Avenue
Room #614
Los Angeles, CA 90024
Training course for certification.

Neurodevelopmental Treatment

NDT

Neuro-Developmental Treatment Association
Executive Secretary
P.O. Box 14613
Chicago, IL 60614
For directory of NDTs.

NDT Association, Inc.
P.O. Box 70
Oak Park, IL 60303
708-386-2454

Dr. Karel & Berta Bobath
c/o Western Cerebral Palsy Center
London, England NW3-5RN

Breastfeeding Support and Education

La Leche League

La Leche League International
916 Minneapolis Avenue

Franklin Park, IL 60131
800-La Leche (9:00 A.M. to 3:00 P.M. CST)

Preparation classes, breastfeeding help, local chapter information, reprints (breast massage and hand expression), books.

Nursing Mothers Counsel

Nursing Mothers Counsel, Inc.
P.O. Box 50063
Palo Alto, CA 94303
415-591-6680

Breastfeeding help; call for local phone number.

Well Start

Well Start
San Diego Lactation Program
P.O. Box 87549
4062 First Avenue
San Diego, CA 92138
619-295-5193
Breastfeeding helpline.

A.S.P.O./Lamaze

American Society for Psychoprophylaxis in Obstetrics, Inc.
800-368-4404
202-857-1128

Certified Lamaze instructors (knowledge of area resources); childbirth educator training programs.

Bookstore:
A.S.P.O./Lamaze Bookstore
2931 S. Sepulveda Blvd.
Suite F
Los Angeles, CA 90064
213-479-8669

ICEA

International Childbirth Education Association, Inc.
Box 20048
Minneapolis, MN 55420
800-823-4815
800-752-4249
Bookcenter: 612-854-8660

Childbirth educators, training programs, educational literature, books.

Childbirth Graphics

Childbirth Graphics, Ltd.
1210 Culvar Road
Rochester, NY 14609-5454
716-482-7940

Request catalogue; many pamphlets available for breastfeeding.

Visual Aids/Written Materials

Geddes Productions (Kitty Frantz, R.N., CPNP)
10546 McVine Avenue
Sunland, CA 91040

Telephone Support for Professionals

Breastfeeding & Human Lactation Study Center
University of Rochester
School of Medicine and Dentistry
Box 777
Rochester, NY 14642
716-275-0088

References

1. Neifert M, Seacat J. A guide to successful breastfeeding. *Contempt Pediatr*. 1986;3:2–14.
2. Klaus M. The frequency of suckling: a neglected but essential ingredient of breastfeeding. *Obstet Gynecol Clin North Am*. 1987;14:623–633.
3. DeCarvalho M, Robertson S, Freedman A, et al. Effects of frequent breast-feeding on early milk production and infant weight gain. *Pediatrics*. 1983;72:307–311.
4. American Academy of Pediatrics policy statement based on task force report: The Promotion of Breastfeeding. *Pediatrics*. 1982;69:654–661.
5. Nutrition Committee of the Canadian Pediatric Society and the Committee on Nutrition of the American Academy of Pediatrics. Breastfeeding: a commentary in celebration of the International Year of the Child, 1979. *Pediatrics*. 1978;62:591–601.

6. Lucas A, Cole T. Breastmilk and neonatal enterocolitis. *Lancet*. 1990;336:1519–23.
7. Lucas A, Marley R, Cole T, et al. Breastmilk and subsequent intelligence quotient in children born preterm. *Lancet*. 1992;339:261–264.
8. Meir P, Anderson G. Breastfeeding behavior of small preterm infants. *MCN*. 1985;396.
9. Ryan A, Rush D, Krieger F, et al. Recent declines in breastfeeding in the United States, 1984–1989. *Pediatrics*. 1991;88:719–727.
10. Lawrence R. Breastfeeding: *A Guide for the Medical Profession*. 3rd ed. St. Louis, Mo: CV Mosbey, 1989.
11. Minchin M. Infant formula: a mass uncontrolled trial in perinatal care. *Birth*. 1987;14:209–218.
12. Newton E. Lactation and its disorders. In: Mitchell G, Bassett L, eds. *The Female Breast and Its Disorders*. Baltimore, Md: Williams and Wilkins; 1990:45–47.
13. Yamauchi Y, Yamanouchi I. The relationship between rooming-in and breastfeeding variables. *Acta Paediatr Scand*. 1990;79:1017–22.
14. Kovar M, Serdula M, Marks J, et al. Review of the epidemiologic evidence for an association between infant feedings and infant health. *Pediatrics*. 1984;74(S):615–638.
15. Lawrence R. Practices and attitudes toward breastfeeding among medical professionals. *Pediatrics*. 1982;70:912–920.
16. Richards M. Breastfeeding and the mother-infant relationship. *Acta Paediatr Scand Suppl*. 1982;299:33.
17. Narayan I, Murthy N, Prakesh K, et al. Randomized controlled trial of effect of raw and holder pasteurized human milk and of formula supplements on incidence of neonatal infection. *Lancet*. 1984;1111–1113.
18. Wright A, Holberg C, Martinez F, et al. Breastfeeding and low respiratory tract illness in the first year of life. *Br Med J*. 1989;299:946–9.
19. Cunningham A, Jelliffe D, Jelliffe E. Breastfeeding and health in the 1980s: a global epidemiologic review. *J Pediatr*. 1991;118:659–66.
20. Ho M, Glass R, Pinsky P, et al. Diarrheal deaths in American Children. *JAMA*. 1988;260:3281–5.
21. Saarinen U. Prolonged breastfeeding an prophylaxis for recurrent otitis media. *Acta Paediatr Scand*. 1982;71:567–571.
22. Schwartzbaun J. An exploratory study of environmental and medical factors potentially related to childhood cancer. *Med Pediatr Onc*. 1991;19(2):115–121.
23. Davis M, Savitz D, Graubard B. Infant feeding and childhood cancer. *Lancet*. 1988;2:365–8.
24. Borch-Johnson K, Mandrup-Paulsen T, Bachau-Christiansen B, et al. Relation between breastfeeding and incidence rates of insulin dependen diabetes mellitus. *Lancet*. 1984;2:1083–86.
25. Mayer E, Hammen R, Gay E, et al. Reduced risk of IDDM among breastfed children: The Colorado IDDM Registry. *Diabetes*. 1988;37:1625–1632.
26. Pisacane A, Grazcano L, Mazzarella G, et al. Breastfeeding and urinary tract infection. *J Pediatric*. 1992;120:87–89.
27. Businco L. Prevention of atopic disease in "at-risk newborns" by prolonged breastfeeding. *Ann Allergy*. 1983;51:691–698.
28. Chandra R, Puri S, Cheem P. Predictive value of cord blood lgE in the development for atopic disease and the role of breastfeeding in its prevention. *Clin Allergy*. 1985;15:517–522.
29. Clandinin M, Chappell J, Van Aerde E. Requirements of newborn infants for long-chain polyunsaturated fatty acids. *Acta Paediatr Scand*. 1989;351(suppl):63–71.
30. Sturman J, Rassin D, Gaull G. Taurine in development. *Life Science*. 1977;21:1–22.
31. Harzer G, Houg M, Dieterich I, et al. Changing patterns of breast milk lipids in the course of lactation and during the day. *Am J Clin Nutrition*. 1983;37:612–621.
32. Lemons J, Maye L, Holl D, et al. Differences in the composition of preterm and term milk during early lactation. *Pediatr Res*. 1982;16:113–117.
33. Bullen J. Bacteriostatic systems in human milk. In: Wilkinson AW, ed. *Immunology of Infant Feeding*. New York: Plenum Press, 1981.
34. Lucas A, Blackburn A, Ansley-Green A, et al. Breast vs. bottle: endocrine responses are different with formula feeding. *Lancet*. 1980;1:1267–1269.
35. Coursin D. Convulsive seizures in infants with pyridoxine deficient diets. *JAMA*. 1954;154:406–408.
36. Spak C, Berg V, Ekstrand J. Renal clearance of fluoride in children and adolescents. *Pediatrics*. 1985;75:575–579.
37. Hahn-Zoric M, Fulconis F, Minoli I, et al. Antibody responses to parenteral and oral vaccines are impaired by conventional and low protein formulas as compared to breastfeeding. *Acta Paediatr Scand*. 1990;79:1137–1142.

38. Kramer M. Do breastfeeding and delayed introduction of solid food protect against subsequent obesity? *J Pediatr.* 1981;98:883–887.
39. Prentice A. Breastfeeding and the older infant. *Acta Paediatr Scan.* 1991;374(suppl):78–88.
40. Morley R, Cole T, Lucas P, et al. Mother's choice to provide breast milk and developmental outcome. *Arch Dis Child.* 1988;63:1382–85.
41. Labbak M. Breastfed babies have better teeth. *Parents.* 1988;1.
42. Klaus M, Kennell J. *Parent-Infant Bonding.* 2nd ed. St. Louis, Mo: CV Mosby; 1982.
43. Aono T, Shioji T, Shoda T, et al. The initiation of human lactation and prolactin response to suckling. *J Clin Endocrinol Metab.* 1977; 44:1101.
44. Huggins K. *The Nursing Mother's Companion.* Rev ed. Boston: The Harvard Common Press; 1990.
45. The Cancer and Steroid Hormone Study of the Center for Disease Control and the National Institute of Child Health Development. The reduction in risk of ovarian cancer associated with oral-contraceptive use. *N Engl J Med.* 1987;316:650–655.
46. Neifert M, Seacat J, Jobe W. Lactation failure due to inadequate glandular development of the breast. *Pediatrics.* 1985;76:823.
47. Day T. Unilateral failure of lactation after breast biopsy. *J Fam Pract.* 1986;23:161–162.
48. King R, Welch J, Martin J, et al. Carcinoma of the breast associated with pregnancy. *Surg Gynecol Obstet.* 1985;160:228.
49. Lincoln D, Paisely A. Neuroendocrine control of milk ejection. *J Reprod Fert.* 1982;65:271.
50. Hauben D, Mohler D. A simple method for the correction of the inverted nipple. *Plast Reconstr Surg.* 1983;71:556.
51. Esperance M. Pain or pleasure: The dilemma of early breastfeeding. *Birth and Fam J.* 1980;7:21.
52. Riordan J. The effectiveness of topical agents in reducing nipple soreness of breastfeeding mothers. *J Hum Lact.* 1985;1:36–40.
53. Worthington-Roberts B. Lactation, human milk, and nutritional considerations. In: Worthington-Roberts B, Vermeersch J, Williams SR, eds. *Nutrition in Pregnancy and Lactation.* 4th ed. St. Louis, Mo: CV Mosby Co; 1988.
54. Van Raaj J, Shank C, Vermaat-Miedema, et al. Energy cost of lactation and energy balances of well-nourished Dutch lactating women: reappraisal of the extra energy requirements of lactation. *Am J Clin Nutr.* 1991;53:612–619.
55. Whichelow M. Success and failure of breastfeeding in relation to energy intake. *Proc Nutr Soc.* 1975;33:62A.
56. Butte N, Garza C, Stuff J, et al. Effect of maternal diet and body composition on nutritional performance. *Am J Clin Nutr.* 1984;39:296.
57. Eggert J, Auerbach K, Rayburn W. Nutrition and lactation. In: Dilts P, Sciarra J, eds. *Gynecology and Obstetrics.* Vol 2, rev ed. Philadelphia, Pa: JB Lippincott Co; 1992:1–14.
58. Clyne P, Kulcaycki A. Human breast milk contains bovine lgG. Relationship to infant colic? *Pediatrics.* 1991;87:439–444.
59. Siegal R. Herbal intoxication: psychoactive effects from herbal cigarettes, teas, and capsules. *JAMA.* 1976;263:473.
60. Committee on Drugs, American Academy of Pediatrics. Transfer of drugs and other chemicals into human milk. *Pediatrics.* 1989;84:924–936.
61. Brazelton T, School M, Robey J. Visual responses in the newborn. *Pediatrics.* 1966;37:284–290.
62. Nylander G, Lindeman R, Helsing E, et al. Unsupplemental breastfeeding in the maternity ward. *Acta Obst Scand.* 1991;70:205–209.
63. Sachdev H, Krishna J, Puri R, et al. Water supplementation in exclusively breastfed infants during summer in the tropics. *Lancet.* 1991;337:929–933.
64. Laughlin H, Clapp-Channing N, Gehlbach S, et al. Early termination of breastfeeding: identifying those at risk. *Pediatrics.* 1985;75:3.
65. Rowland T, Zori R, Lafleur W, et al. Malnutrition and hypernatremic dehydration in breastfed infants. *JAMA.* 1982;247:1016–1017.
66. Ferrris A, McCave L, Allen L, et al. Biological and sociocultural determinate of successful lactation among women in eastern Connecticut. *J Amer Diet Assoc.* 1987;316:21.
67. McDonald M. Neuro-developmental treatment approach (NDT). Remarks at conference on The Development of Normal Infant Suckling Skill. Schaumburg, Ill: UCLA Extension, June 10, 1989.
68. Braun M, Palmer M. A pilot study of oral-motor dysfunction in "atrisk" infants. *Phys Occup Ther in Pediatr.* 1985;5:13–25.
69. Franz K. Clinical aspects of suckling problems: tonic bite vs. tight mouth, tongue problems. Remarks at conference on The Development of Normal Infant Suckling. Schaumburg, Ill: UCLA Extension June 10, 1989.
70. Coulter-McBride M, Coulter-Danner S. Suck-

ing disorders in neurologically impaired infants. *Clin in Perinat.* 1987;14:109–130.
71. Illingsworth R. Sucking and swallowing difficulties in infancy: diagnostic problems of dysphagia. *Arch Dis Childh.* 1969;44:655–668.
72. Marmet C, Shell E. Training neonates to suck correctly. *MCN.* 1984;9:291–297.
73. Hewat R, Ellis D. A comparison of the effectiveness of two methods of nipple care. *Birth.* 1987;14:1.
74. Woolridge M, Baum J, Drewett R. Effect of traditional and of a new nipple shield on sucking patterns and milk flow. *Early Hum Dev.* 1980;4:220–227.
75. Thomsen A, Espersen T, Maigaard S. Course and treatment of milk stasis, non-infectious inflammation of the breast, and infectious mastitis in nursing women. *Am J Obstet Gynecol.* 1984;149:492–495.
76. Isada N, Grossman J. Perinatal infection. In: Gabbe S, Niebyl J, Simpson J, eds. *Obstetrics: Normal and Problem Pregnancies.* New York: Churchill-Livingston; 1986:1038.
77. Conn-Stutt P, Carlson-Bowles B, Morman G. The effects of breast massage on volume and fat content of human milk. *Genesis.* 1988;10:22–25.
78. Gartner L, Lee K. Effect of starvation and milk feedings on intestinal bilirubin absorption. *Pediatr Res.* 1980;14:498.
79. Garnter L, Auerbach K. Breast milk and breastfeeding jaundice. *Acta Pediatr.* 1987;34:249.
80. Briggs G, Freeman R, Yaffe S, eds. Drugs in Pregnancy and Lactation. *Baltimore,* Md: Williams and Wilkins; 1990:468.
81. Cooper D. Antithyroid drugs: to breastfeed or not to breastfeed. *Am J Obstet Gynecol.* 1987;157:234–235.
82. Lamberg G, Llkonen E, Asterlund K, et al. Antithyroid treatment of maternal hyperthyroidism during lactation. *Clin Endocrinol.* 1984;21:81.
83. Mountford P, Coakley A. Guidelines for breastfeeding following maternal radiopharmaceutical administration. *J Nucl Med Commun.* 1986;7:399–401.
84. Schneider H, Anderson C, Cousin D. *Nurtirional Support of Medical Practice.* Hagerstown: Harper & Row, 1977.
85. Kaneko S, Suzucki K, Soto T, et al. The problems of antiepileptic medications in the neonatal period: is breastfeeding advisable? In: Janz D, et al, eds. *Epilepsy, Pregnancy, and the Child.* New York: Raven Press, 1982.
86. Quinn P, Lofvera J. Maternal herpetic breast infection: another hazard of neonatal herpes simplex. *Med J Aust.* 1978;2:411.
87. Morse J, Harrison M, Prowse M. Minimal breastfeeding. *JOGNN.* 1986;15:333–344.
88. Niefert M, Seacat J. Milk yield and prolactin risk with simultaneous breast pumping. Presented at the Annual Meeting of the Ambulatory Pediatric Association. Washington, DC; May 1985.
89. Johnson C. An evaluation of breast pumps currently available on the American market. *Clin Pediatr.* 1983;22:40.
90. Lawman K. *Of Cradles and Careers: A Guide to Reshaping Your Job to Include a Baby in Your Life.* Franklin Park, Ill: La Leche League International, 1984.
91. Stiehm E, Vink P. Transmission of human immuno deficiency virus infections by breastfeeding. *J Pediatr.* 1991;118:410–412.
92. Committee on Infectious Disease. Report of the Committee Red Bood, 21st ed. Elk Grove Village, Ill: Academy of Pediatrics; 1988.
93. Yakushiji T, Watanabe I, Kuwabara K, et al. Postnatal transfer of PCB's from exposed mothers to their babies: influence of breastfeeding. *Arch Environ Health.* 1984;39:368.

17
Female Sexual Problems and the Gynecologist

Domeena C. Renshaw

The age old cloak of ignorance and mystery around female sexuality has yielded to scientific study. Sexual dysfunctions in women are not only understood but treatable by informed physicians today. Lesbian issues are covered in another chapter. Solo and conjoint heterosexual couple management are detailed here.

History

Gynecologists are women's front-line sexual consultants and have been since the formation of the specialty. Although sexual enjoyment is not necessarily associated with parity, the early researchers of sexual problems were male gynecologists in the field of infertility. Robert Dickinson, M.D., of the Cleveland Clinic, a gynecological pioneer, noted how great a role sexual ignorance played in infertility and was the first to do couples' counseling. To educate both husband and wife in anatomy and in coital techniques, he molded wax into models of the genitals, including a sagittal section of the female pelvis. Dickinson also inspired and advised Alfred Kinsey, Ph.D. Kinsey's classic sociological sex research included interviews of 8,000 men in the 1940s and 12,000 women in the 1950s. By 1950, Joseph Wolpe, a Philadelphia psychiatrist, had evolved "reciprocal inhibition," a systematic relaxation technique to desensitize anorgasmic women.

William Masters, M.D., a gynecologist from St. Louis, was also impressed by the many sexual problems of infertile couples.[1] He first studied his own theories about the cause of sexual symptoms by careful laboratory observations. Then, for over 5 years he documented on film the sexual performance of 135 couples (men and women aged 19 to 50) in solo and coital sexual activity.[1] Next Masters developed these findings into a successful technique of brief (2-week) treatment of sexual dysfunctions of both men and women.[2] In 1970 this was a revolutionary breakthrough because no other help was available. Subsequently, many modifications of the Masters-Johnson 14-day sex therapy have been satisfactorily applied in office practice.[3]

Morbidity

How widespread are sexual doubts or symptoms? Sexual problems perplex every boy, girl, teen, and adult. "Am I sexually normal?" is perhaps the most common concern about the sexual self, feelings, fantasies, and behaviors. Few youngsters find answers and information in the home. Most learn haphazardly from peers. Only rare school programs and textbooks provide an accurate sex education beyond explaining the menstrual cycle. Many sexual innacuracies and myths are learned from peers, soap operas, X-rated magazines, cable TV, or videocasettes. A new set of concerns may result: "What's wrong with me?" The sensitive gynecologist can play a valuable role to a confused patient and sort out fact, fantasy, and exaggeration. On the one hand, basic accurate sexual education and medical information are reassuring and may be all that is

FIGURE 17.1. Coital graph.

Male	Female
A Impotence	E Female coitus
B Partial erections	F Arousal, inhibition, pelvic congestion prolonged resolution
C Premature ejaculation	G As F with distraction and loss of arousal
D Average mate coitus	H Prolonged plateau/resolution/no climax
E Sustained male coitus	I As E with multiple orgasms
X Point of ejaculatory inevitability	

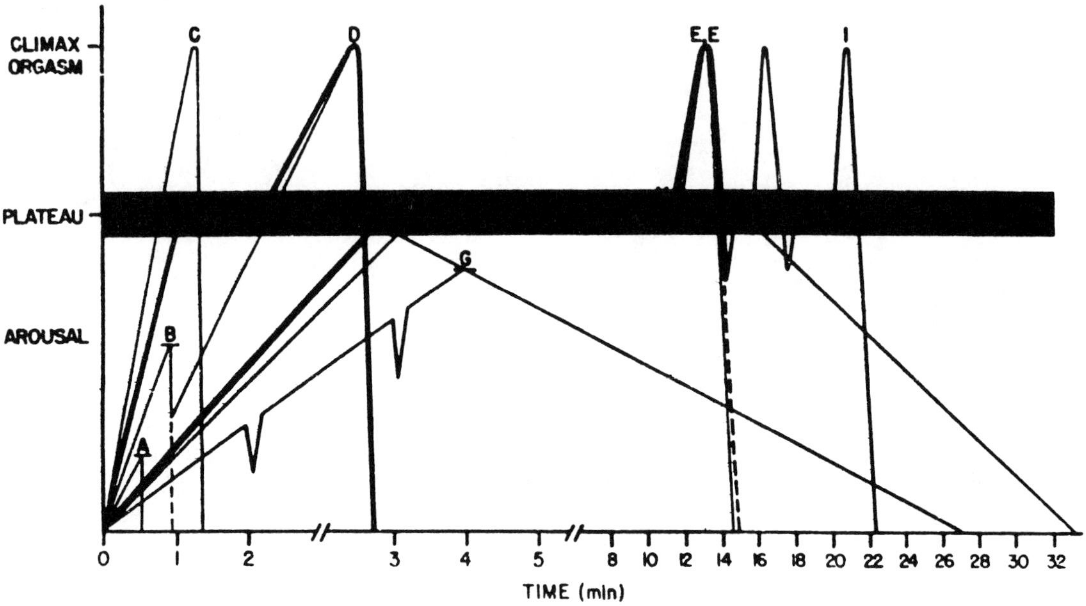

required. On the other hand, sexual dysfunctions may exist in an estimated 50% of marriages. What are they?

Sexual dysfunctions are impaired, incomplete, or absent expressions of normally recurrent sexual desires and responses. When difficulties with pleasurable climatic resolution of appropriate sexual arousal occur, dysfunctions may be accepted as transient or, when there is subjective concern or discomfort, they may become problematic. In some cases, partner dissatisfaction may precipitate awareness of a dysfunction, for example, premature ejaculation, anorgasmia, sexual apathy, or disinterest.

One definition of orgasm, at any age, from in utero to the senium, is a build-up of vasoneuromuscular general and genital tensions. Temperature, heart rate, breathing, and blood pressure all increase to a peak, the penis or clitoris becomes erect, the vagina lubricates, and then there is a discharge of tensions with a few tonic-clonic muscle contractions of all the large and small muscles of the body (including the pubococcygeus encircling the lower vagina). A return to the pre-excitement state then follows, with lowered heart rate, breathing, blood pressure, and total muscle relaxation plus a sense of well-being. The stimulus for orgasm may be the self (masturbation or fantasy), another person (same or opposite sex), or another species (animal).

The phases or stages of the sexual responses are similar for males and females, but the average timing of the cycle is much longer for women (see Fig. 17.1). This information allows the anorgasmic woman to be relaxed and accepting as she learns how her own responses occur. The timing difference between the male and female sexual responses explains why only

20% to 30% of women attain a coital climax. Women have been mislabeled as "frigid" because of a normal physiological difference. This knowledge can be a relief to many couples, since at times a man blames himself for being a poor lover or for having a penis he thinks is too small.

The clitoris is the analog of the penis and has lateral crural roots attached to the pelvic arch were the periosteum is rich in vibratory end organs. These end organs respond well to a pulsating shower or a vibrator, and an intense orgasm may result. Unlike the penis, the clitoris has neither reproductive nor urinary functions; it is there for sexual arousal only. This basic fact is not always realized. A climax always involves both the clitoris and the vagina, whether the stimulus is on the clitoris directly or indirectly through a pulley action of the labial attachments stimulating the clitoris and crurae during the thrusting movements of coitus. The subjective sensations will, of course, differ. With coitus and partner closeness the emotional satisfaction may be optimal. Some women "let go" more intensely when alone. Nonetheless, the objective physiological end result of a climax is total relaxation whatever the mode: coital, oral, manual, or with a vibrator. No one climax is "better or worse," but different. Exploring and understanding her own sexual responses is normal learning and release for every woman, but literally thousands of women need not only permission but direction from the physician to do so. The chart should so reflect: cc: Anorgasmia Rx: relaxed masturbation, recommend alternate days. Return p.r.n. Reading suggested.

All people may be born with similar genital apparatus, but capacities to be aroused and to respond differ. This latter capacity depends on intact brain, spinal cord, and peripheral and autonomic nerves, plus muscles, blood vessels, and end organs. Each woman's personality, early and current life experiences, and response to internal and external erotic stimuli (pleasurable, painful, or conflictual) will influence her response to her partner each day.

Is it possible that a woman in the 1990s is unsure if she is orgasmic? Yes, especially if she expects a different or excessive body response. Her sex education may be marginal. The gynecologist's explicit questions can be educative: During loveplay or foreplay what happens to your breathing? To your heart rate? To your body muscles? Do you feel a change in the clitoris? In the vagina? Do you feel small contractions involuntarily in the whole body and lower vagina, and then total body relaxation? If she is uncertain, suggest she check at home and read one of the self-help books listed at the end of this chapter. Follow up on her next scheduled visit. This approach provides her with accurate medical information and gives dignity to her question.

Do women have orgasms during sleep? Yes; during rapid eye movement (REM) sleep both sexes, from infancy to late age, may have normal sexual arousal and sexual dreams, accompanied by rapid breathing, increased heart rate, clitoral engorgement, vaginal lubrication (erections and occasional nocturnal emissions for men), raised blood pressure, an orgasm, and total muscle relaxation. Sometimes an individual may remember a sexual dream. Unlike the controls exerted during the day, sleep is unconscious and uninhibited. The woman who asks a question about sleep orgasms can be reassured that she is quite normal and that the prognosis is good for waking orgasms.

Sexual questions brought to a gynecologist encompass a range of topics. For example, can a woman have a climax during delivery? This is not common, but it is possible and normal. Can Sexual arousal or climax occur during breastfeeding? This is a normal, undifferentiated nipple reflex orgasmic response that may even occur with a breast pump. Should a woman have intercourse or orgasms during pregnancy? This is normal and healthy unless there is vaginal bleeding or other complications. When should intercourse resume after delivery? A pelvic exam should be done before answering the question to ensure that the genitals have healed. What is the G-spot? This is as yet histologically unsubstantiated as a separate organ or area of prostaticlike tissue and remains controversial. Grafenberg wrote that the area at the trigone of the bladder was erogenous and, when stimulated digitally, caused copious vaginal secretions and orgasms. What if a woman feels depressed and sexually apathetic? A clinical depression causes insomnia, crying, sadness, and lowering of appetites for food, for sex, and for life. If this continues for 6 weeks or more, medication is required. A single bedtime dose of tricyclic antidepressants can be used for at least 6 weeks or until the depression lifts.[4] Used in this way, hypnotics are unnecessary because they have a potential for creating dependency; also, hypnotics can chemically aggravate a depression. What are posthysterectomy sexual changes? If the ovaries remain, then the presurgery sexual pattern should resume after healing. In fact, if there were pregnancy fears before the hysterectomy, sex may improve after. Most patients benefit from reassurance that femininity and desirability will not be altered by a hysterectomy. A woman's vagina remains receptive and elastic. She will gain weight only if she overeats or is underactive. The persistent magazine articles that a woman becomes fat and sexless after the uterus is removed are myths. Why are orgasms painful with an intrauterine device (IUD)? A minority of women have complained of this, which is the result of uterine contractions on the IUD during a climax. Removal of the IUD relieves the symptom. What if a woman is disinterested or sexually apathetic? If no physical cause such as hypothyroid or prolactinemia exists, then check for anorgasmia. If a woman says, "There's nothing in sex for me," avoidance may be a factor. Also ask about masturbatory frequency and if there are other partners (selective sexual apathy). What fantasy is used during loveplay? If the answer is none, then, of course, there will be slow or no arousal.

Sexual expression is the only instinct where deliberate, sustained control or even complete suppression does not result in a threat to the life of the individual, as would happen if breathing, eating, sleeping, elimination, or circulation were cut off. Celibacy can be a normal choice. However, if the celibate person marries, then sexual avoidance may not be tolerated by the partner. Where loss of desire (hypoactive sexual desire) is the presenting complaint, one must ask if celibacy is preferred or if there is selective nonfunction (with one partner but not with another partner) or when alone (masturbation). If the sexual disinterest is global, then physical causes must be excluded. Finally, early sexual trauma or abuse must be questioned and treated.

Mortality

Only when a woman feels desperate about attaining sexual release and/or satisfying her partner may she threaten, attempt, or complete suicide. Her suicidal statement should not be trivialized and will need psychiatric consultation, particularly if divorce is threatened.

Etiology

Physical causes such as an intact hymen, vaginitis, endometriosis, congenital abnormalities, and endocrine abnormalities must always be excluded as possible causes of sexual dysfunction. If all is well physically, consider factors such as sexual ignorance; faulty sexual learning; inhibition; shame about the body; religious orthodoxy; guilt or anxiety about normal sexual fantasy or expression; fears about pregnancy, venereal diseases, AIDS, discovery of an affair, pain, or anticipated pain; chemical factors (alcohol, marijuana, street drugs, needed medications); interpersonal problems; feeling "used"; marked conflict; feeling blamed; resentment or unclear communication patterns; another lover; clinical depression (reduction of all appetites for food, sleep, and sex); deliberate manipulative control of her sexual feelings and responses; and dissociation (not conscious) of sexual feelings due to early sexual trauma. Any of these factors may combine to interfere with pleasurable sexual expression. The wife of an impotent man may doubt her sexual adequacy or his faithfulness and therefore avoid sex.

The Diagnostic and Statistical Manual, 3rd ed. (American Psychiatric Association, Washington DC, 1987, DSM III-R) lists the following sexual desire disorders.

302.71 Hypoactive Sexual Desire Disorder

A. Persistently or recurrently deficient or absent sexual fantasies and desire for sexual activity. The judgment of deficiency or absence is made by the clinician, taking into account factors that affect sexual functioning, such as age, sex, and the context of the person's life.

302.79 Sexual Aversion Disorder

A. Persistent or recurrent extreme aversion to, and avoidance of all, or almost all, genital sexual contact with a sexual partner. Sexual Arousal Disorders

302.72 Female Sexual Arousal Disorder

A. Either (1) or (2):
(1) persistent or recurrent partial or complete failure to attain or maintain the lubrication-swelling response of sexual excitement until completion of the sexual activity
(2) persistent or recurrent lack of a subjective sense of sexual excitement and pleasure in a female during sexual activity.

Orgasm Disorders

302.73 Inhibited Female Orgasm

A. Persistent or recurrent delay in, or absence of, orgasm in a female following a normal sexual excitement phase during sexual activity that the clinician judges to be adequate in focus, intensity, and duration. Some females are able to experience orgasm during noncoital clitoral stimulation but are unable to experience it during coitus in the absence of manual clitoral stimulation. In most of these females, this represents a normal variation of the female sexual response and does not justify the diagnosis of Inhibited Female Orgasm. However, in some of these females, this does represent a psychological inhibition that justifies the diagnosis. This difficult judgment is assisted by a thorough sexual evaluation, which may even require a trial of treatment.

Sexual Pain Disorders

302.76 Dyspareunia

A. Recurrent or persistent genital pain in either a male or a female before, during, or after sexual intercourse.

B. The disturbance is not caused exclusively by lack of lubrication or by vaginismus.

306.51 Vaginismus

A. Recurrent or persistent involuntary spasm of the musculature of the outer third of the vagina that interferes with coitus.

302.70 Sexual Dysfunction Not Otherwise Specified

Sexual dysfunctions that do not meet the criteria for any of the specific sexual dysfunctions.
Examples:

(1) no erotic sensation, or even complete anesthesia, despite normal physiologic components of orgasm
(2) the female analogue of premature ejaculation
(3) genital pain during masturbation

Other Sexual Disorders

302.90 Sexual Disorders Not Otherwise Specified Sexual disorders that are not classifiable in any of the previous categories. In rare instances, this category may be used concurrently with one of the specific diagnoses when both are necessary to explain or describe the clinical disturbance.
Examples:

(1) marked feelings of inadequacy concerning body habitus, size and shape of sex organs, sexual performance, or other traits related to self-imposed standards of masculinity or femininity
(2) distress about a pattern of repeated sexual conquests or other forms of non-paraphilic sexual addiction, involving a succession of people who exist only as things to be used
(3) persistent and marked distress about one's sexual orientation

Primary anorgasmia is never (by hand, mouth, vibrator, or coitus) having experienced an orgasm or climax. *Secondary or situational anorgasmia* is not global, but in some circumstances orgasm can be attained. *Hypoactive sexual desire disorder* may present itself as a variety of statements such as "I'm not interested in sex," "I hate sex," "Sex is a hassle," "I'm sick of sex, it's boring," "I get nothing out of sex, it hurts," or "I'm too exhausted for sex; by the time I've done my office job and the housework, it's just one more job." It's important to address a desire disorder because there is often a dysfunction underneath it. In some of these comments there may be dyspareunia, anorgasmia, or withdrawal from interpersonal conflicts. For fatigue, advice to sleep for 1.5 hours, set the alarm, wake, make love for 30 minutes, and then complete the night's rest may be practical. The medical conditions of Hypothyroidism and hyperprolactinemia should be ruled out.

Dyspareunia describes genital pain on intercourse. It is common but usually temporary, due to an acute local infection. Once the infection heals, the pain subsides. However, dyspareunia may be unrelated to infection; if so, a lack of lubrication or other factors must be sought.

Vaginismus is defined as involuntary spasms of the outer third of the vagina (pubococcygeus muscle) in response to attempts at vaginal penetration. It is estimated that 2% of women have the condition, which can only be diagnosed after physical and pelvic examinations have been done. There must be no physical pathology to account for the vaginal contraction. The muscle is not hypertrophied but hyperresponsive. These spasms prevent not only intercourse but also the physician's examining finger. During the pelvic exam the presence of a tight, intact hymen may explain the difficulty in penetration and the pain. Hymenotomy and healing along with encouragement and reassurance will facilitate final entry and consummation.

Some women with vaginismus admit to being fearful of pregnancy, childbirth, and penetration pain. Many due to early learning, have shame, conflict, and guilt about their genitals. Others insist that they yearn for a child and seek conception. A few will say that sex seems dirty or abhorrent, and some will tell of early sexual abuse or an attempted rape. This may represent an underlying sexual aversion beneath the hypoactive desire. For some women, marriage may have been contracted "for companionship only."

Several other women with vaginismus have admitted to being sexually aroused on a regular basis, have lubricated well, and have accepted, enjoyed, and reciprocated with oral and manual contact to orgasm for the partner, who corroborated the history.

Deliberate contraction of the pubococcygeus muscle for her own or her partner's sexual enhancement has been practiced for centuries in various cultures. In 1952 Arnold Kegel reported that regular exercises of this muscle assisted patients in controlling mild urinary incontinence and also enhanced their orgasmic responses. Kegel's exercises (about 60 successive contractions several times daily) are now part of standard sex therapy to increase vaginal awareness by constricting and relaxing the muscle voluntarily (as if to stop urine flow midstream). This reassures the woman that the control is hers, locally at the lower vagina. Through thinking of these muscles she can instruct herself to relax or contract them.

Occasionally there may be a noninflammatory cause of focal coital pain in the vaginal area, such as at the site of a healed episiotomy or a hymenal remnant that requires excision. Due to pain, or in anticipation of pain, the circular muscles of the outer or lower third of the vagina (pubococcygeus or sphincter vaginae) may contract protectively to prevent penetration. This is usually a transient response related to the actual physical problem and is not considered to be psychogenic vaginismus.

For the practicing gynecologist or family physician, *vaginitis* is one of the most frequently encountered female problems. Acute episodes cause irritation, itching, and vaginal discharge that are distressing to the patient. A variety of causative organisms, often sexually transmitted, can be identified by microscopy and culture. Treatment of choice (often not possible) includes specific medication for both the patient and her partner to prevent recurrence or the emergence of treatment-resistant organisms. Chronic vaginitis can be frustrating to both the clinician and patient after months or years; pain on intercourse (dyspareunia) may be a reported symptom.

The category of *unconsummated marriage* has not found its way into the International Classification of Diseases (ICD) or the DSM III-R but does present itself for treatment. There may be men who are virginal, unable to penetrate or to ejaculate vaginally. There are women with dyspareunia, vaginismus, or other genital problems that preclude coitus. Sometimes both partners have a dysfunction. They may be otherwise compatible and faithful for years before sex therapy is sought.[5]

Sex Therapy

Brief office counseling to manage sexual dysfunctions consists first of a general and sexual history from each partner (Table 17.1) and a physical examination of each partner. Routine urinalysis and blood tests can be done as medically indicated when pathology is noted or discovered. A routine Pap smear is done if needed.

Physical factors (for example, side effects of medications), hormonal changes (hypothyroid, prolactinemia, menopausal atrophic vaginitis), concurrent illness, fear of pain, pregnancy avoidance, and so forth must be ruled out. The physical examination is extended into an educational sexological examination, using a mirror for the woman to view her own genitals in the presence of her partner. Both should be guided by the physician to see and feel the clitoris and vaginal contractions. This experience is of inestimable therapeutic benefit to the couple. It tells them that the condition is known and understood by the physician. Also, with the lights on, the doctor authoritatively gives them permission to look at, learn about, touch, and understand their own genitals. The chart must state "Routine sexological exam done with spouse and nurse."

The physician should then educate the couple. The physician shows and names each aspect of the genitals, for example, "This is a normal-sized soft penis, which is circumcised. These testes are normal and hang asymmetri-cally. This lower outer vaginal muscle is circular and is called the pubococcygeus muscle. Just as you can tell your circular eye, mouth, and rectal muscles to contract and relax voluntarily, you can do the same here. Tighten it. Good. Now loosen or relax it." This is done by the doctor before digital examination. Point out the labia. Take the woman's index finger and place it on her clitoris. When she identifies it, take her partner's index finger and place it on her index finger. Then ask her to remove and guide her partner's finger to roll the clitoris from side to side. Many couples are not in touch with their genitals and benefit greatly from the instruction. (In nonvaginismus cases the pelvic exam and Pap smear follow in partner's presence.) At *no* time is it appropriate to stimulate the patient sexually in the office or on the exam table. This is considered sexual misconduct and may have severe legal repercussions often beyond coverage of malpractice insurance.

Clarify possible causes of the sex symptom if possible. Perceived antecedent factors (for example, shame, guilt, inhibitions, trauma, abuse, painful intercourse, or anticipatory pregnancy or delivery fear) and their corresponding symptoms should be clearly communicated to the couple. This initiates patient understanding preliminary to taking individual responsibility to resolve whatever the conflict.[3,6] A detached, helpless reply of "I don't know why" perpetuates the problem. Invite and answer questions.

Reassurance from the physician must be given when all is healthy. Then tell the couple that if they have motivation to change, all sex symptoms, even vaginismus, are highly reversible through assigned exercises at home. Such therapeutic optimism is well-founded. Muscle spasms can indeed be voluntarily relaxed.

The physician should teach relaxation exercises for dyspareunia or vaginismus. With the woman still on the examining table, in the presence of her partner, instruct her in open-mouth, slow, deep breathing, first inhaling to the count of 4, then exhaling and holding to the count of 4, to relax herself and the vaginal introitus. The physician should insert a single, well-lubricated, gloved examining fingertip as

TABLE 17.1. Explicit sexual history.

How is your sexual functioning? _____
Do you have sexual questions/concerns/worries? _____
Does your partner have a sexual problem? _____
Does your partner blame you or pressure you sexually? _____
How frequently per month do you: try intercourse? _____
How frequently per month do you: have intercourse? _____
How frequently per month do you: have vaginal pain/spasm? _____
Where? (at entrance/mid-vagina/deep inside) _____
Do you use tampons? _____
How frequently per month do you have sexual desire? _____
How frequently per month do you have sexual fantasy? _____
How frequently per month do you have sleep orgasms? _____
How frequently per month do you have waking orgasms? _____ How? _____
How frequently per month do you have oral-genital contact? _____
If uncertain whether orgasmic, answer the following:
What happens to your heart rate? _____
What happens to your breathing? _____
What happens to your clitoris? _____ Throbbing? _____
What happens to your vagina? _____ Lubrication/Contractions? _____
what happens to your body muscles? _____ Tension? _____ Contractions? _____
Relaxation afterwards? _____
How frequently per month do you masturbate each other? _____
How frequently per month do you masturbate alone? _____
How frequently per month do you use a vibrator? _____
How frequently per month do you handshower on the genitals? _____
How have you been best able to climax? _____
Have you had any sexual experiences that were negative or upsetting? _____
Details: _____
Does your partner know? _____
How do you feel about being pregnant? _____ your partner? _____
Do you have worries about pregnancy? _____ or about pain? _____
Have you ever been pregnant? _____
Had a miscarriage? _____ Discussed adoption? _____
Have you been for an infertility work-up? _____
Last pelvic examination: _____ Details: _____
Date of last menstrual period: _____ Any nonmenstrual bleeding? _____
Any vaginal discharge? _____
What books have you read about sex? _____
Do you regard yourself as sexually inhibited? _____ or open? _____
Do you regard your partner as sexually inhibited? _____ or open? _____
What do you see as your greatest sexual problem? _____
Anything else? _____
How can the physician best be of help to you? _____
Further comments: _____

she begins to exhale: "Relax. . . . In to 4 and out to 4 again." The patient is also requested to contract the pubococcygeus very tightly against the examining finger, then to relax and repeat the contraction. This is, of course, a simple illustration of the Kegel exercises. Explain this and tell her to practice at home, using her own fingers in stead of metal dilators, because this provides double sensory value (both fingers and vagina) and there is not only permission but prescription to touch her genitals. During the pelvic exam guide her own finger (lubricated) into the vagina. Comfortable insertion of a childsized or nasal speculum is possible after a gentle one-finger pelvic examination as she slowly breathes out with an open mouth. Occasionally, with a tense and resistant patient, this relaxed pelvic examination with both partners present must be repeated on later visits. Patience, time, and encouragement do help.

Finally, home exercises for all sexual symptoms should be explained. Also suggest that the partners read a sex manual. The couple should be asked to make the time (30 minutes daily) for loveplay, kissing, and body massage, but no breast or genital touching for the first 2 weeks.

For dyspareunia or vaginismus the patient is also to spend 5 to 15 minutes solo, twice daily. The door should be locked and the phone should be off the hook so that she is not disturbed. Sexual fantasy is essential. Relaxed, open-mouthed breathing with simultaneous vaginal insertion first for 2 days with one finger and then for 2 days with two fingers should be continued daily for 2 weeks. In the third, after general and genital foreplay, she can guide her partner's lubricated index finger into her vagina. In the fourth week, after these exercises with foreplay and finger insertion, her partner should lie supine, with her on top. She then inserts his soft penis, lubricated with saliva, KY jelly, or baby oil (which must also be placed on her vaginal opening) into her vagina. She must breathe out slowly to the count of 4. If the penis is erect or semierect, she may be uncomfortable. Therefore, emphasize that the penis must be soft. When a partner says he cannot be soft due to a reflex erection in proximity to her vagina, tell both that this arousal is positive and a compliment. However, prescribe (in the presence of both) that he masturbate and ejaculate first to ensure flaccidity. Then she can mount and insert the soft penis.

In all these tasks, she is told to relax as well as contract the vaginal muscles so that she may get in touch with this voluntary muscle of her body and know how it works. For all these steps, she is active and her partner passive. The trap of "You are hurting me" is thus avoided. She is now to be the active partner, assertive and responsible for finding comfortable, relaxed insertion without blaming him. If the patient is motivated to change, these home instructions can effectively reverse vaginismus of many years. Five or six 30-minute office visits spaced over 12 weeks may be needed.

Masturbatory exercises solo or side by side can attain orgasm for the anorgasmic woman. Some patients and their partners were highly pleased at the relief of the dyspareunia and/or vaginismus and attained pain-free coitus. This was "sufficient pleasure for now," one couple said. Neither wished to pursue orgasmic attainment.

For some women, fear of loss of self-control by having an orgasm may be a stressful thought. The clinician can leave an open door for them to return for further therapy, if and when they are ready.

Clinicians assist persons to attain their optimal sexual pleasure but cannot take them further or faster than they are willing or able to go. It is patient's index of sexual satisfaction, not the physician's, that must always be respected.

Prognosis

For the first time in centuries the outlook for sexual symptom reversal is optimistic. Accurate sex education and the application of home sex-therapy exercises can achieve relaxed, pleasuring, noninvasive assistance. Self-help books can extend the therapy.

Summary

Because sex clinics are few and often overloaded, sex therapy by the practicing gynecologist or family physician merits a diagnostic trial. It can be provided for women patients as solo therapy, in a group therapy instructional setting, or as couple's therapy.[3,4,6,7] There is no subspecialty of sexology.[8] This knowledge has been available for two decades and is waiting for a much wider application than in the 1980s, especially by today's informed gynecologists.

References

1. Masters W, Johnson V. *Human Sexual Response.* Boston, Mass: Little, Brown & Company; 1966.
2. Masters W, Johnson V. Human Sexual Inade-

quacy. Boston, Mass: Little, Brown & Company; 1970.
3. Renshaw DC. Relationship therapy for sex problems. *Comprehensive Therapy.* 1983;9:32–36.
4. Lief H, ed. *Sexual Problems in Medical Practice.* Monroe, WI: American Medical Association; 1981.
5. Renshaw DC. The unconsummated marriage. *MAHS.* 1989;23:74–79.
5a. Renshaw DC. Sexual problems in old age, illness, and disability. *Psychosomatics.* 1981;22:975–985.
6. Renshaw DC. Communication in marriage. *MAHS.* 1983;17:199–220.
7. King B, Camp CJ, Downey AM. *Human Sexuality Today.* Englewood Cliffs, NJ: Prentice Hall; 1991.
8. Rosensweig N, Pearsall FP. *Sex Education for the Health Professional.* New York: Grune & Stratton; 1978.

Suggested Self-Help Reading

1. Barbach LG. *For Yourself (Women).* New York; Doubleday; 1975, 1991.
2. Bloomfield H. *Making Peace with Your Parents.* New York; Random House; 1983.
3. Donovan ME, Ryan WP. *Lover Blocks.* Penguin Books; 1989.
4. Goldberg H. *The Hazards of Being Male.* New York: Nash Pub; 1976.
5. Hendriks H. *How to Get the Love You Want.* New York: Harper & Row; 1988.
6. Maltz W. *The Sexual Healing Journey.* Dunmore, Pa: HarperCollins Pub; 1991.
7. Mason T, Green-Norman V. *Making Love Again.* Chicago: Contemporary Publishers. 1988.
8. Masters W, Johnson V, Kolodny R. *Sex and Human Loving.* Boston: Little, Brown & Company; 1986.
9. Renshaw DC. *Incest: Understanding and Treatment.* Boston: Little, Brown & Company; 1984.
10. Renshaw DC. *Sex Talk for a Safe Child.* Milwaukee, Wis: AMA; 1984. (or Loyola)
11. Visher J, Visher E. *Stepfamilies: Myths and Realities.* Secaucus, NJ: Citadel; 1979.
12. Zilbergeld B. *Male Sexuality.* Boston: Little, Brown & Company; 1978.
13. *Our Bodies, Ourselves: A Book by and for Women.* New York: Simon & Schuster; 1975, 1985.

There are normal differences in how each individual reacts to the same materials. Tell patients to respect these differences and to try to understand how each family, religion, childhood experience, and so on may affect adult sexual feelings.

18
Recurrent Urinary Tract Infections

Mark J. Schacht

Urinary tract infections, particularly recurrent urinary tract infections, are an important problem for any practitioner with a significant female patient population. It has been estimated that between 2% and 5% of all primary care office visits are initiated by urinary tract infection–like symptoms.[1] Others have gone so far as to state that "urinary tract infections rank just behind the common cold as the most frequent infectious disease problem encountered in Western civilization."[2] These infections often are recurrent in nature, and the associated impact on society in terms of cost and morbidity has spurred research into the pathophysiology and treatment of this problem. The insights gained from this research have resulted in a rational and scientifically based approach to recurrent urinary tract infections. This chapter will concern itself with examining this approach, with an emphasis on its practical applications.

Definitions

The starting point of discussion of recurrent urinary tract infections in women is to arrive at an agreement on terminology that allows a common language. Several important terms and their definitions that will appear throughout this paper are listed here. Note that these definitions are widely used and are not particularly controversial.

Bacteriuria. The presence of bacteria in the urine is either due to an infection or to contamination of the specimen. It is usually due to the former when applied within this chapter.

Urinary Tract Infection. Traditionally, a urinary tract infection (UTI) is present if greater that or equal to 10^5 colony-forming units (CFU) of bacteria per mL of urine are present when the urine is cultured. These data were generated in studies performed on both pregnant and nonpregnant females with asymptomatic bacteriuria or acute pyelonephritis utilizing whole void, clean-catch specimens.[3] Recent studies involving women who are symptomatic (experiencing urgency, frequency, and/or dysuria) show that urine cultures in approximately one-third (between 20% and 40%) with eventually confirmed infections grew only 10^2 to 10^4 CFU/mL.[4,5] This means that, in women with symptoms, a midstream urine culture with a count of only 10^2 CFU/mL (that is, only 100) may have significance. Therefore a rigid application of the 10^5 CFU/mL cutoff to define a UTI may lead to significant underdiagnosis of the problem. Conversely, the 10^5 CFU/mL value may lead to an overdiagnosis of urinary tract infection. This is particularly true in women with large numbers of bacteria on their perineum in whom an inadequate sampling technique has been used, resulting in contamination of the specimen. This was identified in one of the original bacteriuria studies where a CFU/mL count of 10^5 or greater on a single culture was found to have a 20% chance of representing contamination.[6] Finally, it cannot be overemphasized that, when utilizing these urine culture values, the data were generated using whole and midstream, clean-catch specimens. If a specimen is obtained by catheterization or bladder aspiration, the presence of any bacteria at all may represent an infection.

Simple Urinary Tract Infection. Simple infections comprise approximately 90% of all UTIs. They occur in an anatomically normal urinary tract and do not result in persistent bacteriuria. Simple urinary tract infections are easily treated and resolved, but they can recur.

Complicated Urinary Tract Infection. A complicated urinary tract infection invloves the upper urinary tract (kidneys) and can result in renal damage. The attendant risk of renal failure makes these infections a significant threat to patient health. The most valuable diagnsotic test to distinguish a complicated UTI is the blood leukocyte count, which will be elevated and show a left shift.[7]

Recurrent Urinary Tract Infection. A recurrent urinary tract infection is a frequent sequential reinfection of the urinary tract, with intervening periods of sterile urine. The exact difinition is somewhat imprecise, particularly with regard to the time elapsed between infections. The reinfections tend to occur in clusters, and the periods of sterile urine are prolonged, usually 6 months or longer.[8] Recurrent urinary tract infections appear to have a distinctive distribution, with approximately 15% of patients experiencing 2 or more UTIs per year while the remaining 85% experience only 0.32 per year.[9] Looking at this in more depth, one study followed 51 infection-prone women in a standardized fashion for a median of 9 years. It was found that the women experienced an average rate of infection of 2.6 per patient-year while not on antimicrobial prophylaxis. However, rates varied widely between individual patients, with the range being 0.3 to 7.6 episodes per year.[10]

It has also been clearly shown in women with recurrent urinary tract infections that the probability of experiencing a recurrence is increased with the number of previous infections and decreased in an inverse proportion to the elapsed time between the first and second infection. In one prospective study, 28.6% of 60 women experiencing their first symptomatic urinary tract infection had a recurrence over the next 18 months, while 82.5% of 106 women who had had previous infections recurred within the same time frame.[11] Data from Stanford University found that a woman with two or more urinary tract infections within less than 6 months had a 66% chance of having another infection within the next 6 months. Significantly, the risk of recurrence remains constant whether the patient is observed, treated with antibiotics (short- or long-term therapy), or treated with antibiotic prophylaxis. Treatment only serves to alter the time frame of recurrence; once it is stopped, the patient returns to her previous susceptibility.[12] Fortunately, there is no association between uncomplicated recurrent UTI and renal scarring, hypotension, or azotemia,[13] and multiple studies have shown that, in otherwise healthy women, recurrent urinary tract infections do not cause serious morbidity often.[14,15]

Persistent Urinary Tract Infection. A persistent urinary tract infection is a sequential reinfection caused by the same orgainsm, without an interventing period of sterile urine. This reflects a failure of therapy to eradicate the infecting organism and indicates the need for further investigation as to the cause of that failure.

Incidence

Lower urinary tract infections are a significant problem in approximately 20% of the female population of the United States. It has been estimated that between 2% and 5% of all office visits to primary care physicians in the United States are prompted by UTI-like symptoms.[1] Bacteriuria has been found to affect approximately 4% of all women by young adulthood, increasing at the rate of an additional 1% to 2% per decade of age; it occurs in 10% of women at menopause and 15% in women over 65 years of age.[16] These numbers translate into the observation that the older a women is, the more likely she is to develop a urinary tract infection and a subsequent recurrence. In light of the aging population, in terms of patient morbidity and cost to society the implications of this must be recognized.

Microbiology

The majority of urinary tract infections in women are from facultative aerobes present in the bowel flora. E. coli accounted for 76% of all positive cultures in a study of 272 females with documented UTIs evaluated in an outpatient setting. Proteus species was the next most common, with 8% of all positive cultures.[17] However, more recent work now suggests that Staphylococcus saprophyticus is the second most common cause of urinary tract infections in young, sexually active women.[18] Other causative agents may include Group B streptococcus, Staphylococcus epidermidis, and Candida.[17]

Pathophysiology

The pathophysiology of urinary tract infections in women is an area into which much successful research has been done over the last few decades. Two concepts have emerged as keys to understanding how a urinary tract infection occurs. The first concept is that of bacterial adherence. The other, which is particularly important to understand recurrent infection, is that of bacterial colonization of the introitus of the vagina with normal bowel flora. The interaction of these two events is though to explain the origin of recurrent urinary tract infections in women.

Bacterial adherence describes the ability of bacteria to bind to an epithelium to which it normally cannot bind. This ability is mediated through a set of bacterial adaptations called pili or fimbriae. These are small, hairlike projections on the surface of the bacteria that attach to specific receptor sites on the patient's epithelium. Type 1 fimbriae bind to mannose or mannoselike receptors on the vaginal epithelium. P-fimbriae are distinct; they do not bind to mannose receptors but bind to epithelial cell memberane glycosphingolipid.[19-22]

Bacterial colonization of the vaginal introitus is of particular significance in understanding recurrent urinary tract infections. Several long-term studies of women with recurrent UTIs noted that bacteriuria is associated with, and usually preceded by, colonization of the vaginal introitus with fecal bacteria. additionally, women with recurrent infections show both a greater incidence and a higher density of vaginal bacterial carriage than do female controls. In other words, women who experience recurrent infections demonstrate more frequent and greater degrees of colonization than women who do not get recurrences, and this colonization seems to precede the development of a true urinary tract infection. Furthermore, women with recurrent UTIs who are between clusters of infection demonstrate a decrease in the degree of colonization.[13] This information leads to a proposed pathophysiology of urinary tract infections, particularly recurrent urinary tract infections.

Bacterial adherence and colonization of the vaginal introitus could be reasonably assumed to interact in the following manner. Bowel flora, usually E. coli, colonize the vaginal introitus by adhering to receptor sites on the vaginal epithelium. This mass of bacteria subsequently adheres to uroepithelial receptors on the urethral mucosa and eventually ascends to the bladder. Within the bladder additional mucosal receptors are present, making eradication of the bacterial innoculum difficult. Once present, the bacteria multiply, resulting in an active infection. However, the factors that allow the bacteria to adhere in some women and not others, and even at some times in an individual and not at other times, remain unclear. The theory is that some biologic change occurs in the epithelium lining the vagina and lower urinary tract, the exact nature of which is not known.[23] This change is indicated by an increased susceptibility to bacterial adherence, mediated by the pili on the bacteria. Interestingly, this increased mucosal susceptibility to bacterial adherence is also seen in the buccal mucosa.[24] The changes at all sites appear to be transient, although, as stated, the factors involved have yet to be identified. Other vaginal factors such as pH, glycogen content, or estrogen content appear to have little effect.[25]

Diagnosis

The classic diagnostic test for a urinary tract infection is the culture and sensitivity performed in a clinical laboratory setting. The serial dilution and pour-plate method is the most accurate,[23] but it is rarely performed due to its technical complexity. Most laboratories utilize calibrated loops to smear agar plates and then incubate the plates and read colony-forming units that are multiplied by the calibration factor. Antibiotic sensitivity testing is then performed. The results reported reflect attainable blood and tissue levels of antibiotic, but not the urinary levels, which may be signficantly higher.

In-office urine culture can also be performed fairly easily. One technique employs a disposable split agar plate and an eye dropper to deliver 0.1 mL of urine to each half of the agar, followed by overnight incubation. The number of colonies can then be estimated and multiplied by 10 to get a final result.[26] An even simpler technique is the dip-slide method in which commercially available agar-coated slides are dipped into the urine and incubated overnight; the colonies found the next morning are multiplied by between 100 and 200 to give an overall count. Obviously, because of the variability of the multiplier, the result produced is not as accurate as with other methods. Actual identification of the causative bacteria is by comparison with photographs, and antibiotic sensitivity testing is not performed.

The urinalysis itself is not a good way to diagnose a urinary tract infection because it requires a bacterial count greater than 30,000 bacteria per mL in either spun or unspun urine, stained or unstained, before bacteria can be seen.[26] However, there are some reports that reagent strips (dipsticks) can be of value in identifying UTIs. Evaluating 100 elderly patients, Evans et al found reagent strips positive for urinary nitrites to be 100% specific and 83% sensitive for urinary tract infections.[27] However, the urine culture and sensitivity remains the best standard.

Treatment

The goal of treatment for female urinary tract infections is to eliminate the infection while at the same time avoiding disruption of the fecal flora, holding down costs, minimizing side effects, and maintaining the normalcy of the bacterial environment at other sites of the body (for example, vaginal). This is a theoretical goal, and it is not usually completely achieved. Multiple antibiotics have been utilized to attain this goal. These include the sulfonamides, ampicillin, amoxicillin, nitrofurantoin, tetracycline, cephalexin, cephradine, trimethoprim-sulfamethoxazole, trimethoprim alone, and quinolones. All work well for uncomplicated UTIs when administered orally. How long these antibiotics should be administered is an area of some controversy. Traditionally, therapy for urinary tract infections has been for 10 to 14 days. Single dose or three-day "short course" therapy has recently become popular and has many advocates. Several studies have compared short-course therapy against 10-day treatment regimens.[28-34] One recent study comparing single-dose and 10-day therapy utilizing trimethoprim-sulfamethoxazole in women with urinary tract infections associated with acute dysuria, urgency, and frequency found that the 10-day treatment yielded a superior cure rate when evaluated at 2 weeks following treatment but that by 6 weeks, this advantage no longer existed. The authors attributed this finding to the inability of single-dose therapy to eradicate the vaginal reservoir of bacteria totally, resulting in a same-strain recurrence by 2 weeks.[35] Currently, the general consensus is that a 3-day course is superior to single-dose treatment.[36] Representative treatment regimens are listed in Table 18.1. If one elects short-course therapy and then the patient experiences a recurrence, the subsequent treatment should be for a full 2 weeks. Also, if short-course therapy is utilized, it is obligatory to obtain a repeat culture 1 to 2 days following completion of treatment to document that the infection has been eradicated.

Antibiotic prophylaxis is utilized in patients with recurrent urinary tract infections to prevent repeat infections. The aim is to suppress the recurrences without selecting for resistant

TABLE 18.1. Sample antibiotic regimens for treatment of urinary tract infections.

Sulfisoxazole: 1 gram p.o. QID for 7 to 14 days
Ampicillin: 500 mg to 1 gram p.o. QID for 7 to 14 days
Amoxicillin: 500 mg to 1 gram p.o. QID for 7 to 14 days
Nitrofurantoin: 100 mg p.o. QID for 7 to 14 days
Nitrofurantoin: 100 mg p.o. QID for 3 days
Tetracycline: 250 to 500 mg p.o. QID for 7 to 14 days
Trimethoprim-sulfamethoxazole: 1 double-strength tablet p.o. BID for 7 to 14 days
Trimethoprim-sulfamethoxazole: 1 double-strength tablet p.o. BID for 3 days
Trimethoprim: 100 mg p.o. BID for 7 to 14 days
Trimethoprim: 100 mg p.o. BID for 3 days
Cephalexin: 250 to 500 mg p.o. QID for 7 to 14 days
Cephalexin: 250 to 500 mg p.o. QID for 3 days
Cephradine: 250 to 500 mg p.o. QID for 7 to 14 days

TABLE 18.2. Sample antibiotic regimens for prophylaxis of recurrent urinary tract infections.

Nitrofurantoin: 100 mg at bedtime
Trimethoprim: 100 mg at bedtime
Trimethoprim/sulfamethoxazole: 1 single-strength tablet every other night at bedtime
Cinoxacin: 250 mg at bedtime
Norfloxacin: 200 mg at bedtime
Cephalexin: 250 mg at bedtime

strains of bacteria in the introital and fecal pathogenetic reservoirs. Medications that are used to achieve this goal include trimethoprim-sulfamethoxazole, introfurantoin, cinozocin, cephalexin, and trimethoprim alone. The prophylactic regimens are more than 90% effective in preventing recurrent infections while being utilized.[37] The patient is usually kept on antibiotic prophylaxis for 6 consecutive months. It is important to note that when removed from the prophylactic regimen, the patients revert to their pretreatment susceptibility to, and rate of recurrence of, urinary tract infections. The exact mechanism by which these regimens actually prevent recurrent infection remains unclear. Some believe it is through maintenance of a low level of antibiotic in the bladder, thereby disrupting bacterial adherence; others believe it is due to some change in the coliform pathogenetic reservoir.[23] Sample regimens are listed in Table 18.2.

Finally, it should again be noted that the antibiotic sensitivities by which therapy is directed may be somewhat misleading because a unique situation in urine culture and sensitivities reporting must be recognized. The sensitivities delivered on a specimen are based on obtainable tissue and blood levels of antibiotic; however, urine concentrations of the drug may run logarithmically higher. This must be kept in mind when choosing an antibiotic because one with fewer side effects and less cost may be just as effective as one with more side effects and greater cost despite the sensitivity report, particularly in uncomplicated UTIs.

Recurrent Urinary Tract Infection in Pregnancy

The prevalence of bacteriuria in pregnant women remains the same as that of nonpregnant women within the same age group, averaging between 4% and 6%.[12] However, the anatomic and physiologic changes accompanying pregnancy make pregnant women with bacteriuria more prone to develop a recurrence of their bacteriuria or urinary tract infection. Factors such as decreased peristalsis of the smooth muscles lining the upper urinary tract, increased urine production due to increases in renal function, and displacement and compression of the bladder by the gravid uterus all contribute to stasis and predispose to recurrent bacteriuria or urinary tract infection. This increased risk was demonstrated in a study of 148 pregnant females with bacteriuria who were treated with placebo medications; 18% developed acute pyelonephritis, 13.5% cleared spontaneously, and an overwhelming 66% experienced persistent bacteriuria.[38] Other authors have demonstrated that even when the bacteriuria/UTI is treated with antibiotics as opposed to placebo, between 16% and 27% of pregnant patients will develop a recurrence later in the pregnancy.[38,39] This bacteriuria is particularly significant in pregnant women because, in general, their incidence of acute pyelonephritis is significantly increased over that of the nonpregnant female population, and, specifically, the incidence in pregnant women with bacteriuria is even greater than that of pregnant women without bacteriuria. For example, the frequency of pyelonephritis in all pregnancies is estimated to be between 1% and 4%.[40] However, in a review of 18 papers concerning bacteriuria and pyelonephritis, Sweet found that the average reported rate of progression to pyelonephritis in pregnant women with bacteriuria to be 28%, and rates as high as 65% have been noted by some authors.[41] The pyelonephritis develops during the third timester in 60% to 75% of these patients, and of these, 10% to 20% experience a recurrent episode just before or just after delivering.[42] Fortunately, aggressive treatment of the bacteriuria can result in a decrease in the rate of pyelonephritis to only 2.9%, roughly the same as that of the general pregnant population.[41]

The larger question that remains is whether bacteriuria/UTI or pyelonephritis during pregnancy has any real effect on the fetus or the mother. In the days prior to antibiotics, pregnant patients with pyelonephritis had a high rate of prematurity and infant mortality. However, despite multiple studies, whether contemporary patients with either bacteriuria/UTI or pyelonephritis, treated with appropriate antibiotics, have the same outcome remains controversial.[41] One recent Australian study does suggest that there is a strong association between low birth weight and UTI during pregnancy and delivery.[43] As long as the issue remains unresolved, the consensus recommendation is to treat both aggressively.[44]

TABLE 18.3. Antibiotics that may be used in the treatment of bacteriuria and pyelonephritis of pregnancy.

May be utilized *anytime* during pregnancy if patient is not allergic:
Cephalosporins
Penicillins
Methenamine mandelate
May be used *under specific cirumstances* during pregnancy:
Short-acting sulfonamides—can be used during first and second trimester
Nitrofuratoin—can be used if patient does not have glucose-6-phosphate dehydrogenase deficiency

TABLE 18.4. Antibioitics to be avoided in the treatment of bacteriuria and pyelonephritis of pregnancy.

Antibiotics to be avoided throughout pregnancy:
Tetracycline—can cause acute maternal decompensation and fetal malformations, fetal tooth malformations and discoloration if used in late pregnancy
Trimethoprim—can cause fetal malformation is used in first trimester
Fluroquinolones—can cause abnormal fetal cartilage and bone development
Erthromycin estolate—can cause cholestatic jaundice in pregnant females
Chloramphenicol—can cause "gray syndrome" (fetal cardiovascular collapse)

Antibiotics to be used with caution during pregnancy:
Aminoglycosides—can cause cranial nerve 7 toxicity in fetus and oto and nephrotoxicity in pregnant females.
Nitrofurantoin—can hemolytic anemia in fetus with glucose-6-phosphate dehydrogenase deficiency
Sulfonamides (long-acting)—can cause fetal hyperbilirubinemia and kernicterus if used in third trimester

Management of Bacteriuria and Urinary Tract Infection in Pregnancy

The screening of pregnant women for bacteriuria or urinary tract infections should begin at the first antenatal visit. Midstream urine cultures should be used; bladder catheterization should be avoided. If bacteriuria is present, it should, as previously noted, be treated aggressively. A repeat culture is necessary after the completion of therapy. If reinfection with the same organism occurs, or if multiple efforts to treat the bacteriuria fail, one must suspect an anatomic abnormality or renal parenchymal infection. These patients will require close follow-up and a complete urologic evaluation after delivery.[45] If the patient develops acute pyelonephritis, parenteral antibiotics should be started immediately. If there is no response to antibiotics in 48 to 72 hours, the patient should have a radiologic evaluation. An abbreviated IVP, with plain film, 15-minute film, and 1-hour film if there is no visualization on the 15-

minute film, is usually sufficient to diagnose anatomic problems.⁴⁶ Ultrasound is usually not of value because of the hydronephrosis that normally accompanies pregnancy.⁴⁴

If bacteriuria is present at the first antenatal visit, there is a high incidence of it recurring later in the pregnancy. This was demonstrated in two previously mentioned studies; in one 148 pregnant women with bacteriuria were treated with placebo medications and 66% had persistence of their bacteriuria,³⁸ and in the other, despite antibiotic therapy, between 16% and 27% of patients developed recurrent bacteriuria later in the pregnancy.³⁹ Conversely, if bacteriuria is not present on the first antenatal visit, it is unlikely to appear later in the pregnancy. Elder found only a 2% incidence of subsequent development of urinary tract infection in 279 pregnant women with negative cultures at their first visit.³⁸ Other authors have supported this finding.⁴⁷

Antibiotics in the Treatment of Bacteriuria and Pyelonephritis in Pregnancy

As has been previously emphasized, bacteriuria during pregnancy, even if asymptomatic, should be treated aggressively. The choice of antibiotics in the gravid patient is crucial. Medications that can be used, and their limitations, are listed in Table 18.3. Antibiotics to be avoided, and the reasons for these recommendations, are listed in Table 18.4. Most data concerning the efficacy of these medications have been generated using the traditional 7- to 10-day treatment regimen.⁴⁸ Some authors, however, have recommended short-course therapy during pregnancy.¹³,⁴⁹ If a short course is utilized, a repeat urine culture should be obtained 1 to 2 days following completion of treatment; this culture must be negative or additional treatment is required. If the patient develops acute pyelonephritis, she should be admitted to the hospital and treated with parenteral antibiotics. If there is no response in 48 to 72 hours, a radiological evaluation as previously outlined should be performed.

References

1. United States Department of Health, Education, and Welfare. Monthly Health Statistics Report. Vol. 24(4) (Suppl 2). National Ambulatory medical Care Survey. May 1973–April 1974. Washington, DC: U.S. Government Printing Office; July 1975.
2. Fair WF. Urinary tract infection: current status. *Seminars in Urology*. 1983;1:91–96.
3. Stamey TA. The diagnosis, localization, and classification of urinary infections. In: Stamey TA. *Pathogenesis and Treatment of Urinary Tract Infections*. Baltimore, Md: The Williams and Wilkins Company; 1980:1–51.
4. Kunz HH, Sieberth, HG, Freiberg J, et al. Zur Bedeutung der Blasenpunktion für den sicheren Nachweis einer Bakteriurie. *Dtsch Med Wochenschr*. 1975;100:2252–2256, 2261–2264.
5. Stamm WE, Wagner KF, Amsel R, et al. Causes of acute urethral syndrome in women. *N Engl J Med*. 1980;303:409–415.
6. Kass EH. The role of asymptomatic bacteriuria in the pathogenesis of pyelonephritis. In: Quinn EL, Kass EH, eds. *Biology of Pyelonephritis*. Boston: Little Brown and Company; 1960:399.
7. Harrison LH. Differential diagnosis of uncomplicated versus complicated urinary tract infection. In: Harrison LH, ed. *Management of Urinary Tract Infections*. Fair Lawn; Medical Publishing Enterprises; 1990:1–6.
8. Kraft JK, Stamey TA. The natural history of symptomatic recurrent bacteriuria in women. *Med*. 1977;56:55–60.
9. Mabeck CE. Treatment of uncomplicated urinary tract infection in nonpregnant women. *Postgrad Med J*. 1972:48:69–75.
10. Stamm WE, McKevitt M, Roberts PL, White NJ. Natural history of recurrent urinary tract infections in women. *Rev Infect Dis*. 1991;13:77–84.
11. Harrison WO, Holmes KK, Belding ME, et al. A prospective evaluation of recurrent urinary tract infection in women. *Clin Res*. 1974;22:125A.
12. Stamey TA. Urinary tract infections in women. In: Stamey TA, ed. *Pathogenesis and Treatment of Urinary Tract Infections*. Baltimore, Md: Williams & Wilkins Company; 1980:122–209.
13. Shortliffe LD, Stamey TA. Urinary infections in adult women. In: Walsh PC, Gittes RF, Permutter AD, Stamey, TA eds. *Campbell's Urology*. Philadelphia, Pa: WB Saunders; 1986:797–830.
14. Ascher AW, Chick S, Radford N, et al. Natural

history of asymptomatic bacteriuria (ASB) in nonpregnant women. In: Brumfitt W, Asscher AW, eds. *Urinary Tract Infection*. London: Oxford University Press; 1973.
15. Freedman LR. Natural history of urinary tract infection in adults. *Kidney Int*. 1975; 8(suppl):96–100.
16. Van Keerbroeck PE. Special considerations in the management of acute urinary tract infection. In: Harrison LH, ed. *Management of Urinary Tract Infections*. Fair Lawn, NJ: Medical Publishing Enterprises; 1990:69–77.
17. Fowler JE. Office bacteriology: techniques and interpretations. *Sem Urol*. 1983;1:97–105.
18. Latham RH, Running K, Stamm WE. Urinary tract infectins young adult women caused by *Staphylococcus saprophyticus*. *JAMA*. 1983; 250:3063–3066.
19. Smith HW. Microbial surfaces in relation to pathogenicity. *Bacteriol Rev*. 1977;41:475–500.
20. Ofek J, Beachey EH. General concepts and principles of bacterial adherence in animals and man. In: Beachey EH, ed. *Bacterial Adherence*. London: Chapman and Hall; 1980:3–29.
21. Savage DC. Survival onmucusal epithelia, epithelial penetration and growth in tissue of pathogenetic bacteria. In: Smith H, Pearce JH eds. *Microbial Pathogenecity in Man and Animals*. London: Cambridge University Press; 1972:25–57.
22. Gibbons R. Position paper. In: Schlesinger D, ed. *Microbiology—1977*. Washington, DC: American Society for Microbiology; 1977:395–406.
23. Sobel J, Kaye D. Urinary tract infections. In: Gillenwater JY, Grayhack JT, Howards SS, Duckett JW, eds. *Adult and Pediatric Urology*. Chicago: Year Book Medical Publishers; 1987:246–302.
24. Schaeffer AJ, Jones JM, Dunn JK. Association of in vitro *Escherichia coli* adherence to vaginal and buccal epithelial cells with susceptibility of women to recurrent urinary tract infections. *N Engl J Med*. 1981;304:1062–1066.
25. Stamey TA, Timothy MM. Studies of introital colonization in women with recurrent urinary tract infections: I. The role of vaginal pH. *J Urol*. 1975;114:261–263.
26. Shortliffe LD, Stamey TA. Infections of the urinary tract: introduction and general principles. In: Walsh PC, Gittes RF, Permutter AD, Stamey, TA, eds. *Campbell's Urology*. Phildelphia, Pa: WB Saunders Company; 1986:738–796.
27. Evans PJ, Leaker BR, McNabb WR, et al. Accuracy of reagent strip testing for urinary tract infection in the elderly. *J R Soc Med*. 1991;841:598–599.
28. Ahlmen, J, Frisen J, Ekbadh G. Experience of three-day trimethoprim therapy for dysuria-frequency in primary health care. *Scand J Infect Dis*. 1982;14:213–216.
29. Sigurdsson JA, Ahlmen J, Berglund L, et al. Three-day treatment of acute lower urinary tract infections in women. *Acta Med Scand*. 1983;213:55–60.
30. Charlton CA, Crowther A, Davies JG, et al. Three-day and ten-day chemotherapy for urinary tract infections in general practice. *Br Med J*. 1976;1:124–126.
31. Fair, WR, Crane DB, Peterson LJ, et al. Three-day treatment of urinary tract infections. *J Urol*. 1980;123:717–720.
32. Harbord RB, Gruneberg RN. Treatment of urinary tract infection with single dose of amoxicillin, cotrimoxazole, or trimethoprim. *Br Med J*. 1981;303:409–415.
33. Ludwig P, Buckwold F, Harding G, et al. Single-dose therapy of acute cystitis in adult females: prospective randomized comparison of four regimens. In: Nelson JD, Grassi C, eds. *Current Chemotherapy and Infectious Disease*. Washington, DC: American Society for Microbiology.
34. Tolkoff-Rubin NE, Wilson ME, Zuromskis P, et al. Single-dose therapy with trimethoprim-sulfamethoxazole for urinary tract infection in women. *Rev Infect Dis*. 1984;4:444–448.
35. Fihn SD, Johnson C, Roberts PL, et al. Trimethoprim-sulfamethoxazole for acute dysuria in women: a single dose or 10-day course. A double-blind, randomized trial. *Ann Int Med*. 1988;108:350–357.
36. Powers RD. New directions in the diagnosis and therapy of urinary tract infections. *Am J Obstet Gynecol*. 1991;164:1387–1389.
37. Stamm WE, Counts GW, Wagner KF, et al. Antimicrobial prophylaxis of recurrent urinary tract infections: a double-blind, placebo-controlled trial. *Ann Int Med*. 1980;92:770–775.
38. Elder HA, Santamarina BA, Smith S, et al. The natural history of asymptomatic bacteriuria during pregnancy: the effect of tetracycline on the clinical course and outcome of pregnancy. *Am J Obstet Gynecol*. 1971;111;41–462.
39. Harris RE. The significance of eradication of bacteriuria during pregnancy. *Obstet Gynecol*. 1979;53:71–73.

40. Kass EH. The role of unsuspected infection in the etiology of prematurity. *Clin Obstet Gynecol.* 1973;16:134–152.
41. Sweet RL. Bacteriuria and pyelonephritis during pregnancy. *Semin Perinatol.* 1977;1:25–40.
42. Cunningham FC, Morris BG, Mickal A. Acute pyelonephritis of pregnancy: a clinical review. *Obstet Gynecol.* 1973;42:112–117.
43. Schultz R, Read AW, Straton JA, et al. Genitourinary tract infections in pregnancy and low birth weight: case-control study in Australian aboriginal women. *Br Med J.* 1991;303:1369–1373.
44. Schaeffer A. Infections of the urinary tract. In: Walsh PC, Retik AB, Stamey TA, Vaughan ED, eds. *Campbell's Urology*. Philadlphia, Pa: WB Saunders; 1992:731–806.
45. Andriole VT, Patterson TF. Epidemiology, natural history, and management of urinary tract infections in pregnancy. *Med Clin North Am.* 1991;75:359–373.
46. Waltzer WE. The urinary tract in pregnancy. *J Urol.* 1981;125:271–276.
47. McFayden IR, Eykyn SJ, Gardner NH, et al. Bacteriuria in pregnancy. *J Obstet Gynecol Br Comm.* 1973;80:385–405.
48. Krieger, JN. Complications and treatment of urinary tract infections during pregnancy. *Urol Clin North Am.* 1986;13:685–693.
49. Chow AW, Jewesson PJ. Pharmokinetics and safety of antimicrobial agents during pregnancy. *Rev Infect Dis.* 1985;7:287–313.

19
Office Evaluation and Treatment of Urinary Incontinence

Peter K. Sand

Epidemiology

Urinary incontinence is a problem of epidemic proportions that produces social embarrassment, isolation, and shame in the affected women. At a recent NIH consensus conference it was estimated that at least 10 million Americans suffer from urinary incontinence and that in the nation at least $10.3 billion is spent annually on this problem.[1] In 1983, the Surgeon General estimated that $8 billion a year was spent on adult diapers for institutionalized patients alone.[2] Current cost estimates, with rising prices of these undergarments and pads and active public marketing of these products, are probably well in excess of this staggering amount.

The exact prevalence of urinary incontinence is unknown and difficult to define when only a minority of affected patients admit to these complaints. Yarnell et al[3] interviewed 1,060 women over the age of 18 in one region of South Wales and found that 45% of these women admitted to having urinary incontinence. Only 9% of the incontinent women in this study had consulted medical professionals about these complaints. In the elderly, those over 75 years of age, the prevalence of incontinence was 59%. These results were similar to the survey of noninstitutionalized elderly people in Washtenaw County, Michigan, by Diokno et al.[4] They interviewed 1,955 people over 59 and found that 37.7% of the women and 18.9% of the men complained of urinary incontinence. Of these incontinent women, 27% complained of stress incontinence and 9% noted urge incontinence, but the majority (55%) had mixed symptoms.

Within institutions for the elderly the problem is even greater. Ouslander et al[5] evaluated 842 patients over the age of 65 in seven nursing homes and found 50% to be incontinent. The NIH consensus statement said that at least $3.3 billion is spent annually caring for incontinence in U.S. nursing homes[1] Many of these people might not need to be in nursing homes if not for their urinary incontinence. Resnick and Yalla found that 41% of 605 institutionalized people had significant urinary incontinence necessitating therapy.[6] Stress incontinence was the primary cause in 21% of the women, but 61% had detrusor hyperreflexia as their primary complaint.[7]

Estimates from the federal government appear to significantly underestimate the problem. The true prevalence of urinary incontinence in the United States is probably at least four times the NIH estimate of 10 million affected Americans.

Etiology of Urinary Incontinence

Adult women may leak urine involuntarily for many different reasons. In a population at the Evanston Continence Center of Northwestern University, more than half of the women evaluated for urinary incontinence have more then one condition causing urinary leakage. The most common condition is *genuine stress*

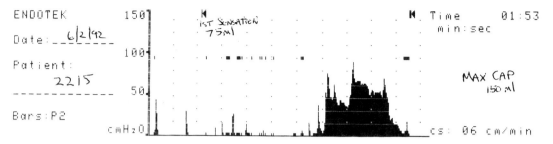

FIGURE 19.1. Detrusor instability occurring during retrograde standing continuous cystometry. The phasic involuntary 40 to 60 cm H₂O detrusor contraction occurs at a bladder volume of 150 mL and is complicated by the patient inappropriately straining during urinary leakage instead of contracting her urethra.

incontinence. This is a condition where involuntary urinary loss occurs due to a rise in intravesical pressure that exceeds intraurethral pressure in the absence of a bladder contraction. This occurs secondary to a rise in intraabdominal pressure. The term *genuine stress incontinence* should be distinguished from the symptom of "stress incontinence," which denotes the patient's statement of involuntary urinary loss during physical exertion. The sign of stress incontinence is noted by the physician with the observation of urinary loss that is synchronous with a rise in intraabdominal pressure. Genuine stress incontinence is found in approximately 75% of all women with urinary incontinence.

Detrusor instability is the second most common reason that adult women leak urine. It is a condition where involuntary bladder contractions occur spontaneously or on provocation during the bladder-filling phase while the patient is trying to inhibit micturition (Fig. 19.1). The bladder contractions are usually phasic in nature and need to be distinguished from "low bladder compliance," a condition where there is a slow and abnormal rise in bladder pressure during filling even at low volumes. Unstable bladder contractions usually are symptomatic; patients complain of "urge incontinence" where they leak urine before reaching the toilet when they feel the urge to go. Or, unstable bladder contractions may be continent, with the patient interpreting the contraction as an urge to void. This contributes to the poor accuracy of symptoms as predictors of the underlying condition(s). Detrusor instability also needs to be distinguished from the other cause of motor urge incontinence, detrusor hyperreflexia.

Detrusor hyperreflexia is the condition where involuntary bladder contractions occur secondary to underlying neurologic disease. Before this diagnosis is made it should be clear to the physician that the involuntary bladder contractions are causally and temporally related to the neurologic condition. Approximately 90% of women with detrusor overactivity and urge incontinence have the idiopathic condition, detrusor instability, while 10% have detrusor hyperreflexia, but this is more prevalent in older institutionalized patients.[7,8] Thus the generic term *neurogenic bladder* used by many to denote the cause of urge incontinence is really a misnomer.

Some patients complaining of urge incontinence may be diagnosed as having *sensory urge incontinence*. This is a poorly understood condition where people who have strong urges to void are overwhelmed by them and actually give up trying to hold back urination, resulting in urinary leakage. This disorder is caused by an increase in afferent sensation from the urethra and stretch receptors in the bladder, which may be caused by infection, trauma, or inflammatory diseases of the bladder and urethra that may overwhelm the usual cortical inhibitory messages that prevent involuntary micturition. The diagnosis of sensory urge incontinence is usually made retrospectively when the cause of increased afferent stimulation has been removed and the patient's urge incontinence symptoms cease.

Other women may have no sensation prior to noting urinary loss and are said to have the symptom of "unconscious incontinence." This may be due to detrusor hyperreflexia from uninhibited urethral relaxation, where involuntary urinary leakage occurs secondary to a sudden drop of the urethral pressure in the absence of a bladder contraction or increased intraabdominal pressure. Sand et al[9] found this in 2% of 534 consecutive incontinent women undergoing multichannel urodynamic testing.

Even other women may leak urine without sensation due to *overflow incontinence*, a condition where there is involuntary urine loss associated with overdistention of the bladder often due to decreased afferent sensation. Many of these patients complain of stress incontinence with or without insensible urinary incontinence. In the absence of an adequate evaluation these patients may be treated with anti-incontinence operations that compound their problems.

The Pathophysiology of Female Urinary Incontinence

Genuine Stress Incontinence

In 1960 Enhorning[10] first proposed inadequate transmission of intraabdominal pressure to the proximal urethra as the cause of genuine stress incontinence. This may be referred to as the extrinsic continence mechanism.[10,11] Complete transmission of this pressure to the urethra is believed to be dependent on the support of the proximal urethra from the surrounding fibrous and muscular supports. While older literature stresses the anterior support of the pubourethral ligaments, DeLancey has shown that primary support comes from the lateral and posterior attachments of the urethra, bladder neck, and vagina.[12,13] Alteration or relaxation of these supports leads to hypermobility of the urethra and reduced abdominal pressure transmission to the proximal urethra. Most believe that pregnancy and vaginal delivery are the primary insults to these supporting structures, leading to urethral hypermobility and stress incontinence.[14–16] Postpartum stress incontinence is probably due to physical damage to the fibromuscular supports of the urethra, diminishing passive transmission of intraabdominal pressure, and to neurologic injury, resulting in diminished active reflex contractions of the skeletal muscle within and around the urethra during increases in intraabdominal pressure.[17,18] Such active increase in urethral pressure actually anticipates the increase in bladder pressure with coughing and straining and therefore must represent an active response of striated muscle.[19,20] This response is lost in women with stress incontinence and is not restored even by successful anti-incontinence operations.[20] Snooks et al[17] and Allen et al[18] have both shown delayed pudendal nerve conduction and neurologic damage with reinnervation patterns in the majority of women undergoing vaginal delivery but not cesarean section, which may account for some of these changes.

The underlying etiology of detrusor overactivity when not due to neurologic injury (detrusor hyperreflexia) is poorly understood. Detrusor hyperreflexia may be caused by any interruption of the neurologic control over micturition, but common causes include multiple sclerosis, cerebral vascular accidents, spinal injuries or tumors, and congenital spinal cord abnormalities.

The idiopathic condition detrusor instability has no known etiology, but several theories may explain some cases. It may be caused by urinary tract infection,[21] anti-incontinence operations,[22] and obstructive phenonemon as in men with benign prostatic hypertrophy. However, in the vast majority of women the cause is obscure but known to increase with age.

Other causes of urinary incontinence include urinary fistulae, diverticula, functional incontinence, and psychogenic incontinence. These causes need to be carefully excluded before treatment is initiated or therapy will never be completely successful.

TABLE 19.1. Comparison of symptoms and urodynamic findings in 218 patients undergoing multichannel urodynamics.*

Symptoms	Genuine stress incontinence	Detrusor instability	GSI & DI	Continent
Stress incontinence ($N = 43$)	25	7	8	3
Urgency and urge incontinence ($N = 13$)	0	10	0	3
Stress and urge incontinence ($N = 132$)	89	13	30	0
Continent ($N = 30$)	0	0	0	30
Total ($N = 218$)	114	30	38	36

*From Sand PK, Hill RC, Ostergard DR. Incontinence history as a predictor of detrusor stability. *Obstet Gynecol.* 1988; 71:257–260. Used by permission.

Initial Office Evaluation of the Incontinent Female

In the past many investigators have tried to establish the etiology of urinary incontinence based on a brief history with a pelvic exam. This is clearly inadequate for most, if not all, incontinent women. But numerous clinical tests to aid the clinician in establishing an accurate diagnosis may be used before initiating treatment.

Urogynecologic History

The urogynecologic history is an essential part of the evaluation of every incontinent female. Although it may not adequately explain the patient's symptoms, it does define them. This then establishes what must be explained by any evaluation. To help clarify the patient's understanding of the problem and to speed the evaluation, patients at the Evanston Continence Center are sent programmed questionnaire to fill out at home before coming to the office. This not only speeds up their evaluation; it also enables the patient to confirm questionable areas and adds to the accuracy of her responses.

With rare exception most authors have found history to be inadequate in establishing an accurate diagnosis.[23–28] Farrar et al[29] found history to be quite accurate. Fifty-four of 56 women (96.4%) complaining of only stress incontinence were found to have stable bladders on cystometry, and 89% of the 110 women who had complaints of urge incontinence were found to have detrusor instability. Hastie and Moisey[30] also concluded that a pure history of stress incontinence was 100% accurate in establishing genuine stress incontinence as the underlying cause of urinary leakage. Other data[23] are quite divergent from this. In a subgroup of 43 women with isolated complaints of stress incontinence, 34.9% (15) were found to have involuntary bladder contractions on cystometry. As Table 19.1 shows, the incontinence history in general was also found to be a poor predictor of the underlying conditions causing urinary leakage. Similarly, Cardozo and Stanton[24] and Webster et al[25] found the history to be inaccurate even in patients with isolated complaints of stress incontinence or urge incontinence. Numerous studies have shown that more than a history is necessary to evaluate the cause of a patient's symptoms. Therefore a basic clinical evaluation including a physical examination with a lumbosacral neurologic evaluation, urine culture, measurement of residual urine, assessment of urethral mobility, demonstration of urinary leakage, and cystometry is invaluable. This basic evaluation is an excellent, cost-effective screening exam for all patients and will be diagnostic in most cases.

In addition to the urologic history, a good general medical history, record of medication and drug usage, and history of prior surgeries should be obtained. Prior anti-incontinence operations may alter the typical prevalence of various types of incontinence.[31] Women who have failed a prior anti-incontinence operation have a higher incidence of detrusor instability.

These questionnaires may be reviewed rapidly before seeing the patient and should be rediscussed with the patient to augment the examiner's understanding of the patient's complaints. This is also a good time to question the patient about fecal incontinence and urinary loss during intercourse, which occurs in as many as 24% of patients and may adversely affect half of all patients' sexual relations.[32] In one study[32] anal incontinence was found to affect 51.1% of women complaining of urinary incontinence. While not always this prevalent, anal incontinence does affect many women with urinary leakage and should be looked for in all incontinent patients.

Patients have also been asked to complete a 24- to 48-hour voiding diary before coming to their appointment. Patients record when they void, the volume voided, when they leak urine, and what they drink. Review of this chart helps to confirm the patient's symptoms but may not correlate well with them. Although many patients with detrusor overactivity void often and in small volumes, clinical overlap with patients without detrusor overactivity prevents the frequency/volume voiding diary from being used as a diagnostic test.[33] It is useful as a quantitive measure of improvement during therapy for detrusor overactivity.[33]

Physical Examination

Careful physical examination is an important part of any work-up and should focus on the pelvic exam and neurologic evaluation of the lumbosacral nerve roots. Assessment to rule out systemic diseases, such as cardiac, thyroid, adrenal, and renal disease that may play a role in incontinence, is also very important.

A careful pelvic examination should first look for signs of genital tract inflammation, infection, and atrophy that may affect the urethra and cause an increase in afferent sensation. This may lead to irritative voiding symptoms such as urgency, urinary frequency, dysuria, and possibly urge incontinence. In addition, heavy vaginal discharge on a sanitary napkin may be confused by some patients as urinary incontinence.

Because the urethra and trigone are estrogen-dependent tissues, estrogen deficiency may contribute to urinary incontinence and dysfunction.[34] Most authors[35-40] agree that estrogen replacement therapy will symptomatically improve many patients, but the mechanism of improvement has been inconsistent in clinical trials; some investigators have shown significant changes in urethral closure pressure and functional length,[35,36,38] while others do not.[39-41] Almost all these studies have shown improvement in pressure transmission ratios that measure the increase in urethral pressure compared to bladder pressure during coughing.[35-38,40]

When these signs of estrogen deficit are noted during the evaluation of incontinent women initiation of vaginal estrogen replacement before any further urodynamic work-up is recommended. The vaginal route of administration is preferred initially because of the excellent absorption of estrogen from these thin atrophic tissues.[42]

After these observations are made, the examiner should look for any signs of urethral deformity such as diverticula or fistula. Vesicovaginal fistulae are most often found just anterior to the vaginal cuff after hysterectomy and will usually leak urine continuously, but urethral fistulae may only cause leakage when urine enters the urethra during episodes of stress incontinence. If a distal urethral sac is seen, it should be palpated to see if it is tender and massaged to see if it empties urine or pus, which may be noted at the urethral meatus or seen on urethroscopy.

Observation of the anterior vaginal wall, posterior wall, and vaginal apex may also be made using the Sim's or bivalve speculum to identify associated genital prolapse. The presence of genital prolapse may have profound effects on lower urinary tract function, often resulting in urinary retention or masking urinary leakage. Posterior rotational descent of the anterior vaginal wall results in the formation of a cystocele, which may be a representation of lateral or paravaginal detachment of the anterior lateral vaginal wall from the arcus tendinous or may represent a central defect due to weakness of the endopelvic connective tissue. Regardless of the etiology, descent of the anterior vaginal wall may result in the urethra or bladder neck folding on itself. This process results in mechanical obstruction or kinking of the urethra that can be identified during urethroscopy and on urethral closure pressure profiles. By examining patients after reducing their prolapse a clinician can identify potential genuine stress incontinence in 60% of women with cystoceles protruding to the vaginal introitus or beyond who do not leak urine with the prolapse unreduced.[43] Stress testing with reduction of the prolapse using a pessary, Sim's speculum, or proctoswabs in the standing position will reveal qualitative information about who has potential genuine stress incontinence, but urethral closure pressure profiles at rest, during Valsalva, and repetitive coughing provide quantitative information about the underlying pathophysiologic effect of the prolapse.[44-46] Many of these women have urinary retention secondary to this mechanical obstruction. Reducing the prolapse during voiding will also enable the examiner to discover whether retention is due merely to the mechanical obstruction or to some other pathologic process. Some women may develop an underactive detrusor or atony over time if the bladder is forced to contract against a kinked urethra chronically. All patients with prolapse to the introitus should be examined with a full bladder after reduction of the prolapse to look for leakage. Care should also be taken to be sure that patients with genital prolapse do not have significant urinary retention. Those with residuals greater than 50 mL should undergo complete evaluation with voiding pressure studies after reduction of the prolapse.

A careful digital exam should be done to assess transvaginal urethral or bladder tenderness and to rule out pelvic pathology. Much attention is often given to uterine leiomyomata causing urinary symptoms, but this is infrequently a cause of incontinence or irritative symptoms. Rectovaginal examination is an important part of this evaluation. Evaluation of possible enterocele dissection along the rectovaginal septum may be appreciated, and anal sphincter tone should be assessed. Rectal prolapse and subtle rectoceles may be appreciated during this exam but often require radiographic evaluation.

The Neurologic Examination

Urinary control in women involves a complex interaction of various reflexes interacting with the autonomic nervous system, which primarily controls bladder and urethral function. These are both modulated by cortical and brain stem centers. Neurologic damage in these areas may cause significant symptomatology.[47] Cerebrovascular disease may cause detrusor hyperreflexia, while peripheral lesions may lead to bladder atony and overflow incontinence. Spinal cord lesions may result in detrusor-sphincter dyssynergia with intermittent voiding patterns and urinary retention due to lack of urethral relaxation during bladder contraction.

Peripheral control is primarily modulated by the pelvic (parasympathetic) and hypogastric (sympathetic) nerves.[48] The parasympathetic system has axons arising in sacral segments S_2–S_4 that release acetylcholine to effect bladder contractions. The sympathetic ganglia originate in the thoracolumbar spine (T_{10}–L_2) and course through the hypogastric nerve to terminate in the parasympathetic ganglia of the bladder wall and in the bladder and urethra, where there are receptors for both alpha and beta fibers.[49] The alpha fibers primarily cause contraction of the urethra and bladder neck while beta receptors cause detrusor relaxation but probably only at low volumes by releasing norepinephrine. In addition, the sympathetic fibers ending on postganglionic parasympathetic neurons modulate and depress parasympathetic transmission. This helps to explain why storage of urine is a passive process. This sympathetic control is also the basis for the pharmacologic success sometimes seen with alpha-adrenergic treatment of detrusor instability.

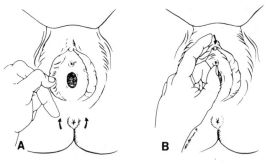

FIGURE 19.2. The anal and clitoral reflexes. Gentle stroking lateral to the anus causes anal sphincter contraction (A). Gentle tapping or pressure on the clitoris causes pelvic floor contraction (B). (From Bhatia NN, Rosenzweig BA. The urologically oriented neurological examination. In: Ostergard DR, Bent AE, eds. *Urogynecology and Urodynamics*. 3rd ed. Baltimore, Md: Williams & Wilkins; 1991:102–114. Used by permission.)

The neurologic exam begins with an assessment of mental status, cranial nerve integrity, and cerebellar control. Muscle strength and the deep tender reflexes should be assessed in the lower extremities to assess the anterior lumbosacral spinal cord. Sacral reflexes are most important and may be evaluated by checking the bulbocavernosus and clitoral reflexes as shown in Figure 19.2. Sensory function may be quickly evaluated from sacral dermatomes L_2 through S_2 by testing for sensation of pinprick around the knee (Fig. 19.3).

If a neurologic deficit is identified or suspected on simple screening exam, electromyography to evaluate skeletal muscle function and innervation or electromyelography to evaluate abnormalities of the autonomic control of the lower urinary tract may be done.

Assessment of Urethral Mobility

Because the underlying pathophysiology of genuine stress incontinence is believed to be urethral hypermobility, an assessment of urethral mobility should be made. This assessment may be made in several ways, the simplest of which is the qualitative visual assessment of distal anterior vaginal wall mobility.

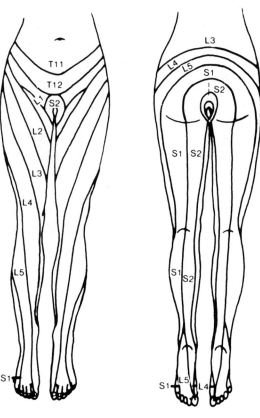

FIGURE 19.3. Lower extremity sensory dermatomes. (From Bhatia NN, Rosenzweig BA). The urologically oriented neurological examination. In: Ostergard DR, Bent AE, eds. *Urogynecology and Urodynamics*. 3rd ed. Baltimore, Md: Williams & Wilkins; 1991:102–114. Used by permission.)

The Q-tip test, as first described by Crystle et al,[50] is an excellent test to document quantitatively the presence of urethral hypermobility. The Q-tip test is a simple, inexpensive way to assess urethral hypermobility quantitatively without any radiation exposure. While the determination of urethral hypermobility on Q-tip testing is a poor predictor of the etiology of a patient's urinary incontinence,[51–54] it remains an accurate and valuable measurement of good urethral support. It is clear that many women develop urethral hypermobility early in their third and fourth decades but remain continent if their intrinsic sphincteric mechanism is capable of

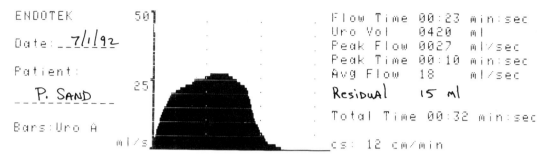

FIGURE 19.4. Normal uroflowmetry study.

compensating for the negative pressure transmission to the urethra created by the urethral hypermobility.[55] It is for this reason that there is tremendous overlap between stress incontinent and continent patients in many[52,53] but not all studies.[54,56] The cotton-tipped applicator is placed at the bladder neck, and then the resting and straining angles are measured several times. The test is arbitrarily said to be positive for urethral hypermobility if the straining angle is greater than 30° to 35°.[51,52,57]

Patients without urethral mobility and stress incontinence have Type III incontinence as described by McGuire,[58] deficient intrinsic sphincter function despite adequate urethral support. These patients are usually managed by periurethral injections, sling procedures, or the placement of artificial sphincters. Straining cystograms and ultrasound studies may also be used to measure urethral hypermobility but are far more expensive than simple Q-tip testing.

Uroflowmetry

Simple uroflowmetry is used as a screening test to detect abnormalities of micturition. It is far more useful in men than in women because of its ability to detect urethral obstruction, a condition that is rare in women. Uroflowmetry is primarily a measure of voiding velocity measured in mL/second. In women, without concerns of physical urethral obstruction, the pattern of the uroflowmetry curve is probably more important than voiding velocity. Figure 19.4 shows a normal uroflowmetry curve. Qualitative assessment of the bell-shaped uninterrupted curve readily identifies it as a normal study. These graphic tracings are made by electronic uroflowmeters. An average flow rate can be measured in a much simpler manner by timing the duration of urine flow and then measuring the volume voided. Simple calculation will reveal the mean flow rate, and the patient may be catheterized to measure the residual urine. Women with residual urine greater than 50 mL need further evaluation of their voiding function with multichannel voiding pressure studies.

Fantl et al[59] showed that while flow rates did not vary significantly with age, parity, weight, height, or menstrual cycle phase, they were directly affected by bladder volume. Like Starling's law for the heart, the bladder volume, within normal ranges, directly affects the efficiency and velocity of bladder emptying.[59,60] Because of this it is difficult to establish normal values for uroflowmetry without using a nomogram. However, in general, for volumes voided in excess of 200 mL, the peak flow rate should be ≥15 mL/second. This is based on Fantl et al,[59] who found an average peak flow rate at this level of 22.6 mL/second with a standard deviation of 7.5 mL/second.

More important than the peak flow rate is the flow pattern. A normal asymptomatic woman will usually void with continuous flow until voiding is completed. Intermittent flow patterns may occur occasionally in normal asymptomatic women but are more common in patients with voiding dysfunction.[59] When a study is abnormal it should be repeated; it may be artificially abnormal because the patient is voiding in an unfamiliar setting.

Intermittent flow patterns and decreased flow rates are most often explained by functional obstruction due to urethral atrophy, inflammation, pain, or fear. Pelvic relaxation with mechanical urethral obstruction is another common cause of abnormal uroflowmetry in women. One of the most common causes of intermittent flow patterns in women is Valsalva voiding, which may be prevalent in women with genuine stress incontinence and decreased urethral resistance. In these women Valsalva is associated with normal to high peak flow rates, but after corrective surgery, where Valsalva will lead to inefficient or absent voiding due to restored normal pressure transmission to the urethra, it can be disastrous. Detrusor sphincter dyssynergia may also cause intermittent flow patterns and retention. Less commonly, underactive and acontractile bladders as causes of decreased flow rates, intermittent flow, and elevated residual urine will be identified. Drugs (alpha-adrenergic and anticholinergics) may also adversely affect voiding function, and their effects may be detected on spontaneous uroflowmetry.

Uroflowmetry is a simple noninvasive test that may act as a screening test for abnormal voiding function. If studies are abnormal or significant suspicion exists of voiding dysfunction, then voiding pressure studies with electromyography should be done.

Endoscopy: Urethroscopy and Cystoscopy

The Robertson urethroscope, with its 0° lens and closed sheath, may be used to assess the urethra and trigone in a retrograde fashion and to observe their function. As subsequent views after instrumentation disturb the exudate that may line the urothelium and manipulation of the scope causes erythema, it is important to be able to visualize the urethra on entry. Urethroscopy may be done using CO_2, H_2O, or saline as the distending media. Although CO_2 is less messy and may offer a better view of the urethra than H_2O or saline, it is more irritating to the bladder and may cause artificial detrusor contractions during filling.[61]

Inspection of the urethra is made with the urethroscope in a retrograde fashion; urethral color, inflammation, and the presence of the normal urethral folds and gland openings may be observed. Pathologic findings such as polyps, cysts, inflammatory fronds, increased exudate, condylomata, and diverticula may all be noted in the urethra. Upon entry into the bladder, the trigone and ureteral orifices may be inspected. The ureteral orifices should be inspected for their appearance and function; the examiner should note normal ejaculation of clear urine from the orifices and good retraction of the ureteral orifices following this.

After inspection of the urethra and trigone the examiner may use the scope to observe the dynamic function of the urethra during filling and with various maneuvers. Robertson[62] has described the technique of dynamic urethroscopy where the clinician is able to observe the bladder neck during filling to diagnose detrusor overactivity and during Valsalva and coughing to diagnose genuine stress incontinence. However, Sand et al.[63] and Scotti et al.[64] have shown that dynamic urethroscopy, although fairly specific, is an insensitive screening test for detrusor instability and genuine stress incontinence.

Cystoscopy may also be used during the routine evaluation of incontinent women but is not advocated in the absence of irritative voiding symptoms or hematuria. In symptomatic patients cystoscopic examination may occasionally reveal signs of chronic cystitis, cystitis cystica or glandularis, interstitial cystitis, polyps, and neoplasms that would not be identified with the urethroscope. These conditions may produce increased afferent sensation leading to sensory urge incontinence and therefore are important to rule out in women complaining of urgency, frequency, and urge incontinence.

TABLE 19.2. Cystometric variables.

I. Filling media
   CO_2
   H_2O
   Saline
   Urine
   Radiographic contrast
II. Temperature of the medium
   Room temperature
   Iced infusions
   Body temperature
III. Method filling
   Orthograde
   Orthograde with diuresis
   Retrograde
      Slow fill (1–10 mL/min)
      Medium Fill (10–100 mL/min)
      Rapid Fill (> 100 mL/min)
IV. Infusion mode
   Continuous
   Intermittent
V. Position
   Supine
   Sitting
   Standing
VI. Provocative maneuvers
   Cough
   Heel bounce
   Position change
   Hand washing
   Valsalva
   Rectal distension
   Running water

Cystometry

Cystometry is a diagnostic test to measure the change in bladder pressure in response to bladder filling. It allows for the evaluation of bladder sensation, compliance, capacity, and control during the storage phase. During a cystometrogram the normal bladder is compliant enough to allow for filling to capacity without any significant increase in bladder pressure, whereas the overactive detrusor will produce involuntary phasic contractions during filling, even at low volumes. Cystometry may be performed in many different ways and may be affected by many different variables (Table 19.2).

"Eyeball cystometry" is a simple, inexpensive, qualitative exam that lends itself well to examination at the bedside in the nonambulatory patient. A simple Foley catheter may be placed transurethrally and the bladder may be progressively filled by pouring sterile water intermittently into an irrigation syringe attached to the catheter 50 mL at a time. The syringe is held about 15 cm above the patient's pubic bone. The patient's first sensation to void and maximum cystometric capacity are recorded, and abnormal (involuntary) bladder contractions are heralded by a rising rather than falling meniscus, often associated with a strong urge to urinate. A rising meniscus may be a reflection of increased abdominal pressure that may be aborted by asking the patient to relax or inspire. Abdominal relaxation may be confirmed by abdominal palpation. For the diagnosis of detrusor instability in geriatric patients, Ouslander et al[65] found this method to have a sensitivity of 75%, specificity of 79%, and a positive predictive value of 85% when compared with multichannel electronic cystometry. This method also allows for the measurement of the residual urine with the initial catheterization. Although not a perfect screening test, its simplicity allows it to be used in situations where no testing would otherwise be possible.

One step up from simple qualitative "eyeball cystometry" is the performance of retrograde, incremental, standing, quantitative cystometry. This can be done with commercially prepared devices (Fig. 19.5), or a water cystometer can be constructed inexpensively by the physician. After the initial pressure is recorded, the bladder is filled in a retrograde fashion, 50 mL at a time, by gravity stopping approximately every minute to measure the bladder pressure. The patient's first desire to void and maximum cystometric capacity are recorded with the corresponding bladder pressure at these volumes. Bladder pressure is recorded by reading the stationary low point of the meniscus each time and after provocation with coughing, heel bouncing, or hand washing. The cystometrogram is considered positive or suspicious for detrusor overactivity if bladder pressure readings rise by more than 15 cm H_2O during filling. This is further confirmed by the presence of involuntary leakage

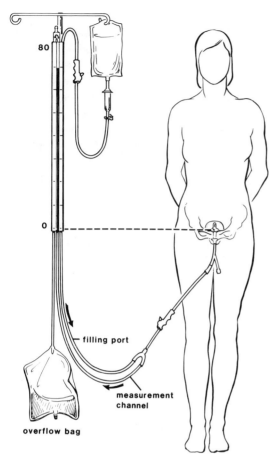

FIGURE 19.5. A simple manometric water cystometer. (From Sand PK, Brubaker LT, Novak T. Simple standing incremental cystometry as a screening method for detrusor instability. *Obstet Gynecol.* 1991;77:453–457. Used by permission.)

with removal of the catheter at maximum cystometric capacity. Normal studies usually show a pressure rise of ≤ 10 cm H_2O during filling, with equivocal studies showing pressure rises between 10 and 15 cm H_2O. Repeating these equivocal studies may improve diagnostic accuracy.[66]

Simple incremental retrograde studies are limited in their sensitivity by missing small increases in bladder pressure that would create urinary leakage in the absence of the obstructing catheter. This may be avoided in some cases by the use of small (8f) catheters without retention balloons that are taped to the patient's leg. These studies are limited in their specificity mainly by the misinterpretation of increased intraabdominal pressure as increased detrusor pressure. This artifact is avoided by the use of subtracted cystometry. A commercial cystometer like that in Figure 19.5 has been found to obtain results that were reproducible in 84% of patients in a prospective trial with a sensitivity of 84.3% and specificity of 69.4% with one cystometrogram.[67] Using two incremental standing cystometrograms increased the sensitivity to 92.3% in 100 consecutive incontinent women when compared to multichannel urethrocystometry. These simple, inexpensive, retrograde cystometers may be used quite accurately in screening incontinent women.[67–69]

Subtracted cystometry may be used to improve upon this diagnostic accuracy in the diagnosis of detrusor overactivity. By being able to measure both abdominal and bladder pressure, one is able to distinguish between abdominal pressure effects and those changes intrinsic to the bladder. Abdominal pressure may be measured through the vagina, through the rectum, or suprapubically. Instantaneous electronic subtraction of bladder minus abdominal pressure allows for continuous recording of the true detrusor pressure. Pressure rises of 15 cm H_2O or more in the true detrusor pressure or phasic rises on continuous cystometry associated with symptoms (urgency, incontinence) are more accurate representations of abnormal bladder compliance or detrusor overactivity. With subtracted cystometry, provocative maneuvers may be utilized more easily with less confusion, and this also improves the diagnostic sensitivity of the test. The most efficient way to perform subtracted cystometry is utilizing an electronic cystometer that records two or three channels of activity. These electronic cystometers typically range in price from $4,000 to $12,000.

When more accuracy or information is needed regarding the bladder's ability to store urine, one may utilize urethrocystometry with electromyography. By also monitoring urethral pressure responses and the electrical activity of the periurethral and intrinsic urethral skeletal muscle this complex study augments the understanding of the detrusor's reaction to bladder filling.

Ambulatory studies that allow for the performance of cystometry during normal orthograde filling from renal urine production may also be used in the patient's own environment, outside the office without a medical audience. These are the most physiologic and sensitive studies available today. They are also very expensive and not readily available in most centers.[70]

The bladder fills incrementally in vivo with urine from the ureters; urine would be the most physiologic media but is very difficult to use during retrograde cystometry. Therefore most investigators use CO_2, water, saline, or radiographic contrast media during cystometry. Although CO_2 has been widely used in the past, many investigators are now replacing it with liquid media.[61,70,71] Compared to liquid meida, carbon dioxide is easy to use and allows for rapid flow rates that are more provocative, and cystometry may easily be repeated rapidly. But CO_2 is an irritating medium that forms carbonic acid when mixed with urine.[61] Water and saline are used by many investigators interchangeably, but saline is probably more physiologic, especially when instilled at body temperature. Little has been described about the accuracy of radiographic solutions as media for cystometry but most of these solutions are nonirritating and probably comparable to saline cystometry. But a contrast medium such as 35% diodone is quite irritating and may be more likely to produce involuntary bladder contractions.[70] Contrast media may be used for cystometry and in combined videocystourethrography studies allow visualization of the bladder during filling.

Cystometry may be done in many positions, but standing studies are the most sensitive. For elderly or infirmed patients standing may be difficult, and the use of a birthing chair may allow for sitting-upright studies. Movement to the standing position with assistance may be provocative in some of these patients.[72]

As listed in Table 19.2 (p. 236) various provocative maneuvers may be used during cystometry to help to elicit involuntary bladder contractions. Although some maneuvers may not be physiologic, it must be recognized that stationary cystometry is unable to reveal involuntary detrusor contractions in all patients with detrusor instability and may actually miss up to 50%.[70] Therefore artificial provocation may be necessary to compensate for the inherent insensitivity of these laboratory urodynamic studies. Recent investigation of provocative maneuvers to elicit detrusor instability by Mayer et al.[73] showed handwashing to be the most provocative. During a cystometry study have the patient cough and heel bounce after every 50 to 100 mL of infusion; reserve handwashing and other forms of provocation until the end of the study if involuntary detrusor contractions have not been recorded.

Despite the use of electronic, continuous, multichannel, subtracted cystometric evaluations with provocation some involuntary detrusor contractions may be missed. Low-pressure detrusor contractions may be mistaken for episodes of stress incontinence in these patients unless the isometric force of these contractions can be measured. Use of stop tests in these ambiguous situations may allow for registration of large isometric contractions that might otherwise go unnoticed.[74] Voluntary interruption of the stream of urine or mechanical obstruction transurethrally with a catheter or transvaginally will close the bladder neck and will allow for measurement of the true isometric pressure generated in the bladder, which improves the sensitivity of cystometry.

Cystometry may also be used to measure compliance or the change in volume associated with a change in pressure ($\Delta v/\Delta p$). Because the bladder is not only composed of active elements (smooth muscle) but also passive elements (collagen and elastin), the evaluation of the filling phase must also evaluate the distensibility or viscoelastic properties of the bladder.[75] If bladder injury from inflammatory disease, radiation, or surgical trauma has occurred, in the absence of phasic contractions cystometry may reveal a low compliance with a slow gradual rise in true detrusor pressure during filling.

The office tests described previously enable physicians to demonstrate urinary leakage and understand its etiology in the majority of patients. However, because of the artificial nature of the testing environment and the inherent lack of sensitivity of some of these tests, not all patients will demonstrate urinary leakage on this screening examination. Those who do not may undergo more complex multichannel urodynamic and radiographic studies that may or may not demonstrate or reproduce their urinary incontinence. An alternative is the use of several simple tests that although not diagnostic are capable of demonstrating involuntary loss of urine.

The Stress Test

Stress testing is a simple clinical test that allows for the objective demonstration of urinary leakage. This simple test is accomplished by having a patient cough, Valsalva, or do anything to increase intraabdominal pressure that will cause her to reproduce her stress incontinence. It is best accomplished with a full bladder, and upright positions are more sensitive than supine investigations. Urinary leakage coincident with increased intraabdominal pressure is indicative of the sign, stress incontinence, and is termed a positive test. However, urinary leakage following a sudden rise in intraabdominal pressure may be a representation of stress-induced detrusor contractions following provocation.[73]

Pad test

Pad testing is a method to document and quantify urinary leakage objectively. Various types of pad tests have been described in the literature, but they all basically use a preweighed sanitary napkin or collecting device to measure urine loss over a period of time during planned or random activities. Twenty-minute, 1-hour, 2-hour, 12-hour, and 24-hour tests have all been used to document and quantify urinary incontinence. The pad test may be a useful adjunct to urodynamic evaluation. Jorgensen et al[76] found that the 1-hour pad test endorsed by the International Continence Society detected nearly twice the number of incontinent women as stress testing and voiding cystourethrography. Because of variable diuresis with no oral fluid load this is consistent with the findings of some investigators[77,78] but not with others.[79,80] It is preferable to use a fixed-bladder-volume 20-minute pad test in clinical trials; this test incorporates all the activities of the standard 1-hour test into a simpler, faster, and reproducible format.[81,82] With rare exceptions (excessive perspiration or vaginal discharge) positive pad tests (>1 g/hour) indicate urinary leakage. In normal subjects mean pad weight increases are less than 1 g/hour.[83,84] When one wants to reproduce the magnitude of a patient's problem reliably, home pad tests over extended periods are best.[85] While the correlation between 24-hour and 48-hour pad tests with the standard 1-hour test is low ($R = 0.35$), the reproducibility of a 24-hour pad test is quite good ($R = 0.96$), and extended testing better correlates with a patient's symptoms.[85]

Complex Urodynamic Evaluation

Initial evaluation with the simple methods described may not be sufficient to define the etiology of an incontinent woman's problem. If some aspect of the patient's symptoms remains unexplained, more sophisticated testing in a urodynamics laboratory is indicated. Other indications include previous failed anti-incontinence operations, women over the age of 60, continuous and/or insensible urine loss, suspicion of neuropathic dysfunction, mixed incontinence symptoms, and women with apparent stress incontinence who do not have evidence of urethral hypermobility.[86,87] The performance of complex multichannel urodynamic and radiographic studies involves expensive equipment that may not always be available to the clinician. Decisions regarding referral in these instances may be difficult.

Urethrocystometry

The performance of urethrocystometry involves the use of pressure-measuring catheters and transducers that are placed in the urethra,

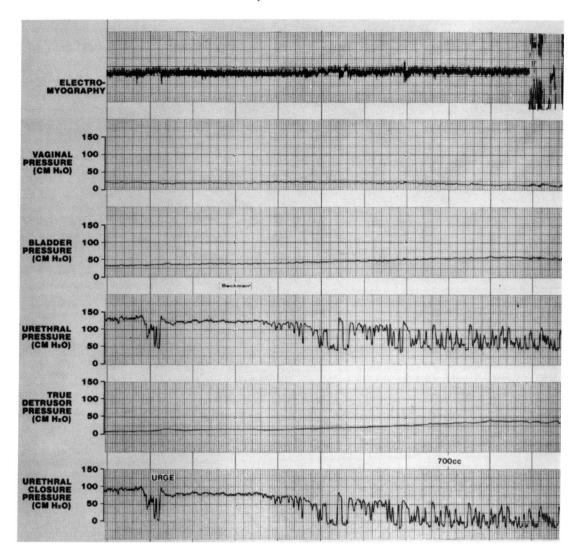

FIGURE 19.6. Detrusor instability on urethrocystometry preceded by marked variation in urethral pressure (urethral instability). (From Sand PK. Evaluation of the incontinent female. *Curr Prob Obstet Gyn Fertil*. 1992;15:105–151. Used by permission.)

bladder, and rectum or vagina to measure intraabdominal pressure.[88,89] The urodynamic equipment is then used to measure not only these three pressures but also the true detrusor pressure from instantaneous subtraction of abdominal from bladder pressure and urethral closure pressure that is calculated by subtraction of bladder pressure from urethral pressure.

Measurement of these five pressures with or without EMG may be accomplished during filling to create a urethrocystometry study (Fig. 19.6).[90] Urethral and bladder pressures may be measured by the same microtransducer catheter that may also be used to fill the bladder through a small infusion channel in the center of the catheter (Fig. 19.7). Patients are positioned in the birthing chair, and a single transducer catheter is placed in either the vagina or rectum to measure the intraabdominal pressure. The dual transducer catheter is then placed in the urethra and positioned so that the proximal microtransducer, which is 6 cm from the catheter tip and distal

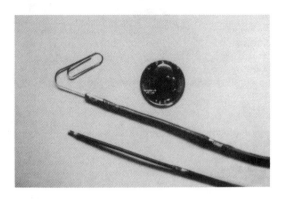

FIGURE 19.7. Dual-channel microtransducer urodynamic catheters with (7F) and without a filling channel (6F). (From Sand PK. Evaluation of the incontinent female. *Curr Prob in Obstet Gyn Fertil.* 1992;15:105–151. Used by permission.)

transducer, is placed at the point of maximal urethral pressure. Filling is then begun in a retrograde fashion either through the lumen of the urethral catheter or via a separate small catheter or pediatric feeding tube.[88] Cystometric parameters and changes are noted as described previously for subtracted cystometry, but urethrocystometry offers the advantage of simultaneous observations of urethral pressure.

Typically, urethral pressure will start to decrease prior to a bladder contraction, whether voluntary or involuntary, and such variations are usually associated with corresponding EMG changes. This urethral instability may be a forerunner of detrusor instability (Fig. 19.6) and often warns of an impending bladder contraction.[91-93] These pressure variations may be independently associated with symptoms of urgency and sensory urge incontinence in some women.[91,92] But some investigators question whether these pressure fluctuations have any significance at all.[94] Exaggerated urethral pressure fluctuations may be directly associated with complete urethral relaxation and may cause urinary incontinence from uninhibited urethral relaxation.[9] Care must be taken to be sure that these urethral pressure variations are not artificial due to patient movement, abdominal pressure changes, or voluntary contraction of the levator ani muscles. Treatment with alpha blockers may relieve symptoms in some of these patients. Pressure variations preceding detrusor instability may be utilized as a marker for biofeedback therapy to inhibit involuntary detrusor contractions. In normal women, filling the bladder will lead to a gradual increase in urethral pressure due to augmentation of skeletal muscle activity associated with a gradual increase in skeletal muscle EMG activity, and it may be triggered by stretch receptors in the bladder trigone.[95]

Urethrocystometry studies in normal volunteers have shown that the average maximum cystometric capacity was 594 mL and the first sensation to void was noted at 32% of capacity.[96] Ninety-five percent of the normal subjects noted first desire to void before filling to 300 mL of H_2O.

Urethral Closure Pressure Profiles

Urethral closure pressure profiles involve the measurement of the urethral pressure all along the length of the urethra. By slowly withdrawing a dual microtransducer catheter through the urethra with a profilometer (automated withdrawal arm) one can measure the urethral pressure at all points along the urethra. This generates a curve that is called a urethral closure pressure profile (Fig. 19.8). The "pressure" in the urethra measured with this system is actually force and is generated by urethral smooth muscle, the periurethral vasculature, the skeletal muscle in the urethra, and the pelvic floor muscles, as well as the inert collagen and elastic tissue in the urethra.[97-101] Damage or attenuation of any of these contributing factors may dramatically alter the urethral closure pressure profile.[97,101]

The curve begins as the proximal or urethral pressure transducer enters the proximal urethra. At this point the urethral pressure is usually greater than bladder pressure, resulting in an increase in closure pressure. In the proximal to mid urethra, above the urogenital diaphragm, the urethral closure pressure profile will reach its peak; then closure pressure will decrease until in the distal urethra, where the closure pressure drops

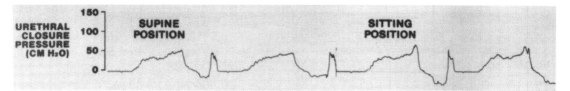

FIGURE 19.8. Resting urethral closure pressure profiles. Note the increase in closure pressure with a more upright (stressful) position in this continent women. (From Sand PK. Evaluation of the incontinent female. *Curr Prob Obstet Gyn Fertil.* 1992;15:105–151. Used by permission.)

below zero because distal urethral pressure is less than bladder pressure. The profile ends when the urethral transducer reaches the urethral meatus, where the urethral pressure tracing falls to zero at atmospheric pressure where it was originally calibrated. The resting profile may be adversely affected by age, surgery, vaginal delivery, and other forms of trauma.[16,55,94,102] Other factors may also affect the resting urethral closure pressure profile, including the position of the patient,[103] the degree of patient relaxation,[98] the microtransducer catheter orientation, and the bladder volume.[99] It is important that these factors all be standardized when resting profiles are done.

Although resting urethral closure pressure profiles in normal women generally have higher closure pressures and functional lengths than in patients with genuine stress incontinence, there is tremendous overlap between these two populations. This makes the resting urethral closure pressure profile a poor diagnostic test for determining who has genuine stress incontinence,[102,105] which has led to the development of dynamic urethral closure pressure profiles during Valsalva and coughing. These tests were designed to help simulate actual physical stresses known to cause incontinence and to observe their effects on the sphincteric unit. While many argue that there is no utility in measuring resting urethral closure pressure, it could be argued that it allows for a graphic illustration of the functional capability of the sphincter and its internal integrity. We[106,107] have shown that women who have urethral closure pressures less than or equal to 20 cm H_2O are at increased risk of failing routine anti-incontinence operations. This finding of a low-pressure urethra appears to be the largest identifiable risk factor for failing anti-incontinence surgery. In these women who appear to have deficient intrinsic sphincteric mechanisms replacement of the proximal urethra above the urogenital diaphragm by traditional retropubic urethropexies and needle suspensions with improvement of urethral pressure transmission does not appear to be enough. While anatomically corrected 97% of the time by these operations, less than 50% of these women are objectively cured of their genuine stress incontinence.[106] These patients with low-pressure urethras also appear to have decreased success with nonoperative therapies.

Dynamic urethral closure pressure profiles may offer the urodynamicist more information about the function of the sphincter urethrae under stress. These profiles may be done at any bladder volume and in any position, but standing with a full bladder is probably the most stressful and sensitive position for this test.[104,108] The Valsalva profile is usually done at a catheter withdrawal speed and recorder paper speed of 5 mm/second. The catheter is withdrawn after the Valsalva maneuver is initiated, and the urethral closure pressure profile is recorded during Valsalva. Leakage of urine in the absence of a bladder contraction during Valsalva (genuine stress incontinence) will be associated with zero closure pressure and functional length by definition (or else the patient would not leak urine). Valsalva appears to generate a more consistent stress when compared to repetitive coughing, but cough profiles appear to be more sensitive and reproducible.[109]

Hilton and Stanton[102] preformed cough profiles in 120 stress incontinent women and 20 asymptomatic women and found that despite bladder neck opening in 25% of the controls, they all had pressure transmission ratios ⩾100% (increase in urethral pressure with coughing/increase in bladder pressure with coughing ×100%). The stress incontinent patients all had deficient pressure transmission ratios.[102] Cough urethral closure pressure profiles allow for the calculation of pressure transmission ratios. Profiles are considered positive for genuine stress incontinence if urinary leakage occurs with pressure equalization. This implies not only that is there negative pressure transmission but also that each negative deflection in closure pressure with coughing descends to or below the zero closure pressure level.

Leak Point Pressure

Recently a new urodynamic test has been described to quantify defective sphincteric function in patients with genuine stress incontinence.[110] The leak point pressure is an indirect measure of urethral resistance to the outflow of urine. It is a simple measurement of the intraabdominal pressure at the time of urinary leakage due to genuine stress incontinence and therefore is a more discriminating measure than the simple grading of the patient's symptoms.

The leak point pressure is measured by a vaginal or anal pressure transducer catheter. The patient is asked to Valsalva with gradually increasing force. The minimum abdominal pressure that causes urinary leakage is the leak point pressure. This test has been found to be quite reproducible and simple to perform and is useful in monitoring therapy for genuine stress incontinence.

Voiding Pressure Studies

Similar to the observations made during urethrocystometry, the same equipment may be used to analyze the emptying phase of the cycle. By using a dual microtransducer catheter or external transducer-catheter system, urethral and bladder pressure can be measured by placing the proximal transducer at the point of maximal urethral pressure. The abdominal pressure may be measured vaginally or rectally; EMG activity can be measured with either needle, wire, patch, or ring electrodes at the anus or urethra.

With this equipment coupled with a uroflowmeter the patient's voiding mechanism and function can be analyzed. Simple uroflowmetry offers only information about voiding speed, pattern, and completeness but normal uroflowmetry may be a representation of many different voiding mechanisms in women. Men may void normally only in one way, by relaxing the urethra and having a bladder contraction, but women may void normally in five different ways. They may void like men by urethral relaxation and detrusor contraction or by urethral relaxation alone. Either of these patterns may be augmented by Valsalva. They may also void by Valsalva alone. And while Valsalva voiding, is normal and more common in women with genuine stress incontinence,[111-113] it may cause prolonged urinary retention following surgery for genuine stress incontinence. In women with Aldridge-type suburethral sling procedures, permanent retention may occur if Valsalva voiding persists.[111]

Rud et al[111] have shown that normal continent women usually void by urethral relaxation (with corresponding decrease in EMG activity) followed by an increase in bladder pressure 3 seconds later. These two events have a cumulative effect of decreasing urethral closure pressure with the onset of urine flow on average 6 seconds after initial urethral relaxation. Failure of urethral relaxation during a detrusor contraction, underactive detrusor contractions (inadequate in time or magnitude), or an acontractile detrusor may be diagnosed as the cause of urinary retention and will predispose patients to overflow incontinence.

Voiding pressure studies may be used to not only analyze voiding dysfunction but also to analyze preoperatively borderline voiding mechanisms that may be decompensated by surgery. These studies allow for the prediction of voiding dysfunction and retention following anti-incontinence operations. Retropubic colposuspensions and needle suspension procedures may increase urethral resistance significantly and in some patients cause prolonged paruresis.[114] Bhatia and Bergman[115] have shown that women with low pressure (<15 cm H_2O) or absent detrusor contractions who Valsalva to void are at a twelve-fold increased risk of requiring prolonged (>7 days) postoperative catheterization.

Videocystourethrography

As mentioned earlier, the combination of cystometry and radiographic techniques has made it possible to study the pressure changes in the lower urinary tract while monitoring its appearance with fluoroscopy or ultrasound. While some find that fluoroscopy adds little to the evaluation,[112] many find it useful. Largely because of its expense and the need for dedicated space to house the fluoroscopic and urodynamic equipment it remains scantly utilized. Anatomic assessment and documentation with corresponding functional analysis of urethral, detrusor, and abdominal pressures clearly makes this an optimal system to analyze urinary tract pathology. The alternative is to evaluate the anatomic changes with endoscopy, pelvic exam, and Q-tip testing, as described previously, separate from the multichannel urodynamic assessment and to integrate the information mentally. The major disadvantages of this technique are the radiation exposure (1 Rad), cost, and need for radiology personnel and space.

Office Treatment of Urinary Incontinence

Although surgical therapy remains the most successful method to treat genuine stress incontinence and mixed incontinence, patients complaining primarily of urge incontinence should be managed with the various methods described here by the gynecologist or the primary care physician on an outpatient basis. Patients with genuine stress incontinence or mixed incontinence may also be managed nonsurgically with a great deal of success. More than 50% of people seen and evaluated at the Evanston Continence Center choose nonsurgical therapy to treat their urinary incontinence initially. Many of these patients are quite satisfied with nonsurgical management and never feel the need to undergo surgical therapy.

Pharmacologic Treatment of Urinary Incontinence

Pharmacotherapy is probably the most widely used, but not always the most successful, method for treating urinary incontinence. Pharmacotherapy is more frequently used as primary treatment of detrusor overactivity but less frequently to treat genuine stress incontinence.

Treatment of Detrusor Overactivity

Both detrusor instability and detrusor hyperreflexia respond to pharmacologic treatment. Five different classes of medications have been used to treat involuntary bladder contractions.[116] The anticholinergics are probably the oldest of these five groups and include propantheline bromide, methantheline bromide, and emperonium bromide (not available in the United States). These agents have been shown to be effective in the treatment of detrusor overactivity by inhibiting peripheral activation of the detrusor muscle through competitive inhibition of acetylcholine receptors at postganglionic sites.[117,118] These medications are limited by their wide range of anticholinergic side effects but may be used effectively in the doses listed in Table 19.3. Side effects of these medications include dry mouth, blurred vision, mydriasis, tachycardia, constipation, urinary retention, and drowsiness.[116] They cannot be used in patients with narrow-angle glaucoma. The side effects may outweigh the benefits of therapy, particularly in older women.[117]

A newer class of agents are the antispasmodic anticholinergic medications, which include dicyclomine hydrochloride, oxybutynin chloride, flavoxate hydrochloride, and hyoscya-

TABLE 19.3. Drugs utilized for the treatment of detrusor overactivity.

Anticholinergics			
Propantheline Bromide	P.O.	13–30 mg	q 4°–6°
Methantheline Bromide	P.O.	50 mg	q 12°–24°
Emeperonium Bromide	I.M.	25–50 mg	
Antispasmodics			
Dicyclomine hydrochloride	P.O.	10–30 mg	q 6°–24°
Oxybutynin chloride	P.O.	2.5–10 mg	q 6°–24°
Flavoxate hydrochloride	P.O.	100–200 mg	q 6°–12°
Hyoscyamine sulfate	P.O.	0.125–0.375 mg	q 6°–12°
Tricyclic antidepressants			
Imipramine hydrochloride	P.O.	25–75 mg	q 8°–24°
Doxepin hydrochloride	P.O.	25–75 mg	q 8°–24°
Sympathomimetics			
Phenylpropanolamine hydrochloride	P.O.	25–75 mg	q 8°–12°
Midodrine	P.O.	25–75 mg	q 8°–12°
Terbutaline	P.O.	2.5–5.0 mg	q 4°
Clenbuterol	P.O.	0.01 mg	q 8°
Calcium channel blockers			
Nifedipine	P.O.	10–20 mg	q 12°–24°
Terodiline	P.O.	25–75 mg	q 24°

mine sulfate (Table 19.3). These agents act both as smooth muscle relaxants in the bladder as well as anticholinergic agents. Oxybutynin and flavoxate also have some local anesthetic properties.[119] These agents are limited by their anticholinergic side effects. Dicyclomine and flavoxate also have additional gastrointestinal side effects that may limit their use.[119] Oxybutynin is probably the most effective of these agents.[120]

The tricyclic antidepressants have also been used successfully to treat detrusor overactivity.[121] The two agents with demonstrated efficacy in this group are imipramine and doxepin. With relief of symptoms in 40% to 50% of patients both appear equal in their efficacy to the anticholinergic and antispasmodic anticholinergic agents. However, because of their long half life, the tricyclics offer the advantage of single daily dosing. These agents also appear to increase bladder outlet resistance to improve patients who have mixed incontinence. In addition to their typical anticholinergic side effects, these agents may be strongly sedating and in rare cases may cause hallucinations or nightmares. These agents appear to block the uptake of serotonin and norepinephrine and also have some antihistaminic and anticholinergic properties, but it is poorly understood how they achieve their effect on the lower urinary tract.[121]

Theoretically, activation of beta receptors in the bladder should facilitate bladder relaxation and increase bladder capacity, but the beta sympathomimetics have not been widely used to treat urge incontinence. Both terbutaline and clenbuterol have been shown to improve the symptoms of urge incontinence, but clinically these have not been very effective.[119] The alpha adrenergic agents such as phenylpropanolamine, ephedrine, and pseudoephedrine have been used in some cases to treat involuntary bladder contractions, presumably by reflex inhibition of the bladder with increased urethral sphincter tone.[122,123] However, these agents are limited in their usage by the side effects of potential elevation of blood pressure, tachycardia, and anxiety. Clinical trials in Europe have shown midodrine to be a promising sympathomimetic agent for the treatment of stress incontinence, and this may also be found to have some reflex detrusor inhibitory effects. Experiments with this agent are currently being done at the Evanston Continence Center.

Calcium channel blockers such as nifedipine and terodiline have shown great promise in the treatment of detrusor overactivity.[124–126]

These agents appear to work primarily by blocking the influx of calcium into the detrusor muscle cells during detrusor contraction. In a small study of women with detrusor instability nifedipine has been shown by Rud et al[124] to reduce bladder contractions and to improve bladder capacity. Prior to its 1991 withdrawal from the market because of concerns about sudden death from cardiac arrhythmias in patients using terodiline, it had been used by more than 3 million people. Clinical trials in the United States have failed to show any significant cardiac arrhythmia but have demonstrated efficacy in women of all ages. The future of this medication, however, remains uncertain at this time.

The Pharmacologic Treatment of Genuine Stress Incontinence

Genuine stress incontinence has been treated pharmacologically with limited success in the past by using alpha adrenergic agents alone or in combination with hormonal therapy in postmenopausal women. Using alpha adrenergic agents such as phenylpropanolamine, ephedrine, pseudoephedrine, and midodrine, investigators have cured 10% to 20% of patients with improvement in up to 50% of women with genuine stress incontinence.[127,128] These agents appear to increase resting urethral tone. Estrogen replacement therapy in postmenopausal women has been shown to increase alpha receptors in the urethra. This has led to a synergistic effect of alpha adrenergic agents and estrogens in some but not all studies.

In addition, as mentioned earlier, tricyclic antidepressants may have activity in the treatment of genuine stress incontinence through central alpha adrenergiclike effects.[121] This action has led to improvement in around 50% of patients using this medication to treat their stress incontinence.

Behavioral Techniques

Behavioral modification by way of exercise, retraining, or biofeedback offers several advantages over pharmacologic therapy in the treatment of urinary incontinence. These therapies are economical and have no side effects. For the treatment of detrusor overactivity, most investigators have found behavior modification and biofeedback to be more successful than pharmacotherapy.[116,129]

Various bladder retraining protocols have been described and influence urinary incontinence from detrusor overactivity.[130] "Prompted voiding" may be used in institutionalized patients to try to have them void at regular intervals to reduce their urinary leakage, but this has little long-term benefit in reversing their underlying condition. "Timed voiding," where patients void on a fixed voiding schedule to try to avoid urge incontinence, is similar, but there is little success in relieving the underlying involuntary bladder contractions. Habit retraining and bladder drill have the advantage of trying to modify the underlying problem to achieve long-term results. With habit retraining, the patient voids at set intervals but can use the bathroom at other times if absolutely necessary. Bladder drill, however, sets a voiding interval that is to be followed strictly with the patient not voiding at other times. As the patient is able to stay dry at each level this voiding interval is gradually increased. During the day the patient voids at set intervals and ignores the sensation of urgency arising from the bladder. Therefore, the bladder is eventually brought back under cortical control. Popularized by Frewen,[131] this technique has been used by numerous investigators with success rates in excess of 75%. This technique requires close surveillance by the physician and an enthusiastic, compliant patient.

Biofeedback has also been used successfully to treat detrusor instability in women.[132,133] By giving patients information about the events prior to involuntary bladder contractions (that is, urethral pressure drop, EMG activity decrease, and so forth) the patient is able to learn to inhibit bladder contractions through contraction of the pelvic floor musculature before urinary leakage begins. About one-third of the patients are totally cured by biofeedback while two-thirds appear to be significantly improved.[132] Biofeedback may also be used along with progressive resistance exercises to treat genuine stress incontinence. Progressive resistance exercises, first popularized by Kegel[134] in 1949, work by teaching the patient to contract the levator muscles in an attempt to strengthen them and to improve the patient's stress incontinence. Kegel initially reported remarkable success treating women with a perineometer (a biofeedback device using a vaginal balloon to measure intravaginal pressure). However, since his initial reports, attempts by investigators to have patients do progressive resistance exercises just with verbal or written instructions have not been nearly successful.

Recently, a resurgence of biofeedback devices including perineometers, EMG devices, and vaginal cones has dramatically improved the success of this therapy. Perineometers use an intravaginal balloon to measure an increase in vaginal pressure created by contraction of the levator muscles around the vagina. However, pressure recordings may be falsely elevated in women who Valsalva instead of contracting their levator musculature, and this is detrimental to their treatment progress. Electromyographic biofeedback devices avoid this problem by measuring not only the EMG activity of the perivaginal musculature but also of the rectus muscles to demonstrate relaxation of the rectus muscles with contraction of levator ani muscles. These devices, however, are quite costly. For this reason, vaginal cones and directed pelvic floor education by physiotherapists are becoming increasingly popular as biofeedback alternatives. Vaginal cones of increasing weight may be placed in the vagina and worn during normal daily activities for fifteen minutes twice a day. Retention of the cone in the vagina implies successful contraction of the levator muscles. Release of the cone into the patient's undergarments suggests that either the patient was not contracting the levator muscles well enough or was straining. If the patient is able to retain the cone successfully, then she is able to move on to the next heaviest cone and therefore receive information regarding her progress. Vaginal cones have been shown to relieve stress incontinence in up to 20% of patients with improvement in 50%.[135]

Pelvic Floor Stimulation

Electrical stimulation of the pelvic floor musculature either with implanted electrodes, transspinal electrodes, or transcutanteous electrodes has been used for over a quarter of a century to treat both genuine stress incontinence and detrusor overactivity. Transvaginal or transanal stimulation appears to be the most popular and simplest form to use; cure rates for stress incontinence were reported in 30% to 60% of women and cure rates for detrusor overactivity were reported in excess of 50%.[136-141] Most of the work with pelvic floor stimulation has been done in Europe, but a few trials in the United States have shown efficacy.[141] These devices (Fig. 19.9) give off low electrical currents that are picked up by the pudendal nerve and directly contract the periurethral musculature; they also reflexively inhibit pelvic nerve activation to prevent involuntary bladder contractions. They have the advantage of being free of side effects, but poor reimbursement from third-party payers has limited their use in this country. The Evanston Continence Center is currently involved in a multicenter placebo controlled trial to test the efficacy of one of these devices. Because of their reported high efficacy in the absence of side effects, it is likely that these devices will become increasingly popular for the treatment of urinary incontinence.

Obstructive Devices

Over the years many gynecologists have recognized that intravaginal devices may be used to resupport the urethrovesical junction or obstruct it to relieve a patient's symptoms of stress incontinence successfully. Tampons have been used by women with mild stress incontinence to relieve leakage, and pessaries of all types have been used to resupport the urethrovesical junction or to obstruct the midurethra to control incontinence. Devices such as the Edwards device have been designed to compress the urethra directly, and recent research with urethral plugs and pads has demonstrated efficacy in blocking the outflow of urine from the urethra.[142] The efficacy and safety of some of these newer devices to

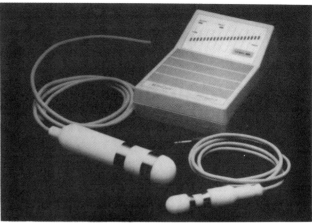

FIGURE 19.9. Two different commercially available pelvic floor stimuation devices in the United States: the Innova device (Empi, Inc.) and Microgyn II device (Hollister, Inc.).

obstruct the urethral outflow of urine remain to be demonstrated.

Hormonal Therapy

As mentioned earlier, estrogen replacement appears to increase alpha receptors in the proximal urethra and trigone.[34] Estrogen also appears to improve smooth muscle tone and appears to have positive effects on the urothelium of the urethra and trigone.[35] Several clinical trials have shown efficacy using estrogen replacement therapy alone to treat stress incontinence.[35,37-40] Improvement in patients with the symptom of urge incontinence is presumed to occur because of correction of the underlying atrophy in the urethra with resolution of sensory urge incontinence.

However, some investigators have failed to show significant improvement in patients using estrogen when compared to placebo.[41] Despite this, most investigators believe that atrophy of the lower urinary tract should be corrected in patients with urinary incontinence. This will sometimes improve or resolve urinary incontinence in elderly women. Due to the good absorption of estrogen, intravaginal administration of estrogen creams initially is preferred; a change to oral or transdermal therapy, because of variable absorption of estrogen when the vaginal epithelium thickens, should follow later. While the exact mechanism of action of estrogen on the lower urinary tract in the treatment of stress incontinence may be unclear, restoration of normal estrogen levels in postmenopausal women with incontinence offers few potential negatives and much to be gained. For this reason estrogen replacement therapy in all patients with urinary incontinence is recommended, unless they have some other direct contraindication (for example, breast cancer or active thromboembolic disease).

Conclusion

Patients presenting to the gynecologist or primary care physician with complaints of urinary incontinence may be effectively evaluated and treated in the office setting. Using a simple core evaluation, with a good history and physical exam, neurologic exam of the lumbosacral nerve roots, urine culture, assessment of the urethral hypermobility, standing cystometrogram, and residual urine check, these patients may be evaluated in approximately 20 minutes with good diagnostic accuracy. Some patients may require more detailed studies and should be referred to urodynamic centers, but the majority may be evaluated and treated by the interested primary care physician with the methods outlined in this chapter.

References

1. *Urinary Incontinence in Adults*. National Institutes of Health Consensus Development Conference Statement 1987;7:1–11.
2. Brazda JF. Washington report. *Nation's Health*. 1983;13:3.
3. Yarnell JWG, Voyle GJ, Richards CJ, et al. TP: The prevalence and severity of urinary incontinence in women. *J Epidemiol Comm Health*. 1981;35:71–74.
4. Diokno AC, Brock BM, Brown MB, et al. Prevalence of urinary incontinence and other urological symptoms in the noninstitutionalized elderly. *J Urol*. 1986;136:1022–1025.
5. Ouslander JG, Kane RL, Abrass IB. Urinary incontinence in elderly nursing home patients. *JAMA*. 1982;248:1194–1198.
6. Resnick NM, Yalla SV. Management of urinary incontinence in the elderly. *N Engl J Med*. 1985;313:800–805.
7. Resnick NM, Yalla SV, Laurino E. The pathophysiology of urinary incontinence among institutionalized elderly persons. *N Engl J Med*. 1989;320:1–7.
8. Ouslander J, Staskin D, Raz S, et al. Clinical versus urodynamic diagnosis in an incontinent geriatric female population. *J Urol*. 1987;137:68–71.
9. Sand PK, Brown LW, Ostergard DR. Uninhibited urethral relaxation: an unusual cause of incontinence. *Obstet Gynecol*. 1986;68:645–648.
10. Enhorning GE. Closing mechanism of the female urethra. *Lancet*. 1960;1:1414.
11. Enhorning GE. A concept of urinary continence. *Urol Int*. 1976;31:3–5.
12. DeLancey JOL. Anatomy and embryology of the lower urinary tract. *Obstet Gynecol Clin N Am*. 1989;16:717–31.
13. DeLancey JOL. Structural aspects of the extrinsic continence mechanism. *Obstet Gynecol*. 1988;72:296–301.
14. Francis WJA: Disturbances of bladder function in relation to pregnancy. *Br J Obstet Gynaecol*. 1960;67:353–366.
15. Francis WJA. The onset of stress incontinence. *Br J Obstet Gynaecol*. 1960;67:899–903.
16. van Geelen JM, Lemmens WAJG, Eskes TKAB, et al. The urethral pressure profile in pregnancy and after delivery in healthy nulliparous women. *Am J Obstet Gynecol*. 1982;144:636–649.
17. Snooks SJ, Swash M, Setchell M, et al. Injurying to innervation of pelvic floor sphincter musculature in childbirth. *Lancet*. 1984;1:546–550.
18. Allen RE, Hosker GL, Smith ARB, et al. Pelvic floor damage and childbirth: a neurophysiological study. *Br J Obstet Gynaecol*. 1990;97:770–779.
19. Constantinou CE, Govan DE. Contribution and timing of transmitted and generated pressure components in the female urethra. In: Zinner NR, Sterling AM, eds. *Female Incontinence*. New York: Alan R Liss; 1981:113–120.
20. Rosenzweig BA, Bhatia NN. Temporal separation of cough-induced urethral and bladder pressure spikes in women with urinary incontinence. *Urology*. 1992;39:165–168.
21. Bhatia NN, Bergman A. Cytometry: unstable bladder and urinary tract infection. *Br J Urol*. 1986;58:134–137.
22. Sand PK, Bowen LW, Ostergard DR, et al. The effect of retropubic urethropexy or detrusor stability. *Obstet Gynecol*. 1988;71:818–822.
23. Sand PK, Hill RC, Ostergard DR. Incontinence history as a predictor of detrusor stability. *Obstet Gynecol*. 1988;71:257–260.
24. Cardozo LD, Stanton SL. Genuine stress incontinence and detrusor instability: a review of 200 patients. *Br J Obstet Gynaecol*. 1980;87:184–190.
25. Webster GD, Sihelnik SA, Stone AR. Female urinary incontinence: the incidence identification, and characteristics of detrusor instability. *Neurourol Urodyn*. 1984;3:235–242.
26. Glezerman M, Glasner M, Rikover M, et al. Evaluation of reliability of history in women complaining of urinary stress incontinence. *Eur J Obstet Gynecol Reprod Biol*. 1986;21:159–164.
27. Ouslander J, Staskin D, Raz S, et al. Clinical versus urodynamic diagnosis in an incontinent geriatric female population. *J Urol*. 1987;137:68–71.
28. LeCoutour X, Jung-Faerber S, Klein P, et al. Female urinary incontinence: comparative value of history and urodynamic investigations. *Eur J Obstet Gynecol Reprod Biol*. 1990;37:279–286.
29. Farrar DJ, Whiteside CG, Osborne JL, et al. A urodynamic analysis of micturition symptoms in the female. *Surg Gynecol Obstet*. 1975;141:875–883.
30. Hastie KJ, Moisey CU. Are urodynamics necessary in female patients presenting with

stress incontinence? *Br J Urol.* 1989;63:155–156.
31. Awad SA, Flood HD, Acker KL. The significance of prior anti-incontinence surgery in women who present with urinary incontinence. *J Urol.* 1988;140–517.
32. Haadem K, Dahlstrom JA, Bengtsson M, et al. Sphincter function in the urethra and anal canal: a comparison in women with manifest urinary incontinence. *Int Urogynecol J.* 1991;2:85–89.
33. Larsson, G, Abrams P, Victor A. The frequency/volume chart in detrusor instability. *Neurourol Urodyn.* 1991;10:533–543.
34. Batra SC, Iosif CS. Female urethra: a target for estrogen action. *J Urol.* 1983;129:418–420.
35. Bhatia NN, Bergman A, Karram MM. Effect of estrogen on urethral function in women with urinary incontinence. *Am J Obstet Gynecol.* 1989;160:176–181.
36. Rud T. The effects of estrogens and gestogens on the urethral pressure profile in urinary continent and stress incontinent women. *Acta Obstet Gynecol Scand.* 1980;59:265–270.
37. Hilton P. The use of intravaginal oestrogen cream in genuine stress incontinence. *Br J Obstet Gynecol.* 1983;90:940–944.
38. Faber P, Heidenreich J. Treatment of stress incontinence with estrogen in postmenopausal women. *Urol Int.* 1977;32:221–223.
39. Walter S, Wolf H, Barlebo H, et al. Urinary incontinence in postmenopausal women treated with estrogens. *Urol Int.* 1978;33:135–143.
40. Karram MM, Yeko TR, Sauer MV, et al. Urodynamic changes following hormonal replacement therapy in women with premature ovarian failure. *Obstet Gynecol.* 1989;74:208–211.
41. Wilson PD, Faragher B, Butler B, et al. Treatment with oral piperazine oestrone sulphate for genuine stress incontinence in postmenopausal women. *Br J Obstet Gynaecol.* 1987;94:568–574.
42. Schiff I, Tulchinsky D, Ryan KJ. Vaginal Absorption of Estrone and 17-beta-estradiol. *Fertil Steril.* 1977;28:1063–1066.
43. Bowen LW, Sand PK, Ostergard DR. Urodynamic effects of a vaginal pessary in women with genital prolapse. Abstract presented at the 34th annual American College of Obstetricians & Gynecologists Meeting, New Orleans, LA, 1986.
44. Fianu S, Kjaeldgaard A, Larsson B. Preoperative screen for latent stress incontinence in women with cystocele. *Neurourol Urodyn.* 1985;4:3–7.
45. Bergman A, Koonings PP, Ballard CA. Predicting postoperative urinary incontinence development in women undergoing operation for genitourinary prolapse. *Am J Obstet Gynecol.* 1988;158:1171–1175.
46. Bump RC, Fantl JA, Hurt WG. The mechanism of urinary continence in women with severe uterovaginal prolapse: results of barrier studies. *Obstet Gynecol.* 1988;3:291–295.
47. Bradley WE, Rockswold GL, Timm GW, et al. Neurourology of micturition. *J Urol.* 1976;115:481–486.
48. Ostergard DR. The neurological control of micturition and integral voiding reflexes. *Obstet Gynecol Surv.* 1979;34:417–423.
49. Raz S. Adrenergic influence on the internal urinary sphincter. *Israel J Med Sci.* 1974;10:608–611.
50. Crystle CD, Charme LS, Copeland WE. Q-tip test in stress urinary incontinence. *Obstet Gynecol.* 1971;38:313–315.
51. Montz FJ, Stanton SL. Q-tip test in female urinary incontinence. *Obstet Gynecol.* 1986;67:258–260.
52. Fantl JA, Hurt WG, Bump RC, et al. Urethral axis and sphincteric function. *Am J Obstet Gyencol.* 1986;155:554–558.
53. Bergman A, McCarthy TA, Ballard CA, et al. Role of the Q-tip test in evaluating stress urinary incontinence. *J Reprod Med.* 1987;32:273–275.
54. Karram MM, Bhatia NN. The Q-tip test: standardization of the technique and its interpretation in women with urinary incontinence. *Obstet Gynecol.* 1988;6:807–811.
55. Rud T. Urethral pressure profile in continent women from childhood to old age. *Acta Obstet Gynecol Scand.* 1980;59:331–335.
56. Walters MD, Diaz K. Q-tip test: a study of continent and incontinent women. *Obstet Gynecol.* 1987;70:208–211.
57. Bergman A, Koonings PP, Ballard CA. Negative Q-tip test as a risk factor for failed incontinence surgery in women. *J Reprod Med.* 1989;34:193–197.
58. McGuire EJ. Urodynamic findings in patients after failure of stress incontinence operations. *Prog Clin Biol Res.* 1981;78:351–356.
59. Fantl JA, Smith PJ, Schneider V, et al. Fluid weight uroflowmetry in women. *Am J Obstet Gynecol.* 1983;145:1017–1024.
60. Drach GW, Ignatoff J, Layton T. Peak urinary flow rate: observations in female subjects

and comparison to male subjects. *J Urol.* 1979;122:215–219.
61. Wein AJ, Hanno PM, Dixon DO, et al. The reproducibility and interpretation of carbon dioxide cystometry. *J Urol.* 1978;120:205–206.
62. Robertson JR. Dynamic urethroscopy. In: Osteryard DR, Bent AE, eds. *Urogynecology and Urodynamics.* Baltimore, Md: Williams & Wilkins;1991:115–121.
63. Sand PK, Hill RC, Ostergard DR. Supine urethroscopic and standing cystometry as screening methods for the detection of detrusor instability. *Obstet Gynecol.* 1987;70:57–60.
64. Scotti RJ, Ostergard DR, Guillaume AA, et al. Predictive value of urethroscopy as compared to urodynamics in the diagnosis of genuine stress incontinence. *J Reprod Med.* 1990;35:772–776.
65. Ouslander J, Leach G, Abelson S, et al. Simple versus multichannel cystometry in the evaluation of bladder function in an incontinent geriatric population. *J Urol.* 1988;140:1482–1486.
66. Brubaker L, Sand PK. Cystometry, urethrocystometry, and videocystourethrography. *Clin Obstet Gynecol.* 1990;33:345–324.
67. Sand PK, Brubaker LT, Novak T. Simple standing incremental cystometry as a screening method for detrusor instability. *Obstet Gynecol.* 1991;77:453–457.
68. Wheeler JS, Niecestro RM, Fredian C, et al. Comparison of a simple cystometer with a multichannel in females with voiding dysfunction. *Int Urogynecol J.* 1991;2:90–93.
69. Sutherst JR, Brown MC. Comparison of a single and multichannel cystometry in diagnosing bladder instability. *B Med J.* 1984;288:1720–1722.
70. Cucchi A. Screening tests for detrusor instability in clinical urodynamics. *Int Urogynecol J.* 1991;2:101–104.
71. Low JA, Mauger GM, Dragovic J. Diagnosis of the unstable detrusor: comparison of an incremental and continuous infusion technique. *Obstet Gynecol.* 1985;65:99–103.
72. Arnold EP. Cystometry-postural effects in incontinent women. *Urol Int.* 1974;29:185–186.
73. Mayer R, Wells T, Brink C, et al. Handwashing in the cystometric evaluation of detrusor instability. *Neuroul Urodyn.* 1991;10:563–569.
74. Frigerio L, Ferrari A, Candiani GB. The significance of the stop test in female urinary incontinence. *Diag Gynecol Obstet.* 1981;3:301–304.
75. Coolsaet B. Bladder compliance and detrusor activity during the collection phase. *Neurourol Urodyn.* 1985;4:263–273.
76. Jorgensen L, Lose F, Andersen JT. One hour pad-weighing test for objective assessment of female urinary incontinence. *Obstet Gynecol.* 1987;69:39–42.
77. Sutherst JR, Brown MC, Richmond D. Analysis of the pattern of urine loss in women with incontinence as measured by weighing perineal pads. *Br J Urol.* 1986;58:273–278.
78. Klarskov P, Hald T. Reproducibility and reliability of urinary incontinence assessment with a 60-minute test. *Scand J Urol Nephrol.* 1984;18:293–298.
79. Lose G, Gammelgaard J, Jorgensen TJ. The one-hour pad-weighing test: reproducibility and the correlation between the test result, the start volume in the bladder, and the diuresis. *Neurourol Urodyn.* 1986;5:17–21.
80. Eadie AS, Glen ES, Rowan D. Assessment of urinary loss over a two-hour test period: a comparison between the urilos recording nappy system and the perineal pad weighing test. *Urogynecologia.* 1985;1:35–37.
81. Lose G, Rosenkilde P, Gammelgaard J, et al. Pad-weighing test performed with standardized bladder volume. *Urology.* 32:78–80.
82. Kinn AC, Larsson B: Pad test with fixed bladder volume in urinary stress incontinence. *Acta Obstet Gynecol Scand.* 1987;66:369–371.
83. Sutherst J, Brown M, Shower M. Assessing the severity of urinary incontinence in women by weighing perineal pads. *Lancet.* 1981;1:1128–1130.
84. Walsh JB, Mills GL. Measurement of urinary loss in elderly incontinent patients. *Lancet.* 1981;1:1130–1131.
85. Thind P, Gerstenberg TC. One-hour ward test vs 24-hour home pad weighing test in the diagnosis of urinary incontinence. *Neurourol Urodyn.* 1991;10:241–245.
86. Horbach NS. Problems in the clinical diagnosis of stress incontinence. *J Reprod Med.* 1990;35:751–757.
87. Bradley WE. Autonomic neuropathy and the genitourinary system. *J Urol.* 1978;119:299–302.
88. Sand PK, Bowen LW, Ostergard DR. The effect of a filling catheter during urodynamics. *Int Urogynecol J.* 1990;1:124–127.
89. McCarthy TA. Validity of rectal pressure

measurements as indication of intra-abdominal pressure changes during urodynamic evaluation. *Urology*. 1982;20:657–660.
90. Asmussen M, Ulmsten U. Simultaneous urethro-cystometry with a new technique. *Scand J Urol Nephrol*. 1976;10:7–11.
91. Kulseng-Hanssen S. Prevalence and pattern of unstable urethral pressure in 174 gynecologic patients referred for urodynamic investigation. *Am J Obstet Gynecol*. 1983;146:895–900.
92. Vereecken RL, Das J. Urethral instability: related to stress and/or urge incontinence? *J Urol*. 1985;134:698–701.
93. Weil A, Miege B, Rottenberg R, et al. Clinical significance of urethral instability. *Obstet Gynecol*. 1986;68:106–110.
94. Tapp AJS, Cardozo LD, Versi E, et al. The prevalence of variation of resting urethral pressure in women and its association with lower urinary tract function. *Br J Urol*. 1988;61:314–317.
95. Kiruluta HG, Downie JW, Awad SA. The continence mechanism: the effect of bladder filling on the urethra. *Invest Urol*. 1981;18:460–465.
96. Abdel-Rahman M, Coulombe A, Devroede G, et al. Urorectodynamic evaluation of healthy volunteers. *Urology*. 1982;19:559–564.
97. Awad SA, Downie JW. Relative contributions of smooth and striated muscles to the canine urethral pressure profile. *Br J Urol*. 1976;48:347–354.
98. Plante P, Susset J. Studies of female urethral pressure profile. I. The normal urethral pressure profile. *J Urol*. 1980;123:64–69.
99. Constantinou CE. Resting and stress urethral pressures as a clinical guide to the mechanism of continence in the female patient. *Urol Clin N Am*. 1985;12:247–258.
100. Weil A, Reyes H, Bischof P, et al. Clinical relevance of urethral stress profile using microtransducers after surgery for stress incontinence in females. *Urol Int*. 1983;38:363–367.
101. Rud T, Andersson KE, Asmussen M, et al. Factors maintaining the intraurethral pressure in women. *Invest Urol*. 1980;17:343–347.
102. Hilton P, Stanton SL. Urethral pressure measurement by microtransducer: the results in symptom-free women and in those with genuine stress incontinence. *Br J Obstet Gynaecol*. 1983;90:919–933.
103. Sorensen S, Knudsen UB, Kirkeby HJ, et al. Urodynamic investigations in healthy fertile females during the menstrual cycle. *Scand J Urol Nephrol Suppl*. 1988;114:28–34.
104. Henriksson L, Ulmsten U, Andersson KE. The effect of changes of posture on the urethral closure pressure in stress-incontinent women. *Scand J Urol Nephrol*. 1977;11:207–210.
105. Versi E, Cardozo L, Studd J, et al. Evaluation of urethral pressure profilometry for the diagnosis of genuine stress incontinence. *World J Urol*. 1986;4:6–9.
106. Sand PK, Bowen LW, Panganiban R, et al. The low pressure urethra as a factor in failed retropubic urethropexy. *Obstet Gynecol*. 1987;69:399–402.
107. Bowen LW, Sand PK, Ostergard DR, et al. Unsuccessful burch retropubic urethropexy: a case-controlled urodynamic study. *Am J Obstet Gynecol*. 1989;160:452–458.
108. Rosenzweig BA, Bhatia NN, Nelson AL. Dynamic urethral pressure profilometry pressure transmission ratio: what do the numbers really mean? *Obstet Gynecol*. 1991;77:586–590.
109. Hilton P. The urethral pressure profile under stress: a comparison of profiles on coughing and straining. *Neurourol Urodyn*. 1983;2:55–62.
110. Ghoniem GM, Roach MB, Lweis VH, et al. The value of leak pressure and bladder compliance in the urodynamic evaluation of meningomyelocele patients. *J Urol*. 1990;144:1140–1142.
111. Rud T, Ulmsten U, Andersson KE. Initiation of voiding in healthy women and those with stress incontinence. *Acta Obstet Gynecol Scand*. 1978;57:457–462.
112. Rud T, Ulmsten U, Westby M. Initiation of micturition: a study of combined urethrocystometry and urethrocystography in healthy and stress incontinent females. *Scand J Urol Nephrol*. 1979;13:259–264.
113. Sjoberg B, Nyman CR. Hydrodynamics of micturition in stress-incontinent women. *Scand J Urol Nephrol*. 1984;16.
114. Bhatia NN, Bergman A. Use of preoperative uroflowmetry and simultaneous urethrocystometry for predicting risk of prolonged postoperative bladder drainage. *Urology*. 1986;28:440–445.
115. Bhatia NN, Bergman A. Urodynamic predictability of voiding following incontinence

surgery. *Obstet Gynecol.* 1984;63:85–91.
116. Sand PK, Brubaker L. Nonsurgical treatment of detrusor overactivity in postmenopausal women. *J Reprod Med.* 1990;35:758–764.
117. Zorzitto ML, Jewett MAS, Fermie GR, et al. Effectiveness of propantheline bromide in treatment of geriatric patients with detrusor instability. *Neurourol Urodyn.* 1986;5:133.
118. Beck RP, Arnusch D, King C. Results in treating 210 patient with detrusor overactivity incontinence of urine. *Am J Obstet Gynecol.* 1976;125:593.
119. Gruneberger A. Treatment of motor urge incontinence with clenbuterol and flavoxate hydrochloride. *Br J Obstet Gynaecol.* 1984;91:275.
120. Cardozo LD, Stanton SL. An objective comparison of the effects of parenterally administered drugs in patients suffering from detrusor instability. *J Urol.* 1979;122:58.
121. Wein AJ. Nonsurgical treatment of lower urinary tract dysfunction. In: Raz S, ed. *Female Urology.* Philadelphia, Pa: WB Saunders; 1983:161–187.
122. Ulmsten U, Andersson KE, Persson CGA. Diagnostic and therapeutic aspects of urge urinary incontinence in women. *Urol Int.* 1977;32:88.
123. Bergman A, Koonings PP, Ballard CA. Detrusor instability: is the bladder the cause or the effect? *J Reprod Med.* 1989;34:834–838.
124. Rud T, Andersson KE, Ulmsten H. Effects of nifedipine in women with unstable bladder. *Urol Int.* 1979;34:421.
125. Ulmsten U, Ekman G, Andersson KE. The effect of terodiline treatment in women with motor urge incontinence. *Am J Obstet Gynecol.* 1985;153:619.
126. Tapp A, Abrams P, Henriksson J, et al. Multicenter study of terodiline in the treatment of detrusor instability: placebo controlled with dose titration. Presented at Eighth Annual Meeting, American Urogynecologic Society, San Francisco, September 1987.
127. Collste L, Lindskog M. Phenylpropanolamine in treatment of female stress urinary incontinence. *Urology.* 1987;30:398–403.
128. Riccabona M. The conservative treatment of stress incontinence in women with midodrine. *Wien Klin Wachr.* 1981;93:163–165.
129. Fantl JA, Hurt WG, Dunn LJ. Detrusor instability syndrome: the use of bladder retaining drills with and without anticholinergics. *Am J Obstet Gynecol.* 1981;140:885.
130. Hadley EC. Bladder training and related therapies for urinary incontinence in older people. *JAMA.* 1986;256:372.
131. Frewen W. Role of bladder training in the treatment of the unstable bladder in the female. *Urol Clin North Am.* 1979;6:273.
132. Cardozo LD, Abrams PD, Stanton SL, et al. Idiopathic bladder instability treated by biofeedback. *Br J Urol.* 1978;50:521.
133. Burgio KL, Whitehead WE, Engel BT. Urinary incontinence in the elderly: bladder-sphincter biofeedback and toileting skills training. *Ann Intern Med.* 1985;103:507.
134. Kegel AH. The physiologic treatment of poor tone and function of the genital muscles and of urinary stress incontinence. *West J Surg Obstet Gynecol.* 1949;57:527–535.
135. Peattie AB, Plevinick S, Stanton SL. Vaginal cones: a conservative method of treating genuine stress incontinence. *Br J Obstet Gynaecol.* 1988;95:1049–1053.
136. Fall M. Electrical pelvic floor stimulation for the control of detrusor instability. *Neurourol Urodyn.* 1985;4:329.
137. Kralj B. Selection of patients for treatment with functional electrical stimulation. *Urogynecologica.* 1985;1:41.
138. Lundstrom S, Fall M, Carlsson CA, et al. The neurophysiological basis of bladder inhibition in response to intravaginal stimulation. *J Urol.* 1983;129:405.
139. Fall M, Erlandson BE, Sundin T, et al. Intravaginal electrical stimulation: Clinical experiments on bladder inhibition. *Scand J Urol Nephrol.* 1978;44(suppl):41.
140. Vereecken RL, Das J, Grisar P. Electrical sphincter stimulation in the treatment of detrusor hyperreflexia of paraplegics. *Neurourol Urodyn.* 1984;3:145.
141. Bent AE, Sand PK, Ostergard DR, et al. Transvaginal electrical stimulation in the treatment of genuine stress incontinence and detrusor instability. *Int Urogyn J.* (in press).
142. Nielsen KK, Kromann-Andersen B, Jacobsen H, Nielsen EM, et al. The urethral plug: a new treatment modality for genuine stress urinary incontinence in women. *J Urol.* 1990;144:1199–1202.

20
Genital Herpes

Timothy L. Sandmann

Introduction

Genital herpes is a common problem encountered by the practicing clinician. The severity of the condition ranges from minimal to severe. Primary occurrences are usually more severe than secondary outbreaks, and the management of the condition should reflect this. Diagnosis is the first step in management, and clinical recognition plays a major role in identifying the symptoms and physical findings. Viral cultures or antigen identification, if possible, to confirm the diagnosis is also important. Patient education, stressing self-recognition and tailoring sexual behavior to curtail transmission, follows closely. Therapy with acyclovir is helpful, especially in decreasing the intensity of the primary episode and, possibly, in subsequent episodes.

Incidence

Genital herpes simplex virus (HSV) is a recurrent sexually transmitted disease (STD). Genital herpes, which has reached epidemic proportions, is one of the most common STDs encountered by the clinician. The Centers for Disease Control reported that there are an estimated 450,000 new cases of genital HSV infection each year in the United States.[1] Herpes is highly contagious, and a large percentage of sexual partners of infected individuals contract the disease.

Etiology

There are two types of HSV: type I (HSV I) and type II (HSV II). HSV I is usually responsible for oral and ophthalmic lesions, and genital herpes is usually a result of HSV II. HSV I can be isolated from genital lesions; the proportions vary, with HSV I causing approximately 3% to 14% of cases, although higher percentages have been found. One report quotes 61% HSV I isolation from genital lesions.[2]

Clinical Findings

Genital HSV is most severe during the primary occurrence. Primary herpes is both a local and systemic disease. The viral incubation period varies from 1 to 26 days following exposure, with the mean being 6 to 7 days. A true primary case usually starts with a prodrome lasting several days. Malaise, fever, headache, local vulvar burning, and vulvar paresthesia are followed by the formation of maculopapular lesions. These lesions become vesicles, then shallow ulcers. The lesions usually are bilateral in the primary episode, cover a large area of the vulva, and demonstrate inguinal adenopathy. The vagina and cervix may be involved to varying degrees. Vaginal and urethral discharge are common. Occasionally, bacterial superinfection occurs. Ulcers can occur for up to 6 weeks but usually heal spontaneously and without scar formation.

Urinary retention can require catherization, when the urethra and bladder are involved. Symptoms are usually resolved in 7 to 14 days. Occasionally a primary pelvic infection is subclinical. Recurrent genital HSV is more commonly a local disease, with symptoms being much less severe. The lesions are usually less extensive than the primary outbreak and are unilateral. The lesions are often preceded by vulvar burning and itching prior to the formation of lesions, but the whole attack usually resolves within 10 days.

Asymptomatic Viral Shedding

Asymptomatic viral shedding of HSV from the genital tract complicates the natural history of the disease and is an important mode of both horizontal and vertical transmission.[3] The frequency of asymptomatic viral shedding is difficult to quantitate. Brock et al demonstrated that approximately 1% of patients with genital HSV will demonstrate asymptomatic viral shedding. This 1% is constantly changing and if sampled frequently, probably all women with recurrent genital HSV infection shed the virus asymptomatically.[3]

Diagnosis

Diagnosis of genital HSV usually centers on the identification of the typical herpetic lesion in the genital area. History of recurrence is helpful, and many individuals experience a prodrome. Painful vesicular genital lesions should be considered HSV until the diagnosis is ruled out. The differential diagnosis includes trauma with secondary infection, syphilis, chancroid lymphogranuloma venereum, herpes zoster virus, carcinoma, and granuloma inguinale.

An accurate diagnosis is important. Clinical suspicion should be confirmed if possible by laboratory testing. Tissue cultures from typical lesions usually exhibit cytopathic changes consistent with HSV within 48 hours of inoculation.[4] Culture is the most widely used method of diagnosis. Immunofluorescent monoclomal antibody staining and enzymeimmunoassays (ELISA) has sensitivity and specificity approaching cell culture.[5,6] Cytology yielding the typical multinucleated cells and eosinophilic inclusion bodies can be obtained from scraping at the base of the vesicles and is rapid but lacks sensitivity. Electronmicroscopy also struggles with sensitivity. Serology tests for detection and titer of immunoglobulin G indicate exposure to HSV but are not reliable as an index to viral reactivation.[4]

Management

Management of genital herpes is twofold. First, the patient needs extensive education regarding the natural history of the disease, risk of recurrence, and asymptomatic viral shedding. Careful counsel regarding the use of condoms and efforts to decrease sexual transmission should be offered. Many patients suffer from the fear of sexual rejection and become sexually withdrawn and depressed. These patients need proper referral and encouragement to seek a support group.

Antiviral therapy is available in the form of acyclovir, which modifies the course of the disease but is not a cure. Acyclovir is available in intravenous, oral, and topical preparations. Treatment of first and recurrent episodes of genital herpes with acyclovir ointment has little relative value compared with oral acyclovir.[7] For severe primary infections intravenous acyclovir (5 mg/kg every 8 hours infused over 1 hour) has been shown to be effective.

Treatment of primary and nonprimary first clinical episodes of genital herpes with 200 mg of acyclovir five times a day for 5 to 10 days significantly reduces viral shedding and shortens lesion healing time.[7] Recurrent episodes of genital HSV treated with 200 mg orally five times a day for 5 days has been shown to reduce the duration of viral shedding by 1 day and to shorten the duration of the episodes by 1 to 2 days.[7]

Patients who have frequent recurrence have

been shown to benefit from continuous suppressive doses of acyclovir administered 400 mg two times per day or 800 mg one time per day. The frequency of recurrence was statistically reduced over placebo.[8,9] Long-term continuous use of acyclovir has been shown to be safe and well tolerated.[7] Suppression of asymptomatic viral shedding has not been demonstrated in patients on suppressive doses of acyclovir.[10]

Management of Pregnant Women with Genital HSV

A detailed discussion of the management of the pregnant patient with genital HSV is out of the realm of this chapter. Since acyclovir does not prevent asymptomatic viral shedding and can cause fetal nephrotoxicity, it should not be used during pregnancy unless the mother has a life-threatening disseminated disease.[7] Vertical transmission from the mother to the neonate is possible and is at highest risk in patients who have a primary genital herpes infection near or at the time of delivery.[7] Women with recurrent disease are at very low risk for vertical transmission, even if virus is being shed at the time of delivery.[11] Since it has been shown that patients with recurrent genital HSV demonstrate sporadic asymptomatic viral shedding, screening cultures do little to predict the likelihood of viral shedding at time of delivery. The American College of Obstetrics and Gynecology, the American Academy of Pediatrics, and the Infectious Disease Society for Obstetrics and Gynecology do not recommend cultures to identify candidates for cesarean delivery.[12-14] A patient with either a history or a partner with a history of genital HSV should undergo a careful history and physical to identify HSV prodrome or active lesions; those without should be allowed to labor.[7]

References

1. Baker DA. Herpes and pregnancy: new management. *Clin Obstet and Gyn*. 1990;33:253–257.
2. Barton IG, Kinghorn GR, Najem S, et al. Incidence of herpes simplex virus types 1 and 2 isolated in patients with herpes genitals in Sheffield. *Br J Vener Dis*. 1982;58:44–47.
3. Brock BV, Selke S, Benedetti J, et al. Frequency of asymptomatic shedding of herpes simplex virus in women with genital herpes. *JAMA*. 1990;263:418–420.
4. Osborne NG, Adelson MD. Herpes simplex and human papillomavirus genital infection: controversy over obstetric management. *Clin Obstet and Gyn*. 1990;33:801–811.
5. Thin RN. Management of genital herpes simplex infections. *Am J Med*. 1988;85(suppl 2A):3–6.
6. Clayton A, Roberts C, Godley M, et al. Herpes simplex virus detection by ELISA: effect of enzyme amplification, nature of lesion sampled and specimen treatment. *J Med Viral*. 1986;20:89–97.
7. Stone KM, Whittingham WL. Treatment of genital herpes. *Per Infec Dis*. 1990;12(suppl 6):610–619.
8. Mostow SR, Mayfield JL, Marr JJ, et al. Suppression of recurrent genital herpes by single daily doses of acyclovir. *Am J Med*. 1988;85(suppl 2A):30–33.
9. Kaplowitz LG, Baker D, Gelb L, et al. Continuous acyclovir treatment of normal adults with frequently recurring genital herpes simplex virus infection. *JAMA*. 1991;265:747–751.
10. Bowman CA, Woolley PD, Herman J, et al. Asymptomatic herpes simplex virus shedding from the genital tract whilst on suppressive doses of oral acyclovir. *Internat J STD & AIDS*. 1990;1:174–177.
11. Prober CG, Sullender WM, Yosukawk LL, et al. Low risk of herpes simplex virus infections in neonates exposed to the virus at the time of vaginal delivery to mothers with recurrent genital herpes simplex virus infections. *N Engl J Med*. 1987;316:240–244.
12. American College of Obstetricians and Gynecologists, ACOG Technical Bulletin No. 122. Washington, DC: American College of Obstetricians and Gynecologists; 1988:1–5.
13. Frigloetto FD, Little GA, eds. *Guidelines for Prenatal Care*. Elk Grove, Ill., and Washington, DC: American Academy of Pediatrics and American College of Obstetricians and Gynecologists, 1988:142–149.
14. Committee on Infectious Disease, American Academy of Pediatrics. *Report of the Committee on Infectious Disease*. 21st ed. Evanston, Ill.: American Academy of Pediatrics; 1988.

21
Dystrophy and Human Papilloma Virus–Associated Disorders of the Vulva

John V. Knaus

A variety of dermatoses affect the vulvar skin. Wide variation in the clinical appearance and symptoms are the hallmark of these skin disorders. This chapter will assist the clinician in making an expedient diagnosis and choosing an effective therapeutic treatment for patients with two of the most common vulvar dermatoses, dystrophy and human papilloma virus–associated disorders.

Recent nomenclature clarification by the International Society for the Study of Vulvar Disease (ISSVD) has simplified the heretofore confusing terminology used in the diagnosis of vulvar dystrophy. The ISSVD replaced the misnomer *vulvar dystrophy* with *nonneoplastic epithelial disorders of skin and mucosa* (Table 21.1). Further, a classification of vulvar intraepithelial neoplasia (VIN) was included (Table 21.2).

TABLE 21.1. ISSVD classification: nonneoplastic epithelial disorders of vulvar skin and mucosa.

Lichen sclerosis
Squamous cell hyperplasia
Other dematoses
　Psoriasis
　Candida
　Condyloma
Mixed disorders of above (note each separately)

TABLE 21.2. ISSVD classification: vulvar intraepithelial neoplasia (VIN).

Squamous VIN
　VIN I: mild dysplasia
　VIN II: moderate dysplasia
　VIN III: severe dysplasia
Nonsquamous VIN
　Paget's disease (intraepithelial)
　Melanoma in situ

Lichen Sclerosis

Vulvar lichen sclerosis, a chronic disorder, typically turns the skin white with a wrinkled or parchment like appearance (Fig. 21.1). In its early stages, lichen sclerosis may be difficult to detect on clinical examination. The patient's symptoms may only be vague but recurrent vulvar pruritus. Usually the patient will have attempted self-treatment with various topical medications without success. As the condition progresses, the characteristic whitish skin change becomes more pronounced and frequently affects the perineum and perianal area. Chronic itching traumatizes the affected area, and superficial ecchymoses, telangectasias, and purpura may be evident on clinical examination. The atrophic component of lichen sclerosis progresses over time, and the morphologic features of the vulva may almost completely disappear (Fig. 21.2). The labia minora may become unapparent on clinical examination. The paraclitorial skin may flatten, completely hiding the clitoris. Contracture of the vulvar skin may ensue, with varying degrees of stenosis of the introitus.

The pathogenesis of lichen sclerosis is

FIGURE 21.1. Characteristic skin wrinkling of lichen sclerosis.

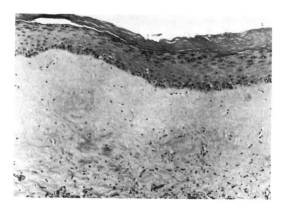

FIGURE 21.3. Microscopic appearance of lichen sclerosis (see text).

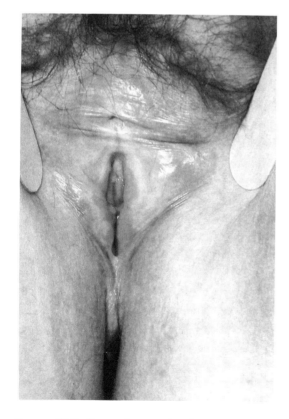

FIGURE 21.2. Loss of normal vulvar morphology from lichen sclerosis.

obscure. The etiology is probably multifactorial. Lichen sclerosis has been variously attributed to environmental factors, trauma, allergy, nutritional deficiency, metabolic abnormalities, psychoneurosis, infection, autoimmunity,[2-4] elevated elastase, decreased 5a-reductase,[6] and the absence of involucrin.[7]

Three microscopic features of lichen sclerosis are classic (Fig. 21.3). At the skin surface, epithelial thinning with flattening or disappearance of the rete pegs occurs. Beneath this is a homogeneous layer of acellular collagen-like tissue. Third, a distinct band of inflammatory cells infiltrate the dermis.

Vulvar skin biopsy is mandatory for an accurate diagnosis of lichen sclerosis, as well as all other vulvar dermatoses. Most experienced gynecologists have, at one time or another, treated a patient with what appeared to be "vulvar dystrophy" on clinical examination and later, in the face of treatment failure, have had the specific diagnosis rendered by a simple skin biopsy. A quick, easy, and effective technique of vulvar skin biopsy is described later in this chapter.

Topical testosterone cream is the primary treatment for lichen sclerosis. Chronic, long-term use of the testosterone is usually necessary for substantial improvement. Most patients are frustrated by the long-term compliance needed for successful treatment of lichen sclerosis. Frequent encouragement from the physician is always required. Treatment results range from ineffective to complete resolution of the dermatoses. Most patients achieve significant relief. Testosterone proprionate 2% must be compounded by the pharmacist. Petrolatum, Complex 15 lotion, or aquaphor are the preferred vehicles for preparing the testosterone. A 2-ounce prescription will last the patient for some time. A small amount of the preparation is massaged into the affected skin twice daily for 2 to 3 months. Significant improvement can rarely be expected for 4 to 6 weeks. Maintenance application of once or twice weekly is usually sufficient for long-term relief. Hydrocortisone (1% or 2.5%) may be used simultaneously with, or added to, the testosterone regimen if severe or persistent itching occurs.

Childhood and adolescent lichen sclerosis is not uncommon. Skin changes are similar to those of the adult. Topical testosterone should not be used in this patient group. Hydrocortisone 1% cream usually provides excellent symptomatic relief. If mild topical steroids do not provide improvement, progesterone ointment can be compounded: 100 mg of progesterone in 1 ounce of vehicle.

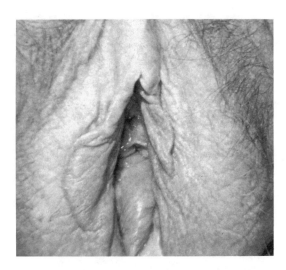

FIGURE 21.4. Thickened whitish vulvar skin characteristic of squamous cell hyperplasia.

Squamous Cell Hyperplasia

Squamous cell hyperplasia, frequently referred to as lichen simplex chronicus, is a chronic skin disorder caused by prolonged contact irritation of the vulvar skin. Repeated scratching provoked by any number of stimuli may initiate squamous cell hyperplasia, which will progress until treated. Chronic contact to an allergen, overvigorous hygiene, and chronic monilia vulvitis are frequent contributors. Squamous cell hyperplasia is a common histologic finding in patients with lichen sclerosis. Formerly termed "mixed dystrophy," proper ISSVD classification now requires both diagnoses to be stated when the conditions coexist.

Squamous cell hyperplasia may involve any skin surfaces of the vulva. Affected skin is usually distinctively thickened, whitish, and well-delineated (Fig. 21.4). Depending on the frequency and severity of the patient's itch/scratch cycle caused by the condition, vulvar skin may appear acutely inflamed, chronically dark red, or in various phases of healing from excoriation. Affected areas may be focal, multicentric, or involve a large confluent portion of vulvar skin.

Vulvar skin biopsy confirms the diagnosis of squamous cell hyperplasia with a characteristic histologic appearance (Fig. 21.5). Epithelial thickening, marked prominence of the rete pegs, and a diffuse dermal inflammatory cell infiltration are present.

Topical corticosteroids are the appropriate treatment for patients with vulvar squamous cell hyperplasia. Triamcinolone acetonide (0.01%), fluocinolone acetonide (0.025% or 0.01%), or similar preparations are applied two to three times daily until symptons abate. Some clinicians advocate a 7:3 preparation of 0.1% betamethasone valerate and crotamiton (EURAX) for patients with severe vulvar pruritus.

Once squamous cell hyperplasia improves or resolves, a maintenance regimen of prn use is prescribed. Prolonged use of the more powerful steroidal preparations (particularly fluorinated compounds) will result in symptomatic

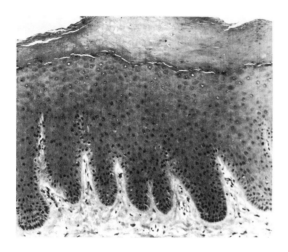

FIGURE 21.5. Microscopic appearance of squamous cell hyperplasia (see text).

atrophy of the treated vulvar skin. Weaker hydrocortisone preparations are ideal for maintenance applications.

Human Papilloma Virus–Associated Vulvar Disorders

Human papilloma virus (HPV) frequently affects the vulvar skin. The Centers for Disease Control estimates that in 1992 lower genital tract HPV infections will be 12 times more common than genital herpes infections. The apparent close association of HPV infection with lower genital tract epithelial neoplasia mandates that the clinician have a clear understanding of and treatment approach for the spectrum of HPV-associated diseases. While this chapter focuses on the vulva, the clinician must assume the presence of HPV at any anogenital site places the entire lower genital tract epithelium at risk and should be examined.

Considerable literature has reported laboratory methods used for identifying specific genital tract HPV viral types. "Experts" have explained the pathogenesis of particular lower genital tract epithelial neoplasia according to viral type, but the issue is far from resolved. Difficulties in under- and oversensitivity and specificity of the various laboratory procedures used in HPV typing have curtailed any clinical applicability for the present. Also, recent studies of populations with known high HPV genital infection rates have failed to develop neoplasia at the frequency expected.[8,9]

Nevertheless, low-risk, intermediate-risk, and high-risk HPV categories for development of neoplasia have been designated. Low-risk HPV types include 6, 11, 42, 43, and 44. The intermediate-risk group includes types 31, 33, 35, 51, and 52. The high-risk HPV types are 16 and 18.

At present, colposcopy of the vulva after vinegar soaking and appropriate (multiple) directed biopsies is the single most effective modality used for the accurate diagnosis of HPV-associated vulvar skin disease.

Limited Vulvar Condyloma

After a biopsy confirms benign histology, limited vulvar condyloma can be appropriately treated by the application of various topical preparations. Podophyllin resin solution (10% to 25% in tincture of benzoin) is applied directly to the lesions. Purified podophyllin extract is now available by prescription for self-application by the patient. Podofilox is applied to the affected area twice daily for 3 days. Four days without treatment follow. This regimen may be repeated. Trichloroacetic acid (85%), which is free of the potential systemic risks of podophyllin resin, delivers equal or greater efficacy. Trichloracetic acid use is safe during pregnancy and can be used on mucosal surfaces, including the vagina and anus.

Refractory and Extensive Vulvar Condyloma

Refractory and extensive vulvar condyloma lesions are best resolved by the aggressive use of a carbon dioxide laser in the hands of a gynecologist with demonstrated expertise in the management of this problem. Enthusiasm for using topical 5-fluorouracil for large lower genital tract condyloma has been tempered by unpredictable results and high failure rates.[10] Nevertheless, some patients may have excellent results from topical 5-fluorouracil application, and its use should remain a treatment alternative. This treatment should be rendered by an individual experienced with this modality.

Vulvar Intraepithelial Neoplasia

For patients with vulvar intraepithelial neoplasia, (VIN), colposcopy of the vulva after vinegar soaking and directed biopsies is mandatory for correct diagnosis, detection of the exact extent of disease, and appropriate choice of treatment. Solitary, clinically evident lesions are occasionally detected by the patient. More often, VIN presents with chronic, intermittent symptoms of vulvar pruritus and irritation. Vulvar colposcopy frequently reveals HPV-associated lesions far in excess of the patient's symptoms or the clinician's suspicions.

Treatment plans should ensure that the patient will have the maximum chance of permanent resolution of her VIN. Treatment must be individualized. While focal VIN I–II may be treated with local excision or carbon dioxide laser ablation, extensive lesions require extended laser ablation in skilled hands. Reports on interferon therapy have given mixed results, and any conclusions await further trials.

Vulvar Vestibulitis

A better understanding of the spectrum of vulvar vestibulitis has occurred in the past few years. Terminology used by various authors for this abnormality includes *idiopathic vulvodynia, vestibular adenitis, vulvar papillomatosis,* and *vulvar vestibulitis syndrome*. It has become clear that chronic HPV infection of the vulvar skin is the source of the symptoms in a large number of patients with this condition.

Patients with vulvar vestilulites may present with severe vulvar burning and tenderness, for which no apparent reason can be detected. Reid et al,[11] have identified four components nearly universally seen in this syndrome. These are described as an irritative acetowhite reaction of the vulvar epithelium attributable to chronic HPV infection, vascular ectasia (inflamed blood vessels) of the vulvar vestibule, painful inflammation of the minor vestibular glands, and deep pain on palpation of the Bartholin's fascia. The first two of these physical findings are thought to represent early, mild manifestations of HPV infection, and the second two a more severe, chronic form. Other authors have suggested a variety of factors associated with the diagnosis of vulvar vestibulitis syndrome. These include chronic recurrent candidiasis, chronic recurrent bacterial vaginosis, application of various topical medications used in treatment attempts, personal hygiene products, and peudendal neuralgia.[12]

Treatment of this condition with topical therapy has a high failure rate. Laser therapy has both advocates and opponents.[13] When conservative treatment is not successful, vestibulectomy with vaginal advancement is reported to have a high rate of success.[14]

FIGURE 21.6. Technique of vulvar skin punch biopsy.

Technique of Vulvar Biopsy

Skin biopsy is indispensable for the accurate diagnosis and initiation of effective treatment of vulvar dermatoses (Fig. 21.6). The affected skin is cleansed with an alcohol wipe or antiseptic solution. A skin wheal is created with local anesthetic. A 3.5 mm dermatology punch (Keyes, Bakers) is simultaneously twisted and pushed into the skin to obtain a core biopsy. The specimen is grasped with tissue forceps, cut from its flimsy subcutaneous tissue attachment, and placed in formalin. The biopsy site receives several minutes of direct pressure before a light dressing is placed. In rare instances, silver nitrate or a single 000 absorbable stitch is required to achieve hemostasis. Healing is excellent. Using this method, multiple biopsies from various areas of the vulva can easily be taken at one time.

References

1. Ridley CM, Frankman O, Jones ISC, et al. New nomenclature for vulvar disease. *J Reprod Med*. 1990;35:483.
2. Meyrick-Thomas RH, Ridley CM, Black MM. The association of lichen sclerosis et atrophicus in autoimmune-related diseases in males. *Br J Dermatol*. 1983;109:661.
3. Harrington CI, Dunsmore IR. An investigation into the incidence of autoimmune disorders in patients with lichen sclerosis et atrophicus. *Br J Dermatol*. 1981;104:563.
4. Meyrick-Thomas RH, Holmes RC, Rowland-Payne CME, et al. Incidence of development of autoimmune diseases in women after the diagnosis of lichen sclerosis. *Br J Dermatol*. 1982;107(suppl 22):29.
5. Douglas CP, Barnes CFJ. Proteslytic enzyme activity measured on extra cellular matrix in vulvar dystrophies. *J Obstet Gynecol*. 1986;6:193.
6. Friedrich EG Jr, DC, Kalva PS. Serum levels of sex hormones in vulvar lichen sclerosis and the effect of topical testosterone. *N Engl J Med*. 1984;310:488.
7. de Oliveira JM, Saleiro V. Involverin expression in vulvar lesions. *J Reprod Med*. 1982;31:828.
8. Crum CP, Burkett BJ. Papilloma-virus and vulvovaginal neoplasia. *J Reprod Med*. 1989;34:5466.
9. de Villiers EM, Wagner D, Wesch H, et al. Human papilloma DNA in women without and with cystological abnormalities. *Gynecol Oncol*. 1992;44:33.
10. Pride GL. Treatment of large lower genital tract condyloma accuminata with topical 5-fluorouracil. *J Reprod Med*. 1990;35:384.
11. Reid R, Greenberg, MD, Daoud Y, et al. Colposcopic findings in women with vulvar pain syndromes. *J Reprod Med*. 1988;33:523.
12. Marinoff SC, Turner ML. Vulvar vestibulitis syndrome: an overview. *Am J Obstet Gynecol*. 1991;165:1228.
13. Shafi, MI, Finn C, Luesley DM, et al. Carbon dioxide laser treatment for vulvar papillomatosis (vulvodynia). *Br J Obstet Gynaecol*. 1990;97:1148.
14. Mann MS, Kaufman RH, Brown A Jr, Adam E. Vulvar vestibulitis: significant clinical variables and treatment outcome. *Obstet Gynecol*. 1992;79:122.

22
Bartholin Duct Cysts and Abscesses: A Rational Approach to Treatment

David W. Doty

Few office or outpatient procedures bring greater or faster relief of pain to the patient than the rapid and appropriate treatment of an acute Bartholin duct abscess. The surgical treatment for these abscesses has evolved from a very complicated (and bloody) vulvar dissection with subsequent removal of the Bartholin gland and a 4- or 5-day hospital stay to today's complete treatment in the office or emergency room. The anatomy and physiology of this somewhat obscure gland, the differential diagnoses a physician should consider, and the treatment options available, taking into account the expense involved and the location where the patient is treated, will be analyzed in this chapter.

Historical Perspectives

The evolution of diagnosis and treatment of various forms of inflammation of the Bartholin gland is interesting; therapies suggested 30 and 40 years ago are now becoming the accepted standard of care. Several contemporary authors have reviewed the first correct anatomical description of these paired vestibular glands by Caspar Bartholin, a Danish anatomist, in 1677.[1] From the time of Bartholin's discovery more than 200 years passed before the function of these glands was clearly understood. In 1967 Word was one of the first to determine that the problem of apparent abscess or cyst formation in the Bartholin gland rests with the duct that connects the gland with the vestibule and not with the gland itself.[2] As a direct result of occlusion of the duct or the ductal ostia, either by infection or physical obstruction, a cyst or abscess forms not in the gland but in the duct of the gland. Green, in his basic text, concludes that these cysts are actually postinflammatory pseudocysts, again as a direct result of ductal occlusion.[3]

Anatomy of the Bartholin Gland

The Bartholin glands, also known as the major vestibular glands, are located deep in the perineum and surface at the introitus in the groove between the hymen and the labia minora. Usually, unless there is some degree of inflammation, the ductal ostia are not easily seen. The Bartholin gland is homologous to the male Cowper gland; its function is to provide continuous lubrication to the epithelium of the vestibule.[4] The Bartholin gland is composed primarily of mucus-secreting glands that eventually transform from glandular columnar cells to ductal stratified squamous epithelium similar to that of the vagina. It is in this transitional zone, as is seen in the cervix, that neoplasms occasionally occur.[5]

Typical Patient Presentation

In 1982 Poma indicated that Bartholin duct cysts or abscesses account for 2% of all new gynecology patients.[6] Patients will usually present with complaints of severe unilateral vulvar and vaginal pain. The subsequent swelling of the Bartholin duct causes difficulty in standing, sitting, or any activity that results in pressure on the perineum. The clinician may discover that it is somewhat difficult to complete an adequate vaginal examination or obtain necessary cervical cultures because of the exquisite point tenderness at the site of the abscess.

A Bartholin duct cyst, indicating an occlusion of the duct without subsequent infection, may also be found at the time of annual gynecologic exam, and the patient may be totally asymptomatic. Often the patient is surprised that the cyst is present and is easily palpable; no treatment is required for the small, asymptomatic cyst. The patient should be made aware of the cyst and its potential for change to an abscess in the future. As a Bartholin cyst enlarges, the patient might report a feeling of vague pressure in the perineum, mild discomfort with intercourse, or pressure with prolonged sitting or standing. When a Bartholin duct abscess begins to form, the patient is then aware of more localized unilateral discomfort. As the infected contents of the duct continue to distend the duct, the patient notes tenderness and erythema over the gland, followed by progressive swelling and edema. Occasionally, edema will spread into the labia minora and majora; the infected duct may actually rupture into the labial space, causing profound distention of the labia as the infection dissects into the vulvar tissue. More commonly the abscess "points" to a location close to the occluded ductal ostia, and without therapeutic intervention the abscess will eventually rupture spontaneously as a direct result of progressive necrosis and disintegration of the duct wall.[2] With rupture of the ductal abscess, the patient notes immediate and complete relief of her discomfort. The clinical aim is to evaluate, diagnose, and treat the symptomatic patient long before spontaneous rupture occurs.

The Etiology of Bartholin Duct Abscess Formation

The beginning of any Bartholin duct cyst or abscess occurs with the occlusion of the duct and the subsequent buildup of the mucinous secretion of the gland. If no infectious material is located within the occluded duct, a Bartholin duct cyst will usually evolve. Conversely, if infection is present, the development of a significant abscess can be quite rapid, usually depending on the original source of the inflammatory process. The development of a Bartholin duct abscess was originally considered to be a primary indicator of a gonococcal infection.[4] Although the clinician should always be suspicious of a gonococcal infection (and the patient should be evaluated fully for all sexually transmitted diseases), it is now known that virtually any bacteria that inhabits the perineum can result in abscess formation. Cultures obtained from the abscess cavity reveal myriad organisms including streptococci, staphylococci, coliforms, and anaerobes.[7] Chlamydial organisms are occasionally recovered from Bartholin duct abscesses, particularly from those patients with concomitant gonoccocal infections.[8] If all abscess contents were fully evaluated by culture, many mixes of organisms would be found. As a direct result of the progressive occlusion of the duct and subsequent anoxia by necrotic debris the predominate bacterial isolates would be anaerobic.[9] The actual identification of specific organisms found in the abscess cavity is not nearly as important as the needed course of therapy, that is, incision of the abscess followed by drainage and the development of a new ductal ostium to the vestibule.

The Evolution of Therapeutic Modalities

The evolution of various treatments for Bartholin duct abscesses has closely followed the level of understanding of the etiology and subsequent formation of the abscess. When it was finally determined that occlusion of the Bartholin duct was the causative agent and that the duct required patency for the gland to be functional, treatments became more successful. Prior to the 1930s the recommended surgical therapy of choice was the total excision of the infected or cystic Bartholin gland; usually no attempt was made to preserve its function. The myriad surgical dilemmas encountered by surgeons while attempting excision of a cystic or infected gland or duct included poorly defined margins of the abscess cavity, hemorrhage, and localized cellulitis, all of which made the dissection tedious and bloody. When it was determined that the gland was important in providing continuous lubrication to the vestibule, attempts were then made not only to provide adequate drainage of the cyst or abscess, thereby relieving the patient's pain, but also to provide a new tract to the perineum and restoration of function.

Word was one of the first gynecologists to recognize the need for a patent outflow tract from the Bartholin gland. His earliest attempts at creating a new ductal tract or fistula to the perineum involved a series of braided linen sutures passed directly from the cyst cavity to the vestibule and back again. The sutures were tied together and left in situ for epithelization.[2] Because of significant patient discomfort, secondary to the presence of a foreign body in the vulva, few patients allowed the sutures to remain long enough to create a fully epithelialized tract. In 1948 Davies reported his clinical success with simple incision and drainage of the cyst or abscess followed by packing the cyst cavity with gauze; the packing was changed twice weekly for a recommended 3-week period.[10] This form of therapy, although much more successful than Word's foreign bodies, was found to be time-consuming for both the patient and the physician.

In 1952 Krieger and Crile proposed using an inflatable bulb-tipped catheter placed within the cyst cavity to create the needed fistula.[11] This technique was found to be far superior to that of gauze packing because the inflated bulb kept the cyst cavity from collapsing while the catheter device remained in situ. Word further refined this specific catheter technology by devising a plugged latex catheter that would eventually bear his name. He listed several advantages of this form of therapy, including that the technique was simple, easily learned, and safe for the patient, that the treatment was one where the patient remained ambulatory, and that postoperative scarring was rare. Word felt that the greatest and most important advantage was that the gland or duct structure was preserved and returned to its normal and physiologic function.[2] Despite sound surgical and physiologic basis, the Word catheter enjoyed little popularity as a first-line therapy until it was rediscovered by Kovar and Scott and others in the early 1980s.[12] In 1987 Yavetz et al described a similar therapy that they termed fistulization, utilizing the usual cyst incision followed by placement of a modified 14F Foley catheter within the cyst cavity.[13]

In 1950 and again in 1960 Jacobson described the technique of marsupialization as a means of treating Bartholin duct abscesses.[14] As with the Word catheter, the objective of this variation was the creation of a new tract from the gland to the vestibule via a new mucocutaneous junction created by the wall of the abscess or cyst and the vestibule or labia. Jacobson's article emphasized the importance of the gland-to-vestibule well-being and some of the significant complications that could be encountered by surgical removal of the gland. It was noted that the excision of the gland was nearly impossible if the abscess had spontaneously ruptured; however, even after spontaneous rupture adequate treatment was easily accomplished by the newly applied procedure of marsupialization. Jacobson stressed the ease of the surgical procedure, its low recurrence rate, and its ability to be performed on an outpatient basis (a rare occurrence in the 1960s). The use of marsupialization was further expanded by Blakey et al in 1966 and Azzan in 1978.[15,16] Azzan indicated that 83% of Bartholin duct abscesses occurred in relatively young patients, between the ages of 20 and 50; he stressed the benefits of outpatient treatment and the rare occurrence of postoperative scarring and dyspareunia in this sexually active age group.

The latest variation, proposed by Cho and others[17] in 1990, is termed the *window operation*; its name is derived from the excision of a small part of the cyst wall prior to suturing, that is, making a window into the cyst cavity. It is felt that the larger opening creates a better ostium to the vestibular surface after closure and subsequent healing. Whether marsupialization or the window operation is utilized, the aim of therapy is the creation of a patent outflow tract that preserves the function of the Bartholin gland and prevents the recurrence of Bartholinitis or duct abscess.[17]

A third school of treatment advocates aspiration of the abscess cavity with a needle or IV catheter, affording the patient instant relief with a minimal amount of incisional pain. In 1982 Poma advocated this approach both as a means of getting better and more specific cultures of the abscess cavity and as being much more cost-effective for the patient.[6] As the study of aspiration continued, less and less importance was placed on the diagnostic outcome of the culture, and routine culturing for sensitivity was eventually dropped.[6] In 1985 Cheetham reported the use of aspiration as a means of differentiating by bacteriologic analysis and identification Bartholin duct abscesses from cysts.[7] If the aspirate appeared to be thick mucus, the mass was judged to be a cyst and no antibiotics were given; if frank pus was noted, the aspirate was cultured and appropriate antibiotics were begun. Despite the finding that no new ductal ostium was created by this technique, a cure rate of 85% was reported.

As newer operative technologies become readily available, new and different therapies will be tried to improve the outcome of the patient with a painful cyst or abscess. The carbon dioxide laser has been shown to be an effective tool in the creation of new duct ostia and for the subsequent drainage of the cyst cavity. Davis reported on a series of patients treated with the carbon dioxide laser in an office setting, using local anesthesia.[18] The laser allowed a 1.5-cm defect or window to be made in the cyst wall without significant postoperative scarring or fibrosis, and no cases of infections or recurrence were reported.[18] The scope of the carbon dioxide laser was further expanded by Lashgari and Keene when the laser was used not only to enter the cyst cavity but also to vaporize the cyst wall or capsule, thus decreasing the chance of recurrence. Although this procedure required general anesthesia, again in an outpatient setting, the usual operating time was less than 20 minutes and the cyst cavity was reduced to one-third its original size prior to treatment.[19] Both these reports stressed that the carbon dioxide laser is preferred for the treatment of Bartholin duct cysts and not for primary abscesses. It has been suggested that the laser is effective in the control of recurrence because of its ability to sterilize the surrounding tissues. Obviously, using the laser is a more expensive modality but one that has its place, especially in the area of recurrent cysts from a previous treatment failure.

Anesthetic Choices

When faced with the patient with acute pain from an infected or overly distended Bartholin duct cyst, the clinician must decide such things as the preferential method of anesthesia, the appropriate form of surgical therapy, and whether antibiotics will be used in the postoperative period. Prior to the 1960s, patients with these abscess cysts were typically admitted for prolonged hospitalizations that followed total excision of the Bartholin gland. It has been well documented that this form of treatment is overzealous and unneeded.

With the advent of more extensive office and outpatient procedures, anesthesia for these procedures has also undergone a significant transformation. In many instances, if the cyst or abscess is simply to be incised, the distention of the duct is significant and anesthesia is rarely required. However, simple incision and drainage (I&D), with a recurrence rate approaching 70%, is not an optimal form of therapy. For that reason, all the recommended procedures require additional manipulation of the cyst wall and therefore additional anesthesia. For an initial incision, topical ethyl chloride sprayed directly onto the cyst will afford adequate levels of anesthesia. If a portion of cyst wall is to be excised, as in marsupialization or the window procedure, local infiltration can then be used more comfortably after the initial treatment of ethyl chloride.

As an alternative to more local forms of anesthesia, Downs and Randall have suggested the revival of the pudendal block for outpatient vestibular procedures, including I&D, marsupialization, placement of the Word catheter, and aspiration of Bartholin duct abscesses or

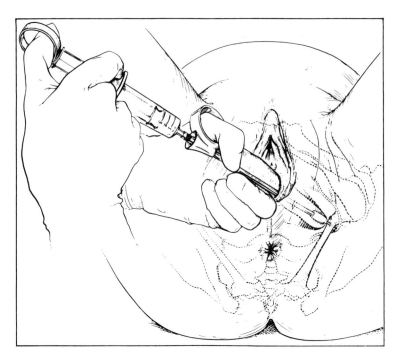

FIGURE 22.1. Local infiltration of the pudendal nerve via the sacrospinous ligament provides excellent unilateral anesthesia to the perineum for the drainage and marsupialization of Bartholin duct abscesses or cysts. (Reprinted by permission of *Williams Obstetrics*, 16th ed.)

cysts.[20] Utilizing 1%–2% xylocaine (without epinephrine), a small wheal is placed in the vaginal mucosa overlying the junction of the ischial spine and the sacrospinous ligament. The pudendal needle, sheathed in the trumpet or needle guard, is then advanced to a level just inferior to the ischial spine, identified by the operator's index finger. Prior to injecting the ligament (and infiltrating the pudendal nerve), aspiration is always attempted to preclude intravascular injection of the selected local anesthetic. One-half of the syringe (usually 5 cc) is injected into the ligament at an angle corresponding with the vaginal axis. The needle is then withdrawn into the trumpet, and the trumpet is repositioned below the ischial spine at an angle parallel to the floor. After the needle is repositioned, an additional 5 cc is injected following aspiration. Because typical abscesses or cysts rarely cross the midline, only one side of the perineum is involved and only one side of the pudendal nerve requires blockade. Down and Randall reported that 90% of their series of patients from the emergency room had adequate anesthesia for outpatient marsupialization.[20] Long considered a form of anesthesia appropriate only for the delivery room; a well-placed pudendal block can make the difference between a simple outpatient procedure in the emergency room or office and a more extensive procedure in the hospital requiring general anesthesia (Fig. 22.1).

It quickly becomes apparent that all abscesses or cysts will not be adequately anesthetized using ethyl chloride, local infiltration, or a pudendal block. A dissecting abscess or an uncooperative patient will require more extensive instrumentation, and a spinal or epidural block or general anesthesia may be required. It is the physician's responsibility to match the individual patient, the abscess or cyst to be treated, and the anesthetic that is appropriate for the selected procedure.

Operative Procedures for Treatment of Bartholin Duct Abscesses

Aspiration

The most quickly performed procedure that provides the patient with relief from pain would be simple aspiration of the cyst or abscess, as advocated by Poma.[6] A large-bore needle, 14 or 16 ga., or an appropriate large-bore IV catheter is prepared with a 10-cc syringe. The perineum is cleansed with Betadine or an other suitable antiseptic. Because of the exquisite tenderness of the vestibule and perineum, a spray form of antiseptic is usually needed. Similarly, topical anesthesia can be accomplished with the use of topical ethyl chloride, sprayed on the surface of the cyst, to a point of frost formation. Because of the short duration of anesthesia, the puncture should be made quickly after application of the ethyl chloride. The puncture is made within the vestibule, inside the labia minora and not in the labial tissue, even if the labia minora is significantly swollen by the cyst. Usually the cyst or abscess will "point" to the location of the original duct orifice. The contents of the cyst can then be aspirated and sent for aerobic and anaerobic culturing. This form of therapy is certainly rapid and cost-effective but does not result in the formation of a new or accessory ductal ostium. For that reason, recurrences are more common following simple aspiration than following marsupialization or other techniques that result in the formation of a new ductal ostium.

Marsupialization

The creation of a new or accessory duct for the Bartholin gland is the primary benefit of the marsupialization, Word catheter, or window procedures. Marsupialization has emerged as the procedure of choice for both treating the acute phase of an abscess and providing the long-term benefit of a new ductal ostium that preserves gland function. The patient is prepared for the procedure by cleansing the perineum and utilizing appropriate anesthesia, either local infiltration, pudendal block, or another regional blockade. Occasionally general anesthesia will provide the best form of anesthesia, especially for the excitable or nervous patient. Friedrich and others have stressed the importance of placing the initial incision within the vestibule, between the hymenal ring and the labia minora. Because the original ductal ostium is located within the confines of the vestibule, attempt to create a new ductal ostium in the same locale; the mucus secretion from the gland will continue to provide moisture to the vestibule.[21] An elliptical incision is made as close to the hymen as possible, well within the vestibule, and not in the labia minora. The incision will usually measure 2 to 3 cm in length and, to allow for a larger window into the cyst cavity, is made as an ellipse. Initial descriptions of marsupialization included a simple straight incision into the cavity, but with a straight incision, the opening is frequently completely closed by the time of the patient's first postoperative visit. The ellipse of the skin is excised and may be sent for pathologic review. The skin opening, now measuring 1 cm in width, may have included an incision into the cyst or abscess, heralded by a large gush of pus or mucus (Fig. 22.2).

An additional oval-shaped piece of cyst wall is excised, and appropriate cultures are

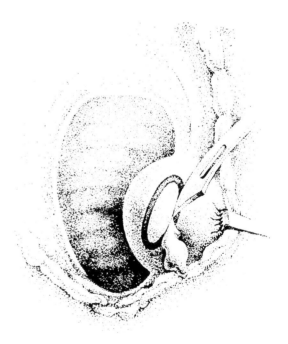

FIGURE 22.2. The incision into the cyst or abscess cavity is marked by a large gush of pus or mucus. An ellipse of skin obtained in marsupialization or the window technique is shown. (Cho et al. *Obstetrics & Gynecology*, I. Vol. 76, No. 5 November 1990. Reprinted by permission.)

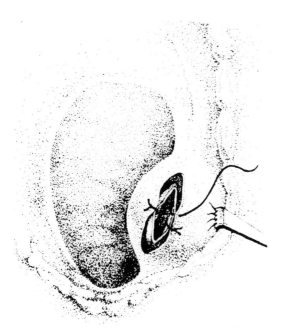 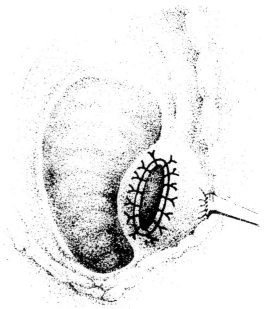

FIGURE 22.3. The cyst/abscess wall is attached, primarily, to the skin edges with absorbable interrupted suture. (Cho et al. *Obstetrics & Gynecology*, I. Vol. 76, No. 5. November 1990. Reprinted by permission.)

FIGURE 22.4. Utilizing interrupted sutures around the cyst cavity creates a large opening that will eventually create a new fistula that connects the gland to the vestibule. (Cho et al. *Obstetrics & Gynecology*, I. Vol. 76, No. 5. November 1990. Reprinted by permission.)

obtained from the cyst cavity. The cavity is then irrigated with warm, sterile saline via a Toomey syringe or other irrigator. With the operator's index finger the cavity is explored for loculations or adhesions. To further decrease the chance of recurrence of infection or abscess formation, it is very important to break up these loculations with blunt digital dissection. To create the desired and new fistulous tract to the vestibule, the edges of the incised cyst wall are approximated to the edge of the vestibular incision with interrupted, absorbable suture, such as 2-0 or 3-0 Vicryl or Dexon; chromic suture may also be used (Fig. 22.3). When the cyst wall is approximated to the epithelium of the vestibular incision, a large opening that will later become the new fistula connecting the gland with the vestibule is noted (Fig. 22.4).

Because of the high incidence of concomitant sexually transmitted diseases and primary Bartholin duct abscesses, the patient should always be evaluated for gonorrhea with both cervical and rectal cultures and Gram stains. Often because of intense perineal pain these cultures cannot be obtained until after the abscess has been evacuated. Postoperatively, the patient is encouraged to utilize sitz baths two to three times daily for 1 week. Antibiotics are usually not needed unless gonorrhea is documented on the Gram stain or is suspicious by history. Because of the common coexistence of chlamydial infections with gonorrhea, treatment would include coverage for both organisms. Amoxicillin (3 gm orally) plus doxycycline (100 mg twice daily) for 7 days provides adequate coverage; other regimens would include ceftriaxone (250 mg IM) plus doxycycline for 1 week. Additionally, the patient with true Bartholinitis (that is, significant edema and extensive erythema or cellulitis with induration of the surrounding tissues) should be placed on broad-spectrum antibiotics. Of the variety of organisms identified by Cheetham,

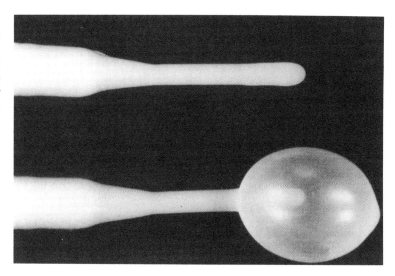

FIGURE 22.5. The Word catheter is placed into the cyst cavity in the uninflated version and following inflation remains in situ to create a permanent fistula. (*Vulvar Disease Therapeutic Procedures*, Chap. 3, Fig. 22, p. 54. Reprinted by permission.)

most are sensitive to penicillin or erythromycin (250 mg four times a day) and metronidazole (400 mg twice daily) for the anaerobic organisms. The patient should refrain from intercourse until she is seen for follow-up care in 1 week. During that time the cyst cavity will have markedly diminished and the window created by the marsupialization will be similarly smaller. The ostium should still be patent; occasionally mucus will be seen coming from the os. The patient should also be reassured that the absorbable sutures will care for themselves but that simple removal of retained sutures at the 2- to 3-week visit is needed on occasion. Follow-up cultures, if needed, should be obtained at the final postoperative visit.

The Word Catheter

The Word catheter was developed to act as a foreign body, around which a permanent fistulous tract would develop between the Bartholin gland and the vestibule. The catheter is made of latex and contains an inflatable bulb, similar to an in-dwelling Foley catheter, that, when inflated, holds the Word catheter in the cyst cavity; the stem of the catheter acts as the foreign body for fistula formation (Fig. 22.5).

Preparation for the insertion of the Word catheter is similar to the procedures for marsupialization. Following appropriate antiseptic and anesthetic preparation, a small incision is made into the cyst cavity. The incision should be small (approximately 5 mm), which allows for insertion of the catheter but also allows the bulb, when inflated, to retain the catheter within the cyst cavity. The bulb is inflated with 2 to 3 cc of sterile saline introduced via a fine 24 or 25 ga. needle. It should be remembered that the Word catheter is not for drainage through the catheter but for the creation of an epithelialized tract that, unfortunately, takes 6 weeks to occur. The vestibular portion of the catheter is tucked into the vaginal canal; intercourse is permissible whenever the initial tenderness of the abscess is gone. If the bulb is overinflated, tenderness will continue until the bulb is deflated. At the end of 6 weeks the bulb is fully deflated and the catheter is removed, leaving a patent tract to the gland.[21]

The Carbon Dioxide Laser

The most recent surgical development for therapy of Bartholin duct cysts utilizes the carbon dioxide laser to make the incision into the cyst cavity, whether for drainage or ablation of the cyst wall. Davis describes using local anesthesia, in the office, at the site of the occluded duct orifice.[18] With the laser operating at a power rating of 800 to 1000 W/cm^2 and a spot size of 1.4 mm, a defect measuring 1.5 cm in diameter is created. Vaporization of the vestibular tissue is created by moving the laser beam in rapid circles until the cyst cavity is entered. Once entered, treatment of the cyst is identical to that of marsupialization, with the exception of suturing the skin edge to the cyst wall. This difference may be beneficial to the patient in long-term follow-up. If excessive suture is used in marsupialization, dyspareunia may occur; this problem is eliminated by the carbon dioxide laser.[18] Lashgari and Keene utilize the laser to make a similar elliptical incision into the cyst cavity and vaporizes the cyst wall.[19] The laser operates at a power density of 9000 W/cm^2 (15 W) and a spot size of 0.4 mm. Once the cyst is entered the capsule of the cyst is vaporized with a defocused beam utilizing a spot size of 2 mm and a power density of 250 W/cm^2 (10 W). Again, no suturing or packing is used, and because this procedure is mostly indicated for cysts and not abscesses antibiotics are rarely needed.[19]

Carcinoma of the Bartholin Gland

Not all neoplasms of the Bartholin gland or duct are cysts or abscesses. Carcinoma of the Bartholin gland is relatively rare, comprising only 2% of all vulvar cancers. Because the duct is noted to have a defined transition zone, cancers are nearly equally spread between squamous cell carcinomas and adenocarcinomas. Unlike patients with cysts or abscesses of the Bartholin duct, who are generally between 20 and 50 years old, patients with carcinoma of the Bartholin gland will be older. Symptoms include bleeding, swelling, and induration over the Bartholin gland, a pattern somewhat different from that of an acute abscess. For that reason, rather than attempt marsupialization, it is suggested that any older patient with a persistent mass in the Bartholin gland undergo complete excision of the gland to rule out a malignant process. Additionally, any patient with a persistent mass or Bartholin duct cyst after marsupialization might also be a candidate for gland excision or biopsy for histologic evaluation.[22]

Summary

The Bartholin gland is extremely important in the maintennance of normal vulvar and vestibular physiology. Occlusion of the duct, rather than infection of the gland itself, causes the Bartholin duct cyst or abscess to occur. Treatment modalities detailed in this discussion have sought to relieve the patient's pain and to return the gland to its normal state via a normal outflow tract or fistula. To prevent unneeded recurrence of abscess formation, creating a new ductal ostium rather than simply draining the abscess, is important. Many of these procedures can easily be performed in the office with local or pudendal anesthesia, saving the patient money while also lessening surgical and anesthetic risk; the ultimate goals of any improved surgical therapy.

References

1. Heah J. Commentary: methods of treatment for cysts and abscesses of Bartholin's gland. *Br J Obstet Gynecol*. 1988;95:321–322.
2. Word B. Office treatment of cyst and abscess of Bartholin's gland duct. *Southern Med J*. 1968;61:514–518.
3. Green Jr. Gynecology: *Essentials of Clinical Practice*. 3rd ed. Boston: Little Brown & Company; 1977:227.
4. Woodruff JD, Friedrich, EG. The vestibule. In: Kaufman RH, Friedrich EG, eds. *Clinical Obstetrics & Gynecology*. Philadelphia, Pa: Harper & Row; 1985:134–141.
5. Friedrich EG. The vulvar vestibule. *J of Repro Med*. 1983;28:773.
6. Poma PA. Bartholin's duct abscess: office management. *Proc Inst Med Chic*. 1982;85–86.
7. Cheetham DR. Bartholin's cyst: Marsupialization or aspiration? *Am J Obstet Gynecol*. 1985;152:569–570.
8. Sweet RL, Gibbs RS. Chlamydial infections. In: *Infectious Diseases of the Female Genital Tract*. Baltimore; Md: Williams & Wilkins, 1985: p. 109.
9. Ledger WJ. Community-acquired gynecologic infections. In: *Infection in the Female*. 2nd ed. Philadelphia, Pa: LEA & Febiger; 1986:134.
10. Davies JW. Bartholin cyst: a simple method for its restoration to function. *Surg Gynecol Obstet*. 1948;86:3/29–3/31.

11. Krieger JS, Crile G. Bartholin's cyst. *Cleve Clin Q*. 1952;19:72–73.
12. Kovar WR, Scott JC. A practical, inexpensive office management of Bartholin's cyst and abscess. *Nebr Med J*. 1983;68:254–255.
13. Yavetz H, Lessing JB, Jaffa AJ, et al. Fistulization: an effective treatment for Bartholin's abscesses and cysts. *Acta Obstet Gynecol Scand*. 1987;66:63–64.
14. Jacobson P. Marsupialization of vulvovaginal (Bartholin) cyst: report of 140 patients with 152 cysts. *Am J Obstet Gynecol*. 1960;79:73–78.
15. Blakey DH, Dewhurst CJ, Tipton RH. The long-term results after marsupialization of Bartholin's cysts and abscesses. *J Obstet Gyneacol Br Commonw* 1966;73:1008–1009.
16. Azzan BB. Bartholin's cyst and abscess: a review of treatment of 53 cases. *Br J Clin Prac*. 1978;32:101–102.
17. Cho JY, Ahn MO, Cha KS. Window operation: an alternative treatment method for Bartholin gland cysts and abscesses. *Obstet Gynecol*. 1990;76:886–888.
18. Davis GD. Management of Bartholin duct cysts with the carbon dioxide laser. 1985;65:279–280.
19. Lashgari M, Keene M. Excision of Bartholin duct cysts using CO_2 laser. 1986;67:735–737.
20. Downs MC, Randall HW. The ambulatory surgical management of Bartholin duct cysts. *J Emer Med*. 1989;7:623–626.
21. Friedrich EG. Therapeutic procedures. In: *Vulvar Disease*. Philadelphia, Pa: WB Saunders Co; 1976:51–54.
22. Dodson MG, O'Leary JA, and Averette HE: Primary carcinoma of Bartholin's gland. *Obstet Gynecol*. 1969;35:578–584.

23
Lesbian Health Issues

Ruth Schwartz

Homosexuality has existed throughout history. Until recently female homosexuality, however, has had little recognition in historical or medical literature for a variety of reasons. Social and cultural changes have affected sexual practices, mores, and public and private acknowledgement of the facts about the existence of homosexuality.

Far more attention is given to male homosexuals than to lesbians. This may reflect the male-dominant culture and a relative lack of importance ascribed to women in general, or it may reflect a greater concern and disapproval of male homosexuality than of lesbianism.

Definitions

Homosexuality is the term first used in the 19th century to indicate a sexual orientation, an attraction for a person of the same sex with one or more of the dimensions of affection, fantasy, or erotic desire. A *homosexual* is one who prefers an erotic, sexual, and usually genital relationship with an individual of the same genital morphology. Like heterosexuals, homosexuals can be celibate, promiscuous, or have moderate sexual activities. There are many varieties of homosexual expression; the underlying motivations that lead to homosexual practices are diverse. The label *homosexual* has suggested a unified grouping that does not exist.[1]

A *lesbian* is a woman with affectional preference for females and, if sexually active, exclusively with females.[2] *Bisexuals* are persons with both homosexual and heterosexual attractions.

A *transvestite* is an individual who attempts a change of behavior and mannerisms to and changes to the clothes of the opposite gender.[3] Cross-dressing occurs in 4% of homosexual females during childhood and adolescence; 8% of adult females cross-dress in response to socially demanding occasions. The majority of female homosexuals prefer "tailored" or masculine-oriented clothes and shun makeup and feminine accessories.[12]

Transsexuals are individuals who feel a discordance or incongruity ("gender dsyphoria") between their anatomic sex and their psychological orientation. This is frequently described as "being trapped in the wrong body." A high percentage of transsexuals have psychological problems, including depression, poor self-esteem, guilt, alcoholism, and attempts at suicide. Some request hormonal therapy and surgical intervention to change their anatomical sex. The diagnosis is predominantly addressed psychologically; in the United States medical or surgical treatment is currently not considered to be appropriate.[4-6] *Ego-deptonic homosexuality* is homosexuality with internal conflict and disturbance, and a dominant wish to change sexual orientation.

Homophobia is a fear of homosexuality. The expanded definition includes negative feelings, attitudes, and behaviors toward homosexual individuals. *Homophilia* is a positive re-

Incidence

The prevalence of homosexuality in the United States had not been fully recognized prior to the survey publication by Kinsey et al[7] in 1948; before that there had been no systematic studies. This survey brought into focus the high prevalence of homosexuality among males of the Western world. Kinsey's report on 5,300 white American males who were extensively interviewed indicated that 37% had some homosexual contact to the point of orgasm. Between the ages 15 and 55, 13% were more homosexual than heterosexual and 4% were exclusively homosexual. Data related to the sexual behavior of 5,914 white American females collected in the late 1940s and early 1950s demonstrated that 28% had some homosexual response, for example, erotic arousal in a homosexual situation.[8] Of these women, 20% had actual homosexual experience and 13% reported orgasm with another woman by the age of 45. In each age group, when compared with males, about half as many women were primarily or exclusively homosexual. At the time of those reports, it was estimated that at least 5% of the general population was homosexual; this estimate triggered additional scientific exploration of male and female homosexuality, including its etiology, personality, psychology, and psychopathology.

Women have become more sexually active, both heterosexually and homosexually, during the second half of the 20th century. A study done in 1979 on 196 medical students in Britain (136 males and 58 females)[9] showed that about 40% of men and a higher proportion of women have been erotically attracted to a member of their own sex at some time. Three percent of both males and females were exclusively homosexual. Differences in this group showed in other aspects of sexuality; 1% of the males and 5% of the females were predominantly homosexual, and 4% of the males and 7% of the females were bisexual.

In 1981 one publication[10] estimated that there are about 10 million lesbians in the United States. Other estimates of 10% to 12% incidence of female homosexuality have been reported.[11] Thus it appears that homosexual attraction was and is not uncommon. In fact, it is much more common than the general society and medical community recognizes, discusses, or accepts.

The "coming out" process, the discovery of gay networks and patronization bars and organizations, is often delayed until lesbians are in their mid 20s or older. Nearly all lesbians sustain a relationship of a year or more, while only 60% of homosexual males were in a relationship that long. Beyond that period, male relationships, once the established, tend to last longer than female relationships.[12]

Homosexual affairs occur among 50% to 70% of women in prisons and are more obvious than in male prisons. Most of the women were not homosexual before imprisonment, and it is likely that the behavior is related to loneliness and lack of affection. Unlike the frequency in male prisons, sexual coercion and group assaults among female prisoners are rare.[13]

Etiology and Associated Factors

The assessment of etiologic and potentially contributory factors of homosexuality is difficult to define due to the limitation of the population base samples. The social stigma related to homosexuality compromises the willingness to cooperate and the verity of the sample. Because only a small minority of homosexuals are open about their homosexuality, the majority are not so categorized on questionnaires. Individuals who are aware of their bisexuality usually do not admit it if questioned directly and would probably provide more meaningful information if anonymity were assured.[10]

Black women are more reluctant than white women to disclose a homosexual experience. Only one-third of known black lesbians surveyed admitted their homosexuality to physicians. Nearly all black homosexual women indicated that they also had heterosexual activities.[14]

Neuroendocrine Factors

Neither male nor female homosexuals have been shown to have inherited an organic or nonpsychogenic factor that influenced sexuality. In one interesting study of homosexuals, 32 women and 38 men showed an increased incidence of left-handedness, which the authors interpreted as providing evidence of prenatal neuroendocrine events contributing to sexual orientation.[15]

There are conflicting findings concerning the levels of sex hormones in homosexual and heterosexual women. A study of the estradiol, progesterone, testosterone, and androstenedione levels at equivalent times during the menstrual cycle showed no significant differences in lesbians with or without previous heterosexual activity when compared with heterosexual controls.[16] Although some lesbians in one study had elevated testosterone levels when compared with heterosexual women, the difference did not reach statistical significance, and the testosterone levels in the homosexual group were within the normal range for females.[17]

An animal study of the prenatal manipulation of sex hormones demonstrated alteration of sexual behavior, but this was also accompanied by change in the genitalia.[3] Prenatal exposure to progestins was associated with no effect on sexual orientation. One study compared a small group of women who had been exposed to DES in utero with their sisters who had not been exposed; an increased incidence of lesbianism in the DES-exposed women was noted.[18] This author's own experience with a large number of DES-exposed women who have been followed over many years has shown no apparent increased incidence of homosexuality and does not substantiate the Ehrhardt et al. study.

The majority of females with congenital adrenal hyperplasia who received corrective endocrine therapy during infancy, early childhood, or later were heterosexual.[3,19] No increased incidence of homosexuality has been demonstrated in children with 5-alpha reductase deficiency. Because of their apparently female genitalia they are reared as girls and begin virilization at puberty, when the increasing testosterone levels override the earlier morphologic consequences of the enzymatic defect. Subsequent to their conversion to defined masculinity, although there are obvious stressful impacts, an increased incidence of homosexuality does not evolve.[3,20]

Parental Factors

Sexual preference in lesbians appears to be independent of parental factors. No typical pattern of parental relationship has been reported for lesbians. This is contrary to the distinct differences in parental-child relationships for gay men when compared with heterosexual men.[12] Some studies of lesbians showed more indifference to parents or the perception of the father as being weak, inadequate, or rejecting. Other studies described the fathers of lesbians as being close-bonding and intimate while the mothers were independent and dominant. All studies have had small cohorts, but the variables did not appear to have an effect on the females' emerging sexual preferences. When compared with heterosexual women lesbians certainly have a higher tendency to feel rejected by their parents.

Adolescent Factors

Adolescent female homosexuality has been considered by many psychiatrists to be the result of a persistent endeavor to deny what the girls regard as a anatomic inferiority. Also, the girl's fear of penetration and of the pain and damage associated with childbearing has been regarded as a factor contributing to homosexuality.[21] Over two-thirds of lesbians gave a history of childhood and adolescent "tomboyishness." In half of these women the characteristic persisted into adolescence. The "tomboy" tends to avoid feminine behavior patterns such as cooking, sewing, and playing house. By contrast, only 16% of heterosexual women felt that they were regarded as "tomboys."[12] Unlike the "sissiness" in boys, "tomboyishness" does not precipitate negative parental or peer reaction.

The onset of homosexual attraction occurs in preadolescence in 80% of lesbians, in 15% it is during adolescence, and in 5% it is during early adulthood. Forty-nine per cent of homosexual women had a romantic attachment to a boy or man; in two-thirds this occurred during adolescence. Sexual fantasies and romantic dreams are less common in homosexual women than in heterosexual women (33% versus 95%). Heterosexual arousal during adolescence occurred in 70% of homosexual women, while homosexual arousal was noted in 5% of heterosexual women.[12]

Psychologic Considerations

In 1973 the American Psychiatric Association (APA) removed *homosexuality per se* from its list of psychiatric disorders. *Ego-deptonic homosexuality* was established as a new classification to categorize those homosexuals who were either in conflict with, disturbed by, or wished to change their sexual orientation. In 1986 the APA eliminated *homosexuality* completely from the official diagnostic nomenclature. It is appreciated that homosexuality does not define the inherent characteristics of an individual and is analogous to heterosexuality in that it merely specifies sexual orientation.[22]

Compared to a few years ago, homosexuals today are regarded differently by the culture and are subject to different societal influences. The stigma has been removed by the savant, and homosexuality is no longer included as psychopathology.

Gender Identity

Attitudes and behaviors associated with femininity are learned and developed in early childhood. By the age of 3 a girl clearly understands her gender.[11]

Studies on gender identity, gender-role behavior, and sexual orientation have predominantly been carried out on males. There are differences in early gender conformity in some lesbians, but, in general, female homosexuals manifest psychological characteristics that have a similarity and commonality with other women. This pertains less frequently to gay men when compared with heterosexual men.[12]

Traditional gender roles are less common in lesbian and gay couples than in heterosexual couples. The allocation of roles may be based on pragmatic conditions (income, skills, or work schedules). The "masculine" role is not universally regarded as being dominant or more valued, and, as a corollary, femininity is no longer equated with an "inferior" role. The dominant role in a homosexual relationship is often assumed by the individual who is basically more assertive or aggressive, while the passive role falls to the partner who is softer and more subdued. The latter may have equal status in the relationship. The majority of lesbians interchange sex roles.[23]

Women are more likely than men to form close passionate attachment, even without physical contact. They prefer a lasting, faithful relationship. Homoerotic fantasies usually precede physical contact by women. More lesbians (26%) than homosexual men marry, but divorce or separation often ensues after a year or two. Some (5%) lesbians may marry a homosexual male for convenience. The casual sexual adventures of homosexual males are relatively uncommon with lesbians. Lesbians have more social contacts with the heterosexual feminine world than homosexual men do. To "confirm" their own identity as "a man" heterosexual men tend to reject homosexual men.

Gynecologic Care

It is very probable that a gynecologist's practice will include the care of lesbian patients, and some of these patients will be pregnant. There appears to be a lack of awareness of the sexual orientation of women on the part of many gynecologists. In 1979 Good[24] surveyed 110 gynecologists in Florida. The identification of lesbianism by one-third of physicians was done solely on clinical judgment and observation. About half the gynecologists indicated that they had not treated any lesbians, and the rest had treated three or fewer. Yet, in one county with 120 gynecologists, there were about 300,000 women (aged 20 to 55). Assuming an incidence of 4% female homosexuality, there were at least 12,000 lesbians in that population base, and they should be well represented in a busy practice. This suggests that either homosexual women avoid gynecologic care because of fears and concerns or that their sexuality is unrecognized.

Physician and Patient Attitudes: The Question of Bias

A study on physicians' attitudes toward homosexuality based on assessment of the members of a California medical society[25] indicated that 37% were homophilic (favorable attitudes), 40% were neutral, and 23% were homophobic (unfavorable attitudes). Female physicians tended to be more homophobic than male physicians, who tended to be more neutral. The younger the physician, the greater the percentage of homophilic attitude expressed. About one-third of obstetricians and gynecologists, family practitioners, surgeons, and orthopedists were homophobic. When asked how they felt about treating homosexual patients, 39% acknowledged being somewhat or unequivocally uncomfortable. This contradicts the basic ethics inculcated in medical training that physicians provide unbiased care for all patients, setting aside personal prejudice. Although physicians are reluctant to acknowledge *personal* prejudice in patient care, they are aware of prejudice in other physicians.[26] Many people are unaware of their own biases.

How do lesbians and female bisexuals feel about their physicians? When questioned about how open they were with their physicians about sexual orientation, 49% of lesbians were explicitly open and 1% presumed that the physician knew; (63% of homosexual men shared the information). One-third of all homosexuals thought that their providers were unsupportive or hostile; women were more likely to feel a prejudicial attitude.[27,28] Another study showed that 25% of lesbians had "a poor experience" with their gynecologists, 50% a neutral experience, and 25% an "adequate" experience.[29]

Medical Provider Approach

The ability to diagnose and treat health care problems, especially those with psychologic components, is dependent on an awareness of accurate information regarding lifestyle, including smoking, substance abuse, exercise, or sexual orientation. Yet there is still controversy about how homosexuals should be treated by society and by physicians.

To provide holistic and comprehensive care, especially in obstetrics and gynecology, rapport and trust must be established between the physician and the homosexual female patient. Requisites for obtaining adequate information about the patient are acceptance and understanding on the part of the physician and gaining the patient's trust.

A routine sexual history may alleviate difficulties for both physician and patient and may result in a more meaningful and mutually satisfactory relationship. The history should begin with generalized questions; if it becomes apparent that the patient is in the office due to a type of sexual activity, proceed to more specific areas. Over time, when the physician-patient trust grows, further questions and discussion about the satisfaction with the patient's lifestyle or needs should ensue.

Certain suggestions and caveats that pertain to all doctor-patient dialogues are particularly applicable to the management of lesbian patients.

1. Physicians should try to be aware of their own feelings, concerns, and attitudes so that any personal biases can be set aside.
2. A place to record sexual orientation should be included on the medical history form. The options should include "married, divorced, separated, widowed, single, and living with." Although a physician does not need to know the sexual orientation of every patient, in certain situations the following information would be very helpful: the patient's sexual history

current sexual behavior, support system, existence of a significant other, and previous diseases or disorders.
3. Recognition of a partner is as important with a lesbian patient as it is with other patients.
4. During dialogue, gender-neutral pronouns should be used with new patients. For example, when inquiring about sexual activity use *lover* or *partner* rather than *he* or *boyfriend*.
5. Language is very important; problems in language occur when terminology is unclear or has been associated with negative stereotypes. Many prefer the words *gay* or *lesbian* rather than *homosexual*, which some consider to be offensive. Using appropriate language will increase "visibility" of lesbians and homosexuals.[1,30]
6. Physicians should recognize that most lesbians are primarily or exclusively homosexual, but some are bisexual.
7. Physicians should not assume that all sexually active women are heterosexual or that a woman who is not having heterosexual intercourse is celibate.
8. Promoting birth control is inappropriate for a patient who is exclusively homosexual. Lesbians resent the assumption that all patients are heterosexual, especially when physicians empirically promote birth control.
9. The routine question, "Do you have intercourse?" is frequently used and may preclude further communication. By contrast, "Tell me about your sexual activity" presented in the same tone as "Tell me about your menstrual period" will enhance communication especially for the patient who has some anxieties or who needs more information and communication about her sexual practice.
10. Physicians should not assume that there is an abnormality in the patient's biology or endocrine system. It is no more likely for a lesbian to have an endocrine or genetic problem than any other woman, and the lesbian should be investigated for the same indications as heterosexual patients.
11. If the patient has psychological problems and needs referral for psychotherapy, one should not assume that all therapists are unbiased. Psychologists and psychiatrists vary widely in their adherence to the APA standards of unbiased practice.[31]
12. Confidentiality of the patient's sexual orientation should be maintained as with all patients (for example, do not share information of bisexuality or other affairs with the patient's partner unless specific permission is given by the patient). The physician should have the patient's approval before recording sexual orientation in her record.

Little attention has been focused on the health of homosexual women, which reflects on the deficiency of information in scientific literature.[23,24,29]

Abnormal Pap Tests

Forty-six percent of lesbians had yearly Pap tests; 73% had biyearly Pap tests. Many lesbians do not believe that it is necessary to have routine Pap tests. Considering that a history of heterosexual intercourse has been noted in 95% of bisexual women and 77% of lesbians, the potential risk is underestimated.[32] Abnormal changes of genital squamous epithelium as consequence of sexually transmitted HPV may take months or years to develop. Also emphasizing the role of routine Pap smears for lesbians is that 5% of cervical cancers are adenocarcinoma, unrelated to sexually transmitted disease.

The incidence of abnormal Pap tests is less common among lesbians than among heterosexual women. In homosexual females only 1% to 2.9% had a history of abnormal Pap smears. Homosexual women with cervical intraepithelial neoplasia (CIN) had a history of prior heterosexual intercourse, confirming that the risk of CIN among lesbians is related to their exposure to heterosexual coitus.[32,33]

Vaginal Infections

The frequency of gynecologic infections is lower in homosexual women than in heterosexual women. Bisexual women are more likely than lesbians to report a history of gynecologic infections, including trichomonas, candidiacies, and gonorrhea. The risk of transmission of trichomonas to the female partner is unknown; the possibility of digital transmission of the organism has been raised with no documentation. Evidence of sexual transmission of bacterial organisms is conflicting. Bacterial vaginosis does occur in women who have never had coitus.[34] Since it is unknown if orogenital or direct genital contact can transmit disease, it would be prudent to advise patients to avoid sexual contact until the infection is cured. Routine treatment of the partner is not recommended.

The only infection that has been reported with greater frequency in lesbians than the general population is herpes genitalis, which can be transmitted by direct contact with either oral or genital lesions during female sexual activity.[33]

Sexual expression used by lesbians include kissing, manual and oral stimulation of genitalia, breast stimulation, and direct genital to genital contact. These methods are also used by heterosexual couples. The sexual behavior of female homosexuals is associated with a significant lower risk of infection than that of heterosexuals. Therefore screening for sexually transmitted disease should be individualized; those with recent heterosexual encounters are at greater risk than those who are exclusively homosexual.

HIV Infection

National surveillance[35] assessed behavioral risks for HIV infection in lesbian women (0.8% of HIV-positive female are lesbians). Of the 79 HIV-positive lesbians 95% were intravenous drug users (IVDU). Of 103 bisexuals reviewed 79% were IVDU, 16% had male sex partners of risk or with infection, and 4% had had blood transfusions. There is a known risk of female-to-male HIV transmission via vaginal secretions. Transmission of the HIV virus through lesbian sexual contact has not been well documented. It is possible that the virus could be transmitted by lesbian partners via vaginal secretions or bleeding. Transmucous membrane contamination potentially leads to HIV transmission.

Cystitis

Cystitis occurs in about 15% of women in their reproductive years and is more common in homosexual women who have had heterosexual intercourse than in homosexual women who have not had heterosexual intercourse.

Endometriosis

The incidence of endometriosis is speculative. Accepting a 3% to 4% incidence[36] of endometriosis, the rate is thought to be similar in all women. Irregular or abnormal menses, dysmenorrhea, and abdominal pain also occur with the same incidence in heterosexual and homosexual women.

Pregnancy

In contrast to male homosexuals, lesbians commonly have an interest in babies. Strong maternal feelings and lesbianism coexist in many women. Many feel deprived because of the absence of a family home with children. As with traditional mothers, maternal identity appears to be a central part of the lesbian mother's self-concept and is a more important factor than race, socioeconomic status, or sexual orientation.

Surveys on middle-class homosexual women showed that 12% of white[32] and 34% of black homosexual women had children.[14] About 19% of the whites had pregnancy terminations. Spontaneous abortion rates are similar for bisexuals and lesbians (20%) but slightly higher than the estimated rate of 15% for the general population. The ratio of elective abortions to live births (1.2:1.0) was slightly less for homosexual women for unmarried heterosexual women (1.5:1.0).

Some lesbians marry in order to have children. Others have intercourse with a cooperative or an unsuspecting male partner, which is not acceptable to some lesbians. Artificial insemination may be used, with self-obtained donors or with purchased frozen semen that the women self-insert; others have artificial insemination in a medical facility. Adoption is difficult, especially for a single woman, but is successful for some.

Lesbian Motherhood

The fear of losing custody of their children, especially if divorce occurs, often causes homosexual married women to conceal their sexual orientation. In contested custody cases involving a lesbian, the judicial system is very likely to give custody to the father.

The problems encountered by lesbians having children include instability of the lesbian family unit, the absence of a role model for a male child, a preference of lesbians for daughters, problems in dealing with children's wishes and their concerns about their father, the few established role models of lesbian mothers, the need for a support system, and the disapproval of their physicians and society.

Physical Abuse

The incidence of violence against homosexuals increased 15% in 1991. Eighty percent of the attackers are adolescents. Victims are often unwilling to report the nature of the attack to police because police sometimes verbally and physically assault gay men and lesbians (a 29% increase in police abuse of male and female homosexuals was reported in 1991). Victims may also be concerned that their sexual identity will become public information. In 1991, 600 cases of abuse followed through the courts; only 2 resulted in convictions. Patients may also be afraid to tell treating physicians how the episode occurred.[37]

Psychological and Psychiatric Problems

The American Psychiatric Association has stated that it "is not abnormal or unhealthy to be homosexual." What is unhealthy and sometimes the source of stress and sickness is homophobia in society, which can lead to lowering of a homosexual's self-esteem, suicide, and violent attacks on lesbians.[2] The patient's emotional well-being is a concern for all in the health case field.

Many situations create stress for homosexual patients, including policies that exclude, segregate, or demean a person because of sexual orientation. In the past, these policies also included ethnic origin, race, sex, creed, age, and socioeconomic status. All the latter are now illegal in businesses and in some social organizations but persist in government organizations. All military organizations screen people for homosexual contacts to exclude them from service. In 1991, 926 uniformed personnel were dismissed from the service because of homosexuality.[38] Gay rights advocates contend that the military tracks down homosexuals with "homophobic zeal." The U.S. Congress is considering a repeal of the exclusion. The U.S. Immigration and Naturalization Service bans homosexuals from immigration. Many religious groups ban homosexuals, as do the Boy Scouts and Girl Scouts. In many regions Planned Parenthood will not see patients unless they are seeking birth control. The American Medical Association advises doctors to at least mention that therapies for reversal of sexual orientation exist.[1]

There is bias in language, psychological theories, and interventions that psychotherapists make.[31] Heterosexual bias is the cultural belief that values heterosexuality as superior to and/or more "natural" than homosexuality. This contributes to emotional stress that homosexuals face.

Research and generalizations are often made concerning psychotherapy of the homosexual person and an assumption of "deviant" behavior. Most studies and research on homosexuals have been limited to those in psychiatric hospitals, clinical or private therapy, prison populations, those involved in homophilic organizations, or those who visit known homosexual social sites (such as certain bars or parks). This population represents only a small minority of homosexuals.

There are differences in psychological adjustment between lesbians and heterosexual females, but both are equally well adjusted.[39] Social rather than clinical factors distinguished a group of college-educated female homosexual women from a matched group of heterosexual women.[22] There was no difference in the "Neurotic Scale," although lesbians were higher on extrovertism. Lesbians also scored higher on civil rights attitudes, more socialistic or political and economic attitudes, more radical on attitudes toward law and justice, and more liberal on attitudes concerning women's rights. The drop-out rate from college is higher among lesbians than among heterosexual women.

A review of empirical studies indicated that the results of studies of adjustment in lesbians were not different from those of heterosexuals, although some studies showed more submissiveness and higher goal-direction and self-acceptance.[22] Homosexual adults who accept their homosexuality do not regret their sexual orientation, function effectively both sexually and socially, and are no more distressed psychologically than heterosexual women.

Sexual Dysfunction

Sexual dysfunction occurs in some lesbians and bisexual women. Some homosexual women desire a change of sexual orientation because of dissatisfaction; a number of psychotherapeutic approaches are used to expedite this. Change in sexual orientation is a controversial issue among therapists. There is no unanimity as to whether it is to a patient's advantage to encourage behavioral changes if the patient requests it.

Alcoholism and Drug Abuse

Homosexual women are at higher risk for alcohol and drug abuse[40] than heterosexual women due to psychosocial variables such as stress levels and the cultural importance of bar settings. The incidence of alcoholism is about three times that of the general population[41]; however, this is not accompanied by higher rates of heavy abuse. The use of alcohol and drugs by homosexual women declines substantially with age.

Behavior Patterns

Lesbians have a greater tendency to drink, to be cross-gendered in self-assessment, and to drop out of college. There is a greater prevalence of psychiatric disorders, especially with alcohol abuse. The degree of functional disability as a result of these disorders does not make lesbians significantly more disabled than heterosexual women, and there is little impairment of function. Any person who shows a dominant pattern of frequent sexual activities with many partners who are strangers present with evidence of shallow narcissistic, impersonal, often compulsively driven genital-oriented rather than person-oriented sex, and is almost always regarded as pathological.[12,42] Depression is not uncommon with lesbians, especially in the early time of their "coming-out" process.

Summary

The significant incidence of female homosexuality means that it is highly probable that the practicing gynecologist will provide care for women of that group. Therefore it is imperative that gynecologists be knowledgeable about the homosexuals and be appreciative of the sensitivity of the dialogue with the homosexual patient. The taking of a history should allow for these patients. Sought-after advice should be based on available data related to psychologic or societal issues as well as to specific gynecologic problems. It is to be emphasized that lesbianism should not be labeled as psychopathology and that, as health-care providers for women, gynecologists should intensify data acquisition concerning homosexual females so that care is improved.

References

1. Council on Scientific Affairs, American Medical Society. Health care needs of a homosexual population *JAMA*. 1982;248:736–739.
2. Simpkin RJ. Lesbians face unique health care problems. *Can Med Assoc J*. 1991;145:1620–1623.
3. Meyer-Bahlburg HFL. Hormones and psychosexual differentiation: implications for the management of intersexuality, homosexuality and transexuality. *Clin End Metab*. 1982;11:681–701.
4. Gebhard PH, Johnson AB. *The Kinsey Data: Marginal Tabulations of the 1938–1963 Interviews conducted by the Institute for Sex Research*. Philadelphia, Pa: WB Saunders: 1979.

5. Ferber M. *Human Sexuality: Psychosexual Effects of Disease.* New York: Macmillan Pub Co.; 1985.
6. Kolodny R. Masters WH. Johnson VE. *Textbook of Sexual Medicine.* Boston, Mass: Little, Brown & Co.; 1979.
7. Kinsey AC, Pomeroy WB, Martin CE. *Sexual Behavior in the Human Male.* Philadelphia, Pa: WB Saunders; 1948.
8. Kinsey AC, Gebhard PH. *Sexual Behavior in the Human Female.* Philadelphia; Pa: WB Saunders; 1953.
9. McConaghy N, Armstrong MS, Birrell PC, Buhrich N. The incidence of bisexual feeling and opposite sex behavior in medical students. *J Nervous & Mental Disease.* 1979;167:685–688.
10. Gartrell N. Lesbian as single female. *Am J Psychotherapy.* 1981;35:502–517.
11. Stein TS, Cohen CJ. *Contemporary Perspectives on Psychotherapy with Lesbians and Gay Men.* New York: Plenum Publishing Corp; 1986.
12. Saghir MR, Robins S. *Male and Female Homosexuality: A Comprehensive Investigation.* Baltimore, Md: Williams & Wilkins; 1973.
13. Diamont L. ed. *Male & Female Homosexuality: Psychologic Approaches.* The Series in Clinical and Community Psychology. Washington, DC: Hemisphere Publ Corp; 1987.
14. Cochran SD, Mays VM. Disclosure of sexual preference to physicians by black lesbian and bisexual women *West J Med.* 1988;149:616–619.
15. McCormick CM, Witelson SF, Kingstone E. Left-handedness in homosexual men and women: neuroendocrine implications. *Psychoneuroendos* 1990;15:69–75.
16. Dancy CP. Sexual orientation in women: an investigation of hormonal and personality variables. *Biol Psychol* 1990;30:251–264.
17. Gartrell NK, Loriaux DL, Chase TN. Plasma testosterone in homosexual and heterosexual women. *Am J Psychiatry.* 1977;134:117–119.
18. Ehrhardt AA, Meyer-Bahlburg HF, Rosen LR. et al. Sexual orientation after prenatal exposure to exogenous estrogen. *Arch of Sex Behavior.* 1985;14:57–77.
19. Ehrhardt AA, Evers K, Money J. Influence of androgen and some aspects of sexually dimorphic behavior in women with the late-treated adrenogenital syndrome. *J Hopkins Med J.* 1968;123:115–122.
20. Imperto-McGinley J, Peterson RE, Gauter T, Sturia E. Androgens and the evolution of male gender identity among male pseudohermaphrodites with 5a-reductase deficiency. *New Engl Med.* 1979;300:1233–1237.
21. Mills JK. The psychoanalytic perspective of adolescent homosexuality: a review. *Adolescence.* 1990;25:913–922.
22. Diamont L. An investigation of a relationship between liberalism and lesbianism. Presented at American Psychology Association. 1977.
23. Kenyon FE. Homosexuality in gynaecologic care. *Clin Obstet Gyanaec.* 1980;7:363–386.
24. Good RS. The gynecologist and the lesbian. *Clin Obstet Gynecol.* 1976;19:473–482.
25. Mathews WC, Booth MW, Turner JD, Kessler MA. Physicians' attitudes toward homosexuality: survey of a California medical society. *West J Med.* 1986;14:106–110.
26. Pauley IB, Goldstein SC. Physicians attitudes in treating homosexuals. *Med Asp Human Sexual.* 1970;4:26–45.
27. Dardick L, Grady KE. Openness between gay persons and health professionals. *Ann Int Med.* 1980;93:115–119.
28. Smith EM, Johnson SR, Guenther SM. Health care attitudes and experiences during gynecologic care among lesbians and bisexuals. *Am J Public Health.* 1985;75:1085–1087.
29. Johnson SR, Guenther SM, Laub D, et al. Factors influencing lesbian gynecologic care: a preliminary study. *Am J Obstet Gynecol.* 1981;140:20–28.
30. Avoiding heterosexual bias in language. *Am Psychol.* 1991;46:973–974.
31. Garnets L, Hancock KA, Cochran SD, Goodchilds J, Peplau LA. Issues in psychotherapy with lesbians and gay men: a survey of psychologists. *Am Psycholog.* 1991;46:964–972.
32. Johnson SR, Smith EM, Guenther SM. Comparison of health care problems between lesbians and bisexual women: a survey of 2,345 women. *J Reprod Med.* 1987;32:805–811.
33. Robertson P, Schachter J. Failure to identify venereal disease in a lesbian population. *Sex Trans Dis.* 1981;8:75–76.
34. Sweet RL, Gibbs RS. *Infectious Diseases of the Female Genital Tract.* Baltimore, Md: Williams & Wilkins; 1990.
35. Chu SY, Buehler JW, Fleming PL, Berkelman RL. Epidemiology of reported cases of AIDS in lesbians, United States 1980–1989. *Am J Pub Health.* 1990;80:1380–1381.
36. Schenken RS. *Endometriosis: Contemprorary Concept in Clinical Management.* Philadelphia, Pa: JB Lippincott Co; 1989.
37. Cotton P. Attacks on homosexual persons may

be increasing, but many "bashings" still aren't reported to police. *JAMA*. 1992;267:2999–3000.
38. Lancaster J. Navy presses relentless search for gays. *Washington Post Weekly Edition*. 1992;9:31.
39. Siegelman M. Adjustment of homosexual and heterosexual women: A cross-national replication. *Arch Sex Behavior*. 1979;8:121–125.
40. McKirnan DJ, Peterson PL. Alcohol and drug abuse among homosexual men and women: epidemiology and population characteristics. *Addic Behavior*. 1989;14:545–553.
41. Hall JM. Alcoholism in lesbians: developmental, symbolic interactionist, and critical perspectives. *Health Care for Women Int*. 1990;11:89–107.
42. West DJ. *Homosexuality Re-examined*. Minneapolis, Minn: University of Minnesota Press; 1977.

24
Hirsutism and Virilism

Ian S. Tummon

What is Hirsutism?

Hirsutism is increased terminal hair growth in normal and abnormal areas that is considered to be excessive in that individual's racial and cultural setting (Fig. 24.1). A common but distressing benign condition, hirsutism is most often caused by an increase in androgen production from the adrenal cortex or overies or by increased peripheral conversion to androgen. A predisposition may also be present, with increased sensitivity of the follicles to normal levels of circulating androgens.

When and Whom to Treat?

As with growth-hormone therapy for short children without growth-hormone deficiency, demands for treatment for hirsutism are likely to increase.[1] Unrealistic female appearances are often used in commercial advertising, and such images may induce consumer demand for treatment of borderline hirsutism. The decision of when and whom to treat can be difficult because the diagnosis of hirsutism is semiobjective at best.

Hirsutism May Not Be Caused by Hyperandrogenism

Not all hirsutism is due to increased androgen. Increased hair growth with minimal coarsening of the hair shaft is termed hypertrichosis. Fine, downy hair on the limbs is a characteristic of anorexia nervosa and other weight-related amenorrheas. This hair growth is not associated with any abnormality of testosterone. Thyroid disorders are rarely associated with hirsutism, but an exception is juvenile hypothyroidism. In this condition the hair is more hypertrichosis in type. Medications that produce hypertrichosis are due to an unknown cause. Aside effect of some drugs is an abnormal increase in hair growth. Cyclosporine not only carries with it the general complications of immunosuppression, such as increased susceptibility to opportunistic infections or malignancy, but also hirsutism.[2] Minoxidol, administered for refractory hypertension, is now marketed to treat male-pattern baldness. Other offending medications that may cause hirsutism include phenytoin, danazol diazoxide, and streptomycin.

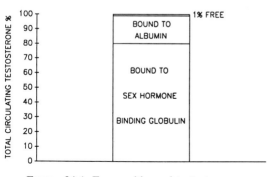

FIGURE 24.1. Free and bound testosterone.

Distinguish Virilism from Hirsutism

Hirsutism must be distinguished from virilism. Hirsutism *may* be the first indication of a malignancy. Suspicion of a serious underlying problem will be heightened by a rapid onset of symptoms and by signs of masculinization, such as increased laryngeal size, deepening of the voice, clitoromegaly, temporal hair recession, and increased muscle mass as well as hirsutism, acne, and menstrual disturbances. Androgen-secreting neoplasms of the ovary and adrenal are both, fortunately, rare causes of hirsutism. A rapid progression of hirsutism may help to differentiate a benign process from a neoplastic one.

Gradual worsening of hirsutism over many years suggests a functional disorder, whereas rapid appearance of symptoms suggests an underlying ovarian or adrenal neoplasm. If plasma testosterone levels exceed 7 nmol/L (200 ng/dL) despite a normal pelvic exam, the diagnosis may be an androgen-producing tumor. Careful history may reveal a source of exogenous androgen and thus avoid wasteful and invasive investigations and treatments.[3] However, diagnosis may only be possible through surgical exploration. Intraoperative ovarian vein androgen measurements can help to establish a histopathologic diagnosis in cases of microscopic virilizing ovarian neoplasms.[4] Surgery should not, however, be undertaken unless an adrenal source has been excluded. Computerized tomography or magnetic resonance imaging of the adrenalglands to identify tumors and adenomas has replaced many hormonal tests previously used as diagnostic aids. These techniques are sensitive but should not be routinely employed when evaluating patients whose only problem is hirsutism. In contrast, imaging of the adrenalglands *should* be employed to evaluate virilized hirsute patients. Adrenal adenomas or carcinomas that secrete androgens are particularly unusual causes of hirsutism and virilization. Adrenal adenomas or carcinomas may secrete androgens or cortisol. Testosterone levels are usually very high, often above 1,750 nmol/L (50,000 ng/dL). Androgen secretion is usually fixed and unresponsive. Women with Cushing's syndrome may have associated acne and hirsutism but rarely have signs of virilization.

Pathophysiology

The transition from vellus to terminal hair is dependent on androgen. The underlying endocrine etiology of hirsutism is either increased androgen production or increased sensitivity to androgens. The ovaries and adrenal glands each contribute 25% of the testosterone pool in the reproductive years. The remaining 50% is formed by peripheral conversions of preandrogens. Circulating testosterone is present in two forms: bound, which is biologically inactive, and unbound, which is biologically active. These forms of circulating testosterone are determined by the presence of sex-hormone-binding globulin that is formed in the liver. Under normal conditions, 99% of testosterone circulates in the bound form and less than 1% constitutes the unbound fraction (Fig. 24.2). In cases of increased testosterone production, however, there is suppression of sex-hormone-binding globulin, resulting in an increase in the unbound testosterone fraction. The presence of excess unbound testosterone leads to hirsutism. Intracellulary testosterone is converted to dihydrotestosterone, the active compound in the follicle.

The number of hair follicles in human skin is fixed at birth. It differs among ethnic groups but remains the same between sexes within the same race. Hair coloration also plays a part, since in women of fair complexion slighter degrees of increased hair growth may cause relatively more distress.

Chronic anovulation accounts for 75% of hirsutism cases. The pathogenesis of chronic anovulation is multifactorial, and chronic anovulation does not represent a specific endocrinopathy.[5] A self-perpetuating sequence of events may arise either at a central or ovarian level. A tonic luteinizing hormone increase is characteristic but not invariably found. Estradiol levels are too low to stimulate positive feedback of luteinizing hormone and

vellus hair

terminal hair

FIGURE 24.2. Vellus and terminal hair.

too high to permit adequate follicle-stimulating hormone increase. Hyperprolactinemia may be the initiating factor for chronic anovulation.

Hyperandrogenicity is also linked to insulin resistance. This is a rather complex situation. Insulin growing factor-I may be the main regulator of sex-hormone-binding globulin. By interaction with insulin growth factor-I receptors, insulin carries on its inhibitory activity on sex-hormone-binding globulin.[6] Higher insulin concentrations in anovulatory compared with ovulatory women with hyperandrogenemia may indicate that insulin resistance in the ovaries contributes to the mechanism of anovulation.[7]

Mild hirsutism often occurs following the final cessation of menses. There are usually no abnormal ovarian or adrenal findings. The ovarian stroma continues to synthesize testosterone and androstenedione at a normal premenopausal rate. Sex-hormone-binding globulin levels are low due to a lack of estrogen stimulation, and free testosterone levels may thus be elevated; these factors can account for mild hirsutism.

Clinical Assessment

Although hisrutism is distressing, the problem is more often cosmetic than life-threatening. Most androgen-secreting ovarian tumors are derived from the sex-cord mesenchyme. Sertoli-Leydig tumors or arrhenoblastomas are among the most common virilizing ovarian tumors. Granulosa cell tumors, although usually associated with estrogen production, have also been reported as causing hirsutism. Postmenopausal hirsutism has been associated with thecomas. Lipid cell tumors or luteomas are also commonly associated with androgenic manifestations. If imaging suggests an adrenal or ovarian tumor, then surgical exploration may be indicated. In women with virilization, pelvic exploration should be considered when unilateral adnexal enlargement is encountered, even if serum testosterone is only mildly elevated.

Hyperandrogenism and hirsutism are associated with lipid profiles that are risk factors for arterial disease. Thus, successful treatment of hyperandrogenism offers potential for reducing cardiovascular disease.[8]

Biochemical Evaluation

The source of excess androgen may be either the ovaries or the adrenal glands, and distinguishing between these sources may be difficult. Total serum testosterone concentration gives an inaccurate reflection of androgen availability.[9] Serum-free testosterone concentration is a more sensitive indicator of hyperandrogenism because sex-hormone-binding globulin is often reduced in association with hirsutism. Rittmaster and Thompson have found the ovaries to be the major source of circulating testosterone and androstenedione in women with chronic anovulation.[10] In women with hirsutism the ovaries were the major source of androgen secretion.

Adrenocorticotrophic hormone (ACTH) testing has been advocated to identify adult patients with congenital virilizing adrenal hyperplasia. The defect is most commonly due to partial deficiency of 21-hydroxylase. Baseline levels of precursor steroids for the defective enzyme are frequently normal. A simple ACTH test has been described to detect individuals with partial enzymatic deficiency. Hassiakos and co-workers determined the prevalence of late-onset congenital adrenal hyperplasia in a group of hyperandrogenic women presenting with menstrual disturbances and/or infertility. In this study, 17% of women showed evidence of partial 21-hydroxylase deficiency, and 6% showed evidence of 3 β-hydroxysteroid dehydrogenase deficiency. Neither basal hormonal levels nor clinical characteristics distinguished women with adrenal hyperplasia from other hyperandrogenic women. Azziz and co-workers[12] studied hyperandrogenic patients with ACTH stimulation; only 0.8% were found to suffer from presumed 11 beta-hydroxylase deficiency late-onset adrenal hyperplasia. They concluded that a systemic search for this deficiency in hyperandrogenism is probably unwarranted.[12] Siegel and co-workers also used ACTH stimulation to study women with hirsutism.[13] They measured progesterone, 17-hydroxypregnenolone, 17-hydroxyprogesterone, dehydroepiandrosterone, androstenedione, 11-deoxycortisol, and cortisol. Sixty percent of subjects demonstrated subtle defects in adrenal steroidogenesis. No correlation between the score for hirsutism and basal or ACTH-stimulated hormone levels was found. There was no significant correlation between the basal plasma dehydroepiandrosterone sulphate levels and the hormonal response to ACTH, nor were the basal levels of hormones predictive of the levels after ACTH stimulation. They reasoned that a substantial proportion of women with hirsutism have mild defects in adrenal steroidogenesis, revealed by an ACTH stimulation test, that are indicative of late-onset (nonclassic) congenital adrenal hyperplasia. Measurements of basal steroid levels were not helpful in differentiating among the causes of increased androgen production in such patients and may be misleading. This work has not been replicated, and the accuracy of the diagnostic criteria for maturity onset adrenal hyperplasia remains uncertain.

Weight Reduction

Obese women with chronic anovulation are more likely than lean women to have hirsutism with chronic anovulation. Sex-hormone-binding globulin concentrations are lower in obese women and are inversely related to insulin. Insulin has a direct inhibitory action on sex-hormone-binding globulin secretion. Weight reduction of more than 5% is associated with an improved biochemical profile and restoration of fertility.

Mechanical Therapy

Simple cosmetic measures are usually required to control the effects of hirsutism. These measures include bleaching, plucking, using depilatory creams, waxing, shaving, and undergoing electrolysis.[15]

Pharmacotherapy

Estrogens

Traditionally, estrogen-progestin combined oral contraceptives that suppress luteinizing hormone have had a role in controlling hirsutism. Luteinizing hormone stimulates the theca interna within the ovarian stroma to produce testosterone. The choice of progestin is very important for the treatment of hirsutism. A progestin with the least androgenic potential is preferred. Although biopotency studies of progestins equivalencies are difficult, ethynodiol diacetate as the progestin is usually preferred.[16]

Antiandrogens

One approach in the medical treatment of hirsutism lies in a group of drugs that block androgen effect at the hair follicle, the antiandrogens. An antiandrogen is a substance that acts by preventing androgens from expressing their activity at target sites.

Combined Estrogens and Low-Androgen-Potency Progestins

Porcile and Gallardo[17] studied monthly versus bimonthly treatment with oral contraceptives containing desogestrel in the maintenance of the remission of hirsutism. They found that monthly treatment kept hirsutism in remission and that testosterone and free testosterone levels decreased. Bimonthly treatment was equally effective keeping hirsutism in remission, although testosterone levels did not decrease. Bimonthly treatment was not better than monthly treatment in relation to changes in plasma lipids.[17] They also compared desogestrel with cyproterone acetate in oral contraceptives and found no differences between groups. They concluded that treatment should be prolonged to give best results.[18] Similarly, Prelevic and co-workers found favorable changes in hormonal profiles and hirsutism using a low-dose estrogen-antiandrogen combination (35 μgm of ethinyl estradiol and 2 mg of cyproterone acetate, Diane-35).[19]

Sprionolactone

Spironolactone also has mild antiandrogenic activity. A number of its side effects such as gynecomastia, decreased libido, and impotence initially suggested interference with androgen action. Several explanations of this have been given, including peripheral conversion of testosterone to estradiol and competitive binding of dihydrotestosterone or its receptor. Low-dose spironolactone improves hirsutism in a majority of hirsute women, irrespective of age, severity or duration of hirsutism, menstrual status, or serum hormone levels. Favorable responses were associated with increased severity of hirsutism ($P = 0.04$) and lower serum dehydroepiandrosterone sulfate levels ($P = 0.05$).[20] In a randomized study, effects of cyproterone acetate and spironolactone on hair growth were compared. Subjects completed 6 months of therapy with 100 mg per day of cyproterone acetate or 100 mg per day of spironolactone, both with concomitant estrogen therapy.[21] The reverse sequential regimen of ethinyl estradiol (30 μg daily) was given on days 5 to 25 and cyproterone on days 5 to 14. Both treatments reduced hair diameter, and neither treatment demonstrated an advantage over the other. In another randomized trial, Prezelj and co-workers found no improvement using combination therapy with dexamethasone and spironolactone over spironolactone alone.[22] In a double-blind placebo-controlled study of the efficacy of 100 mg daily of spironolactone in idiopathic hirsutism, McLellan and co-workers demonstrated no objective benefit from spironolactone treatment; subjective improvement was noted.[23]

Cimetidine

Cimetidine, a histamine receptor type 2 blocker, has been suggested as a possible treatment of hirsutism. In a prospective, randomized, controlled trial, 1.5 gm a day of cimetidine was given to hirsute women for 3 months. As measured by two assessment methods there was no significant change in the hair growth rate in either group. The treatment period is too short to make a final judgment.[24]

Cyproterone

Cyproterone acetate may be best considered as a nonandrogenic progestin.[25] Due to its delayed onset of action, it is most commonly administered with ethinyl estradiol in the standard reverse combination. Side effects consist of weight gain, fatigue, decrease of libido, and, rarely, galactorrhea. In a well-designed double-blind study, Barth and co-workers demonstrated the surprising finding that in the therapy of hirsute women 2 mg of cyproterone acetate daily appears to be as effective as higher doses.[26] McKenna concluded that the beneficial effect may have been overestimated.[25]

Flutamide

In a small sample, treatment with flutamide for only 3 months favorably influenced hirsutism in women. More study of this drug for treatment of hirsutism seems warranted.[27]

Ketoconazole

Ketoconazole, an imidazole derivative that inhibits steroid synthesis by reducing side-chain cleavage of cytochrome P450-dependent pathways, has been evaluated in the treatment of hirsutism with mixed results. In a randomized, placebo-controlled, double-blind crossover study design, Ferriman-Gallwey scores[28] were significantly reduced with ketoconazole (600 mg a day) compared with placebo.[29] In contrast, Venturoli and co-workers found a benefit in only 14 of 42 women treated with 400 mg a day; side effects and complications were significant.[30]

5-α-Reductase Inhibitors

The steroid 5-α-reductase enzyme is responsible for the formation of dihydrotestosterone from testosterone. Although specific inhibitors are not yet generally available for human use, it is expected that they will become available within the next several years. These compounds hold promise for treatment of resistant acne and idiopathic hirsutism. Progesterone and the 19-nor derivatives inhibit 5-α-reductase activity at high doses, whereas medroxyprogesterone acetate does not.[31] Therefore the progestin component may expand the usefulness of oral contraceptives in the treatment of hirsutism.

Adrenal Suppression

If hyperandrogenemia is due to maturity onset adrenal hyperplasia, then exogenous glucocorticoid is the treatment of choice. There is no consensus of what is and what is not a significant abnormality. Results of ACTH stimulation need to be interpreted with caution. It is still to be determined whether excessive 17-OH-progesterone responses to ACTH represent heterozygotes for 21-hydroxylase deficiency or an acquired adrenal dysfunction secondary to chronic anovulation.[32] If maturity-onset adrenal hyperplasia is diagnosed, then treatment with prednisone (5 mg) or dexamethasone (0.25 mg daily) is recommended.

Ovarian Suppression

Selective ovarian suppression with chronic administration of gonadotropin-releasing hormone agonists has been employed for cases of hirsutism resistant to first-line therapy.[33] However, treatment is expensive, costing more than $300 per month. Bone demineralization is also a major concern.[34] Estrogen add-back therapy has been proposed as a mechanism to prevent bone loss. It has been confidently predicted that such complications are likely to be effectively managed by estrogen replacement therapy,[35] but confirmatory data are lacking. Some contrary data suggest that the customary doses of estrogen add-back do not protect against bony demineralization.[36] Although gonadotropin-releasing hormone analogue therapy successfully suppresses gonadotropin secretion and excess ovarian androgen production, it does not break the cycle of disordered hormonal patterns that perpetuates the syndrome.[37]

Cautions and Limitations

In treating hirsutism, it is important that any treatment be safe. Rarely is it necessary to extend pharmacotherapy beyond the use of a combination of estrogens and antiandrogens. Glucocorticoids should only be used with a diagnosis of maturity onset adrenal hyperplasia.[9] Chronic administration of gonadotropin-releasing hormone agonists alone is limited by loss of trabecular bone. Enthusiasm for the polypharmacy of ovarian suppression with agonists and add-back estrogen should be tempered until the question of its effects on bone is settled.[36]

Regardless of the therapy utilized, it is often difficult to assess the success of treatment. Subjective assessment is notoriously unreliable, but the semiquantitative method of Ferriman and Gallway, which allows grading of different areas of the body, may be of value.[28] Quantitative assessment by shaving the area and weighing the hair is also of some value, but cellular debris may account for a large portion of the total weight. Whatever method is used, counseling should emphasize that improvement is often slow and that long-term treatment may be required.

Therapeutic Nihilism or Thrifty Good Sense?

Helfer and co-workers have questioned the routine measurement of batteries of hormone measurements.[38] They ask if it is cost effective to measure prolactin, luteinizing hormone, follicle-stimulating hormone, dehydroepiandrosterone, androstenedione, testosterone, and testosterone's biologically active fractions in the evaluation of hirsute women. Apparently, it is not cost-effective. Erhmann and Rosenfield term standard empirical treatment of hirsutism as "nihilism."[39] They advocate a 5-day, low-dose dexamethasone suppression test to select those most benefiting from ACTH testing. Given the cost constraints, perhaps this approach is simply good sense. However, hirsutism must be distinguished from virilism. If there is any question of virilism, hormonal investigation is mandatory.

References

1. Judd SJ, Carter JN. The changing face of hirsutism. *Med J Aust.* 1992;156:148.
2. Min DI, Monaco AP. Complications associated with immunosuppressive therapy and their management. *Pharmacotherapy.* 1991;11:119S–125S.
3. Parker LU, Bergfeld WF. Virilization secondary to topical testosterone. *Cleve Clin J Med.* 1991;58:43–46.
4. DeFreitas A, Dudzinski MR, LaRocque JC, Coddington CC. Ovarian vein sampling in rapidly progressing virilization: a case report. *J Reprod Med.* 1991;36:546–548.
5. Abdel-Gadir A, Khatim MS, Mowafi RS, Alnaser HM, Alzaid HG, Shaw RW. Polycystic ovaries: do these represent a specific endocrinopathy? *Br J Obstet Gynaecol.* 1991;98:300–305.
6. Pugeat M, Crave JC, Elmidani M, et al. Pathophysiology of sex hormone binding globulin (SHBG): relation to insulin. *J Steroid Biochem Mol Biol.* 1991;40:841–849.
7. Sharp PS, Kiddy DS, Reed MJ, Anyaoku V, Johnston DG, Franks S. Correlation of plasma insulin and insulin-like growth factor-I with indices of androgen transport and metabolism in women with polycystic ovary syndrome. *Clin Endocrinol Oxf.* 1991;35:253–257.
8. Wild RA, Grubb B, Hartz A, Van Nort JJ, Bachman W, Bartholomew M. Clinical signs of androgen excess as risk factors for coronary artery disease. *Fertil Steril.* 1990;54:255–259.
9. Conway GS, Jacobs HS. Hirsutism *BMJ.* 1990;301:619–620.
10. Rittmaster RS, Thompson DL. Effect of leuprolide and dexamethasone on hair growth and hormone levels in hirsute women: the relative importance of the ovary and the adrenal in the pathogenesis of hirsutism. *J Clin Endocrinol Metab.* 1990;70:1096–1102.
11. Hassiakos DK, Toner JP, Jones GS, Jones HW Jr. Late-onset congenital adrenal hyperplasia in a group of hyperandrogenic women. *Arch Gynecol Obstet.* 1991;249:165–171.
12. Azziz R, Boots LR, Parker CR Jr, Bradley E Jr, Zacur HA. 11 betahydroxylase deficiency in hyperandrogenism. *Fertil Steril.* 1991;55:733–741.
13. Siegel SF, Finegold DN, Lanes R, Lee PA. ACTH stimulation tests and plasma dehydroepiandrosterone sulfate levels in women with hirsutism. *N Engl J Med.* 1990;323:849–854.
14. Franks S, Kiddy D, Sharp P, et al. Obesity and polycystic ovary syndrome. *Ann N Y Acad Sci.* 1991;626:201–206.
15. Wagner FR Jr. Physical methods for the management of hirsutism. *Cutis.* 1990;45:319–321, 325–326.
16. Ehrmann DA, Rosenfield RL. Clinical review 10: an endocrinologic approach to the patient with hirsutism. *J Clin Endocrinol Metab.* 1990;71:1–4.
17. Porcile A, Gallardo E. Oral contraceptive containing desogestrel in the maintenance of the remission of hirsutism: monthly versus bimonthly treatment. *Contraception.* 1991;44:533–540.
18. Porcile A, Gallardo E. Long-term treatment of hirsutism: desogestrel compared with cyproterone acetate in orald contraceptives. *Fertil Steril.* 1991;55:877–881.

19. Prelevic GM, Wurzburger MI, Balint-Peric L, Puzigaca Z. Effects of a low-dose estrogen-antiandrogen combination (Diane-35) on clinical signs of androgenization, hormone profile and ovarian size in patients with polycystic ovary syndrome. *Gynecol Endocrinol.* 1989; 3:269–280.
20. Crosby PD, Rittmaster RS. Predictors of clinical response in hirsute women treated with spironolactone. *Fertil Steril.* 1991;55:1076–1081.
21. O'Brien RC, Cooper ME, Murray RM, Seeman E, Thomas AK, Jerums G. Comparison of sequential cyproterone acetate/estrogen versus spironolactone/oral contraceptive in the treatment of hirsutism. *J Clin Endocrinol Metab.* 1991;72:1008–1013.
22. Prezelj J, Kocijancic A, Andolsek L. Dexamethasone and spironolactone in the treatment of non-tumorous hyperandrogenism. *Gynecol Endocrinol.* 1989;3:281–288.
23. McLellan AR, Rentoul J, MacKie R, McInnes GT. Lack of effect of spironolactone on hair shaft diameter in hirsute females. *Postgrad Med J.* 1989;65:459–462.
24. Lissak A, Sorokin Y, Calderon I, Dirnfeld M, Lioz H, Abramovici H. Treatment of hirsutism with cimetidine: a prospective randomized controlled trial. *Fertil Steril.* 1989;51:247–250.
25. McKenna TJ. The use of anti-androgens in the treatment of hirsutism. *Clin Endocrinol Oxf.* 1991;35:1–3.
26. Barth JH, Cherry CA, Wojnarowska F, Dawber RP. Cyproterone acetate for severe hirsutism: results of a double-blind dose ranging study. *Clin-Endocrinol Oxf.* 1991;35:5–10.
27. Marcondes JA, Minnani SL, Luthold WW, Wajchenberg BL, Samojlik E, Kirschner MA. Treatment of hirsutism in women with flutamide. *Fertil Steril.* 1992;57:543–547.
28. Ferriman D, Gallwey JD. Clinical assessment of body hair growth in women. *J Endocrinol Metab.* 1961;21:1440.
29. Akalin S. Effects of ketoconazole in hirsute women. *Acta Endocrinol Copenh.* 1991;124; 19–22.
30. Venturoli S, Fabbri R, Dal-Prato L, et al. Ketoconazole therapy for women with acne and/or hirsutism. *J Clin Endocrinol Metab.* 1990; 71:335–339.
31. Cassidenti DL, Paulson RJ, Serafini P, Stanczyk FZ, Lobo RA. Effects of sex steroids on skin 5 alpha-reductase activity in vitro. *Obstet Gynecol.* 1991;78;103–107.
32. Dewailly D. ACTH stimulation tests in women with hirsutism. *N Engl J Med.* 1991;324;564–565.
33. Andreyko JL, Monroe SE, Jaffe RB. Treatment of hirsutism with a gonadotrophin releasing hormone agonist (naferelin). *J Clin Endocrinol Metab.* 1986;63:1031–1035.
34. Tummon IS, Pepping ME, Binor Z, Radwanska E, Dmowski WP. Unchanged bone mineral density during ovarian suppression with GnRH-a or danazol for endometriosis. *Fertil Steril.* 1988;49:792–796.
35. Adashi EY. Potential utility of gonadotropin-releasing hormone agonists in the management of ovarian hyperandrogenism. *Fertil Steril.* 1990;53:765–779.
36. Sugimoto AK, Nisker JA, Hodsman AB. Long-term gonadotropin-releasing hormone agonist with standard postmenopausal estrogen replacement failed to prevent vertebral bone loss in premenopausal women. *Fertil Steril.* In press.
37. Shaw RW. Use of nafarelin to investigate the pathophysiology of the polycystic ovarian syndrome. *J Reprod Med.* 1989;34(suppl 12): 1039–1043.
38. Helfer EL, Miller JL, Rose LI. Cost effectiveness of routine gonadotropin and androgen measurements in hirsute women. *Am J Med Sci.* 1990;299:94–97.
39. Ehrmann DA, Rosenfield RL. Hirsutism: beyond the steroidogenic block. *N Engl J Med.* 1990;323:909–911.

25
Rape

Teresita M. Hogan

Introduction

The crime of sexual assault has reached epidemic proportions. It is estimated that one in every six women is raped during her lifetime.[1,2] Sexual assault incidence has increased four times as fast as the overall crime rate in the United States. A woman is sexually assaulted once every 6 minutes.[3,4] Although these numbers are impressive, physicians often do not believe that this crime will affect their patients or themselves. For example, in a practice with 500 patients, 84 of these women are likely to be victims of rape. Physicians should have the skills to manage a condition that could strike 15% of their patients.

Any office practitioner can effectively treat victims of sexual assault; all that is required is a little willingess. However, three common reasons exist for resistance to treat rape victims. First is the mistaken belief that the crime will not strike patients, even though rape occurs in all communities. Second, the fear of being dragged into court exists. However, this is rarely the case; only 4% of sexual assault cases ever progress to the prosecution phase.[5] Third is the intimidation of properly collecting legal evidence. A goal of this chapter is to provide information to help physicians perform this skill with ease.

Of women who are raped, only 10% to 25% ever report the crime.[6] Therefore many victims of rape will present under the guise of a routine exam or some other "shielding" complaint, and by doing so they are denying themselves the benefit of a complete rape exam. Even if legal evidence can no longer be collected, the medical care needed by these women will be woefully incomplete if a history of sexual assault is not elicited.

A physician must therefore maintain an appropriate level of suspicion. When women present with possible rape-induced problems—fear and anxiety, sexually transmitted diseases, sexual dysfunction, difficulty in interpersonal relationships, psychomatic complaints, sleep disturbances, or unexplained gastrointestinal symptoms—carefully investigate rape as a possible cause.[7] Some physicians believe that the majority of rape victims will offer a shielding complaint rather than openly disclose the sexual assault. Often the most difficult task in the treatment of a rape victim is simply eliciting the rape history. Women may initially deny a rape experience but later disclose it when trust in the physician has been attained.[8]

Once the disclosure or suspicion of sexual assault has been made, evaluation and treatment can begin. The sexual assault exam consists of three main parts: diagnosis and treatment of all mental and physical injuries; prevention of sexually transmitted diseases and pregnancy; and provision of legally admissible physical evidence for prosecution.

The first two sections are routine for any health care provider and require only minor adjustments in focus and documentation. Due to time elapsed since the assault or refusal of the victim to prosecute, the evidence exam, a relatively simple task, will often not be necessary.

To facilitate the care of sexual assault victims, several materials should be in stocked in an office:

1. *The sexual assault evidence collection kit.* Every state has its own variation, and some major cities have a different kit. These kits are readily obtained from the local emergency department or police at a nominal charge.
2. *Handouts.* Handouts for victims of violent crime and survivors of sexual assault put the patient in touch with myriad services to aid in emotional and psychological recovery. These handouts are also easily obtained through the local emergency department, rape crisis center, or women's clinic.
3. *Rape victim advocates.* Most communities of greater than 100,000 people have support services for sexual assault victims that often respond immediately to help the victim during examination and police questioning. If personal attendance is not possible, there is usually a hot-line number to call. The numbers for these services can be obtained from the emergency department or from local telephone listings.
4. *Locked cabinet.* An office must have a locked cabinet to store the collected evidence kit and to maintain a legal chain of evidence. It must be clear that no one could have tampered with the kit until it is picked up by law enforcement personnel.
5. *Woods lamp.* A hand-held Woods lamp is necessary for detection of dried semen stains.
6. *Acid phosphatase test strips.* These test strips change color when exposed to the acid phosphatase enzyme. Because the enzyme deteriorates with time, it is best detected during the presenting physical exam. Strips are available in most rape kits or through the manufacturer.

History

As with all patient encounters, the evaluation begins with a thorough history, which may be obtained by any trained individual. For legal reasons documentation is critical. First, the name of the person obtaining the history should be noted. Second, nothing should be documented as fact unless directly witnessed by the examiner. For example, the chart should not read, "The patient was raped by John Smith." Rather, it should read, "The patient reports that she was raped by John Smith." No conclusions should be documented. For example, the final diagnosis should not be "rape" but "reported rape." To be technically correct, since rape is a legal conclusion not yet established, the correct terminology is "reported sexual assault." Although details of a sexual assault history may seem obvious, the critical points that must be documented are included on the following list.

Critical History

1. Begin with the statement, preferably in quotes, made by the patient to report the sexual assault. Note especially if there was or was not vaginal, oral, or anal penetration. Were any objects other than the penis, including hands, tongue, or other materials, inserted into the victim. Also note if the attacker did or did not ejaculate.
2. Was the patient assaulted with any weapon? If so, have the patient describe it.
3. Was the patient restrained in any way? Again, detail how this occurred.
4. Was the patient threatened in any way either verbally or physically? If possible, quote the threat as she relates it. For example, the patient states, "He said he was going to kill me."
5. Where did the rape occur? Note details of the rape site: basement flooring, parking lot (oily or dirty), wooded area, bedroom flooring. Attempt to corroborate this with physical findings on the patient or on her clothing.
6. When did the rape occur? If the time is not known, try to pinpoint it as finely as possible (day or night, dusk, halfway to or from a known location).
7. Since the attack has the patient:
   a. urinated?

b. defecated?
   c. bathed or washed hands or face?
   d. douched?
   e. changed clothes or dusted off clothes?
   f. smoked, gargled, drank any fluids, or ate any foods?
   g. altered her physical state in a way that may affect evidence?

   Obviously, document either positive or negative responses to these questions.

 8. Was the patient injured in any way? Ask specifically the location of any blows, the manner of choking, the way in which the reported injury was inflicted. Attempt to corroborate the description with physical findings on the patient's body during the physical exam. Ask about any unreported markings discovered during the exam and relate these to the history.
 9. How many assailants were involved?
10. Can the patient identify any assailant, even if only by physical description?
11. Did the patient ever lose consciousness? How did this occur? If there is a period that she cannot remember, what events immediately preceded and followed this time gap?
12. When was the patient's last consensual sexual encounter?

These 12 points are a blend of medical and legal needs and as such are unfamiliar to most medical personnel. All are very logical and necessary. Think of them as relating to the first part of the rape exam: diagnosis and treatment of mental and physical injuries. They are the history of the injury itself. Several other historical points must also be elicited that relate to the second portion of the exam: prevention of sexually transmitted diseases and pregnancy.

Supplemental History

1. Last menstrual period and use of contraception
2. History of sexually transmitted disease
3. Risk factors for sexually transmitted diseases and human immunodeficiency virus (HIV)/Hepatitis
4. Symptoms of current sexually transmitted diseases
5. Symptoms of current HIV or AIDS infection
6. Symptoms of current hepatitis infection

Psychological History: Rape Trauma Syndrome

The last area of focus during the history is that of the rape trauma syndrome. The rape trauma syndrome is a behavioral posttraumatic stress disorder following sexual assault.[9] According to the Diagnostic and Statistical Manual of Mental Disorders (DSM-III), this syndrome is defined as follows:

The development of characteristic symptoms after the experiencing of a psychologically traumatic event or events outside the range of human experience usually considered to be normal. The characteristic symptoms involve re-experiencing the traumatic event, numbing of responsiveness to or involvement with the external world, and a variety of other autonomic dysphonic or cognitive symptoms.[10]

Posttraumatic symptoms may occur immediately or may begin after several weeks. Symptoms may last many years. The psychological injury from the assault is responsible for a wide range of behaviors. Thus the examiner should not judge that the patient has not been raped because her demeanor is not as expected. In one typical form of the syndrome, patients remain calm and composed with little outward emotional display.[9] Mood swings are common and, depending on the timing since the attack, the same victim may display contradictory behaviors.[7] This is often witnessed during follow-up examinations.

Rape trauma syndrome is admissible as evidence in court only if the evidence presented shows the typical reactions to rape and does not draw any conclusions.[11] Therefore the examiner must be very specific in the documentation of the patient's behavior, demeanor, and appearance during the interaction. However, document only the victim's actions ("She was crying"), not personal interpretations of that action ("She was upset").

It is appropriate to question the patient as to her feelings about the attack and how she is coping. In-depth questioning of the victim's emotional state may lead to counseling referral and may speed her recovery. This vital need for the mental health of the patient should not be ignored. Survivors of sexual assault feel most comfortable with physicians who project personal concern while exhibiting professional objectivity.[12] Most important, be sympathetic and nonjudgmental.

The Physical Exam

The physical examination should be all-inclusive, focusing on trauma inflicted during the attack, corroborative physical evidence, prevention of sexually transmitted disease, and pregnancy.

Because the physical exam itself is such a vital part of evidence collection, the evidence collection protocol is detailed. Sections of the protocol should be added or eliminated as appropriate to each individual patient.

Associated Physical Injuries

Statistics on associated physical injuries during sexual assault vary.[13,14] Fortunately, physical injury is not an inevitable consequence of rape. The largest studies show that approximately 50% of victims sustain no apparent physical injury.[15] Cases involving a gun had the least number of victims injured, while cases involving a knife or club correlated with a much higher incidence of injury. This is perhaps a result of the level of intimidation of the victim as well as the cruelty of the assailant. If the attacker is known to the victim, the chance of nongenital injury is less than if the attacker was a stranger. But the incidence of genital injuries is similar in both groups. Overall, about 40% of victims sustain nongenital and 16% sustain genital injuries.[16,17] In all studies the injuries sustained by rape victims are predominately minor and require little or no treatment. Nongenital injuries generally consist of abrasions, lacerations, rope marks, bruises, or burns. Approximately 4% of victims sustain injuries requiring hospitalization. The majority of these are beatings involving severe closed head injuries, intraabdominal injury, or fractures of ribs and long bones. Genital injuries consist primarily of vulvar contusions and hymenal and vaginal lacerations.[15]

Although people tend to be more believing and more sympathetic of the victim with obvious external trauma, this is biased and conclusionary. The correlation of physical injury to rape depends on myriad factors: the emotional state of the attacker, the weapon used, the likelihood of discovery, the setting of the attack, the number of assailants, the intimidation of the victim, and so on. Therefore, physicians must be careful of their own mind-set and judgments because they do not know what actually occurred.

Prevention of Sexually Transmitted Diseases

In most states the law requires that specific laboratory evaluation be done on victims of sexual assault. Most physicians would routinely perform these tests, but some states mandate specific testing. The following tests are required during the initial visit:

1. Serologic test for syphilis. (In addition, obtain blood for ABO typing as needed for legal evidence during this blood draw.)
2. Culture for sexually transmitted diseases from all areas indicated by the assault. Only chlamydial cultures, not antigen detection tests, are recommended.[18]
3. Wet mount for trichomonas vaginalis and clue cells.
4. Urinalysis to identify sperm, trichomonas, or fungus
5. Pregnancy testing by serum βHCG

Risk of Disease Transmission

The risk of transmission of infectious diseases in sexual assault is extremely variable. The risk is based on several factors. One is the infectivity of the assailant. Since this is often not known at the time of the exam, a physician cannot reassure the patient on these grounds. If the assailant is known, mandatory testing for the benefit of the victim is a legal minefield and is best left to law enforcement officials. Risk is also determined by the routes of exposure to body fluid and the presence or absence of ejaculation and its volume. The duration and frequency of contact also affects infectivity. Sites of assault also affect infectivity, and other sites may harbor insidious infections. Therefore, if the history is at all questionable, culture the rectum and the mouth even though the victim did not indicate penetration at these sites.

Anal assault and oral assault each occur in approximately 20% of all rapes.[19] The existence of trauma may contribute to a higher risk of infection, especially for blood-borne infections such as hepatitis B and HIV.[20]

The only positive thought about disease transmission is that studies have shown a fairly low prevalence of sexually transmitted disease in victims of rape. Perhaps this is due to the high rates of sexual dysfunction by assailants.[21] Prevalence of gonorrhea ranges from 2% to 13%, while syphilis prevalence is less than 3%.[17,22,-24]

The risk of acquiring chlamydia trachomatis infection is similar to that of acquiring gonorrhea. The risk of transmission of a viral sexually transmitted disease may be less than that of a bacterial infection due to the episodic nature of viral shedding.[18] Gastrointestinal sexually transmitted diseases have been well documented in male homosexuals and may theoretically develop from sexual assault.[25] Therefore, sexual assault victims may be at risk for giardia lamblia and entamoeba histolytica gastroenteritis.

The prevalence of sexually transmitted diseases in rape victims is identical to that of nonvictims in the same clinical area. Unless the patient has recently moved, the prevalence is the same as that of the usual patient population.[26] The assault itself puts the victim at higher risk for sexually transmitted diseases and HIV because the assailant engages in high-risk behavior. Therefore the incidence of disease rises in rape victims.

One study shows that the estimated risk of acquiring a sexually transmitted disease as a result of rape is 5% to 10%.[27] Other more optimistic studies report development of gonorrhea in 3% to 4% of cases and syphilis in 0.1% of cases.[28] The low-risk figure can be very reassuring for the patient.

Human Immunodeficiency Virus and Hepatitis B in Sexual Assault

Much has been written about HIV infection as a result of rape and sexual abuse.[20,26,29,30] By the time a rape victim has sought medical attention, she has certainly thought about contracting HIV from the assault. A physician will *not* protect the patient's best interest by avoiding the subject. It is the physician's responsibility to ensure prompt diagnosis and to counsel the patient on barrier contraception to prevent further spread of the disease. In fact, until it is known that the patient has *not* contracted HIV or another sexually transmitted disease, failure to counsel the victim on abstinence or barrier contraception will leave the physician legally liable. For example, if a patient's partner develops HIV and the chart does not detail that "patient was advised to the risks of passing on any possible infection" or words to that effect, liability can be an issue. HIV testing should be offered to establish a baseline, and an accurate risk history should be documented. If initial HIV testing is negative, follow-up testing is recommended at 12 weeks, 6 months, and 1 year.[25] If the assailant is known to be HIV positive, tests of the victim should be repeated in 24 weeks unless prompted by earlier symptoms suggestive of infection.[31] Data suggest that in 95% of patients antibodies develop within 6 months of HIV exposure.[31]

The use of prophylaxes has been suggested using various anti-HIV substances from vinegar and nonoxynol-9 intravaginally to oral zidovudine.[32] It should be noted that no data presently exist to support the use of prophylactic antiviral medications.[18]

Where the assailant is known to have a high risk of hepatitis B, hepatitis B gammaglobulin should be offered.

Treatment of Sexually Transmitted Diseases

Treatment should certainly be given for any infections identified at the time of initial evaluation. The victim should also be treated for any infection known to be harbored by the assailant. A presumptive diagnosis of gonorrhea may be based on the observation of Gram-negative intracellular diplococci.

Debate exists on administration of routine prophylactic therapy. Generally, if the patient is reliable and will not be lost to follow-up, she should have the option of immediate treatment or treatment pending laboratory evaluation. If patient follow-up is at all questionable, however, immediate prophylaxis should be administered.

Prophylactic therapy should be effective against gonorrhea, chlamydia, and incubating syphilis.[33] Ceftriaxone (250 mg IM) is recommended and will provide coverage for all strains of *N. gonorrhoeae* as well as incubating syphilis. If the victim is allergic to cephalosporins, use spectinomycin (2.0 gm IM). This should be supplemented with chlamydial coverage by:

1. 100 mg of doxycycline by mouth twice a day for 7 days,
2. 500 mg of tetracycline by mouth four times a day for 7 days, or
3. erythromycin if the patient is pregnant.

Patients receiving prophylactic therapy do not require repeated culture. If therapy is not given, the cultures should be repeated in 7 to 14 days.

Serology for syphilis should be rechecked in untreated victims after 8 to 12 weeks, hepatitis B serology should be repeated in untreated patients after 12 weeks, and HIV serology should be repeated in all victims after 12 weeks. Since the risk of bacterial vaginosis and trichomona infection is unknown, prophylaxis against these infections is not recommended.[25] Therefore patients should be re-examined for these pathogens after 8 to 12 weeks. There is no known effective prophylaxis for genital herpes or genital warts.[34]

Prevention of Pregnancy after Sexual Assault

Pregnancy as a result of rape occurs in about 5% of fertile female victims.[27] In addition to the rape and subsequent fear of infection, the victim may have to face the possibility of an unwanted pregnancy and the ethical and moral dilemma of preventing or terminating such a pregnancy. These subjects are also difficult for the concerned health care giver, but the law has again dictated what the health care provider must do, limiting available options.

The physician must inform the female rape victim of the availability of pregnancy prophylaxis. Although the physician is not obligated to administer this if it is against his or her religious beliefs, appropriate referral must be offered to the victim so that she may utilize prophylaxis if she so chooses. The health care provider and the institution are required to inform the patient of the options and to refer her to providers who do not find the treatment morally objectionable. Therefore, if the victim has been subjected to vaginal penetration and is in the fertile portion of her menstrual cycle and is not currently using an active form of birth control, she must be made aware of the options for pregnancy prophylaxis. The physician is legally liable for failure to do so.[35] in a California case a woman filed a malpractice suit after a Catholic hospital failed to recommend postrape pregnancy prophylaxis. Although the patient did not become pregnant as a result of the sexual assault, the court found that her civil rights were violated because "medical patients have a right to self-determination in their treatment [which] must prevail over the moral and religious convictions of the hospital."[36]

First a beta-HCG subunit assay should be done to rule out any preexisting pregnancy. If the patient is not pregnant, prophylaxis may be achieved with 25 mg of diethystilbestrol by mouth twice a day for 5 days or 2 tablets of ethinylestradiol-norgestrel (Ovral) by mouth initially followed by 2 tablets in 12 hours. Antiemetics should be offered for the probable side effect of nausea and vomiting.

Patients must also understand the 1% failure rate of any pregnancy prophylaxis and the teratogenicity effect of all postcoital medications. Since these facts have obvious legal implications, any practitioner offering pregnancy prophylaxis should document explanation of these risks in the chart. Pregnancy prophylaxis is only effective if treatment can be initiated within 72 hours of the exposure.

The Evidence Exam

The purpose of the evidence exam is to corroborate that sexual contact occurred, to corroborate that this contact was nonvolitional (proof of force or coercion), and to identify the perpetrator.

Proof of Sexual Contact

Obviously not every sexual encounter leaves proof of its occurrence. The likelihood that proof exists is affected by many factors, including the time between assault and evidence collection, the occurrence of ejaculation, and the physical state and number of assailants (use of condoms, prior vasectomy, sexual dysfunction). Multiple factors in the victim are also important. Has the victim bathed, douched, urinated, or defecated since the event? Did the victim change clothes or otherwise alter her physical appearance since the attack? Did the patient have consenting intercourse in the 72 hours prior to the exam? Some of these factors have already been established during the history portion of the exam.

Several different markers exist that can prove recent sexual contact, such as the presence of motile and nonmotile sperm and increased levels of acid phosphatase (ACP), an enzyme present in high levels in seminal fluid. Recovery of these markers from a victim's vagina, mouth, anus, or skin is considered conclusive legal evidence of sexual contact.

Evidence Collection

The presence of motile and nonmotile sperm and prostatic acid phosphatase are proof of sexual contact and can pinpoint the timing since the attack and help to identify the assailant.

Motile sperm indicate recent sexual intercourse. Sperm have been documented to survive in the vaginal cavity for up to 3 days.[37] Sperm normally lose their motility within 30 minutes, but some can remain motile for up to 12 hours. Thus the finding of motile sperm attests to an assault within the last 12 hours. This is particularly important if the victim had voluntary intercourse within 72 hours prior to the assault.[38]

The Pap smear technique is the best for securing evidence of sperm and can be used to demonstrate the presence of nonmotile sperm beyond 72 hours from the time of the assault.[39] Sperm have been found in the cervix up to 17 days after coitus; however, they are rarely found after 72 hours.[40]

Other semen-specific markers exist. A major seminal plasma glycoprotein is p30, which is found in high concentrations in seminal fluid and is male specific. It follows a predictable pattern of postcoital decline that can help pinpoint the timing of the assault. The presence of p30 puts an approximate 48-hour time limit on the assault.[41]

ACP is present in both vaginal secretions and in seminal fluid. However, its concentrations are markedly higher in semen. Quantitative ACP assays are excellent indicators of recent intercourse because ACP deteriorates rapidly in the vagina. The presence of menses does not interfere with ACP measurement.[41,42]

Semen contains three genetic markers in high enough concentrations to permit genetic typing of the individual. This allows for the differentation of a random pair of individuals with 90% accuracy.[37] These three markers, which can be obtained from vaginal or anal washings, are the peptidase A enzyme (Pep-A), the ABO group antigens, and phosphoglucomutase (PGM).

Timing is critical in obtaining evidence samples. Therefore the physical exam should not be delayed for police or for rape crisis workers. The victim should know that just 1 hour could result in the loss of evidence. Negative results are the norm in cases when there has been more than a 5-hour delay between the assault and evidence collection.[6] If negative results are used against the victim in court, this short time interval for positive evidence collection should be noted.

The most productive source of evidence is

the vagina, and the test most likely to result in confirmation of recent intercourse is ACP. Levels can remain positive in the vagina for as long as 36 hours. Use of the Pap technique for sperm can yield positive results longer than any other sperm recovery method and should be performed routinely. The second most productive site is the skin. ACP levels may be positive on the skin for 8 to 12 hours and have been noted up to 39 hours after sexual assault.[6]

Proof of Force or Coercion

In documenting the patient's emotional state, the physician should note only direct observations, not interpretations. This can verify emotional trauma and corroborate the rape trauma syndrome. Details of trauma are appropriate in the physical exam. Documentation of the patient's overall mental state is also important. Mental retardation, immaturity, mental illness, and the use of alcohol or drugs all may affect the patient's ability to give consent for sexual intercourse.

Note the patient's age, physical development, stature, and apparent age. When statutory rape is at issue, these facts are important.

Documentation of physical injuries is critical, and the most powerful form of documentation is a photograph. If bruising and other external trauma are physically impressive, suggest photographing the injuries to strengthen the victim's case. Police evidence technicians are available for this task. If the victim is unwilling to submit to photographs, then the examiner should describe any trauma as completely as possible. Measure and note the size of the wounds. To note the exact location of all injuries, it may be appropriate to draw them on a body template or circle them on a stick figure. Detail the age of the injuries (scabbing, infection, color of bruises). Is the timing of the injuries consistent with the reported timing of the attack? Try to correlate the physical injuries with the patient's description of how they occurred.

Collect clothing damaged or soiled during the assault. Place each item in an individual paper bag and seal. Do not use plastic bags because mildew can destroy evidence. Collect any debris from the patient's hair or body and document from where and how it was obtained; seal in an individual paper bag.

Evidence Collection: Identification of the Assailant

Physicians may recover several identifying markers during the rape exam that can serve to convict the assailant:

1. Public hair combings. This may supply hair from the assailant. Hair follicles provide legally binding evidence as individual as a fingerprint. Any form of hair follicle, public hair or hair from the assailant's chest, arms, head, or beard, will serve the purpose. The shaft of hair itself can be compared with the hair of the reported assailant and can also serve as evidence.
2. ABO blood group identification. Obviously a person's blood can be typed from blood smears, either fresh or dry, recovered from the victim's body or clothing. However, 80% of all human beings are secretors (meaning they secrete large amounts of ABO blood group substances into their other body fluids). Secretors can be blood typed from any and all of their bodily fluids.
3. DNA pattern. DNA "fingerprinting" of body fluids from either blood or seminal fluid can be done easily. It can also be done from skin cells that are sometimes found under the fingernails of the victim.
4. Saliva. Saliva from the assailant's oral contact or from spitting onto the victim can help to identify him in court.

Details of the Physical Exam

If a patient were to present immediately after an attack, the following details should be part of any complete evidence collection exam. Sections of the exam can be eliminated as appropriate.

1. Have the patient disrobe while standing over a clean sheet. This way any debris falling from her clothing may be collected and labeled. Place each article of clothing in a

separate paper bag. Label each bag and seal it either with staples or tape so that it cannot be opened without damaging the bag. Head hair combings are appropriate at this time; place fallen hairs in an envelope with the comb. Fifteen to twenty plucked head hairs with the follicles intact should be placed in a separate envelope. The patient should be allowed to pluck her own hair, but someone must witness the procedure for legal verification.

2. Perform a head-to-toe physical exam.
   a. Collect any debris from the patient's hair or skin. Dried secretions or dirt may be gently scraped with a scalpel edge or tongue blade into an envelope or specimen container. Everything should be labeled and sealed. Note where it was found and how it was removed from the patient.
   b. Inspect for external trauma. Document by photography, drawing, or physical description. Inspect the victim's back and buttocks for trauma from lying on a hard surface. Check for broken fingernails. Obtain fingernail scrapings and place in two separate envelopes, one for each hand. Take clippings from each fingernail and include them with the scraping specimens.
   c. Dried semen appears as a lightly crusted, flaky area. Because semen stains fluoresce brightly, check for any sign of fluoresence by using a Woods light passed over the victim's skin. Check the perineal area carefully for dried semen. Collect a portion of the dried semen by scraping as described above. Also remove some dried semen with a saline-moistened swab. Air-dry two slides from each site for police use. (Do not coverslip police evidence slides.) Test the semen for prostatic acid phosphatase. Commercial tapes that change from blue to dark purple when exposed to the enzyme are available; some rape kits contain these test strips.

3. The oral cavity should receive special attention. Inspect the mouth for signs of trauma. Check the frenulum for tears. Prepare two air-dried slides from swabbing the oral cavity. (Do not coverslip police evidence slides.) Sperm may collect in the areas where the gums meet the teeth and between the lower wisdom teeth and the cheek. Immediately upon completion of the physical, examine wet mounts for motile or immotile sperm and trichomonas. Because they will not be police evidence, these may be coverslipped, only the report of the findings is evidence. Test fluid from recesses of the oral cavity for presence of acid phosphatase. Saliva should be collected for blood group antigen testing. Ask the victim to deposit saliva on a filter paper or gauze pad and allow the specimen to air dry, then seal in an envelope. Next, obtain a gonorrhea culture from the throat.

4. Perform a pelvic examination. On external examination note any trauma to the perineum, buttocks, or inner thighs that may have been missed prior to placing the patient in stirrups. In the lithotomy position again reexamine with the Woods light for any dried semen. Obtain any newly discovered specimens as instructed in step 2c. Note injuries, dirt, and foreign material. Collect foreign materials in an envelope. Obtain public hair combings and place them in an envelope with the comb. Allow the patient to pluck her own pubic hairs for evidence. Use a warm-water-moistened speculum and avoid the use of any lubricants. Note any hymenal or vaginal trauma. Swab the vagina and prepare two air-dried smears. Aspirate the vaginal pool with 5 to 10 mL of saline and place in a test tube. Prepare a wet mount for sperm and trichomonas, which should be examined immediately after the physical exam. Test for presence of acid phosphatase. Obtain a Pap smear for the presence of sperm. Culture for gonorrhea and chlamydia. Perform a bimanual exam.

Toluidine blue is a nuclear stain commonly used in gynecologic examination to detect vulvar cancer. Normal vulvar skin contains no nuclei and will not bind the dye. When lacerations or abrasions expose the deeper dermis these tissues uptake the toluidine blue. Studies have shown an increase in the numbers of vaginal injuries from 1 in 24 rape victims to 14 in 24 after application of toluidine.[43]

The use of colposcopy in the sexual assault evidence exam has been advocated to increase the yield of positive physical findings. Colposcopic magnification offers an advantage that may dramatically increase positive examination findings to about 87%.[44] In this study, 87% of victims had gynecologic trauma noted under colposcopic magnification. Predominant injuries were seen between 3, 6, and 9 o'clock on the posterior fourchette. Injuries of mounting trauma included lacerations, abrasions, ecchymosis, and swelling. Abrasions to the medial labia minora were common as were ecchymosis of the hymen and periurethral structures. Colposcopy obviously revealed more injuries than noted on nonmagnified exam, and it was also superior to examination with toluidine blue.

5. Inspect the rectum and anus for trauma and foreign materials. Check for blood, signs of semen, or lubricant. Collect two air-dried slides. If rectal intercourse occurred, a rectal washing may be obtained by inserting the hub of a 10-mL syringe filled with saline. Aspirate and collect in a test tube for evidence. Test for sperm and acid phosphatase. Obtain a gonorrhea culture. Perform a digital examination and check for foreign bodies, fissures, and internal trauma.

The physical evidence exam can be difficult for the already victimized patient. It can be made easier if the patient understands that this is being done to ensure her health and safety. Always explain the exam in detail to the patient before she begins to disrobe. When she understands the logic begind the procedures the ordeal can be easier. If the victim is unsure of involving police, make it clear that the best chance for obtaining evidence is the present. The evidence kit can be collected and held for up to 2 weeks before turning it over to the police. One option is to collect the evidence and hold it while the victim decides.

Chain of Evidence

The law is very specific about what evidence is and is not admissible in court. Ensure that the evidence is collected properly and stored according to the letter of the law or it will be suppressed.

From the time the evidence kit is opened until it is sealed, all materials in the kit must be watched by someone on staff. The patient, family, friends, or crisis workers may never be left alone with the evidence kit. All materials must be air-dried before sealing. All materials must be sealed in their containers so that opening would necessarily damage the container. All materials must be properly labeled so that they can be identified as to the source, location, and person obtaining the sample.

The evidence kit should be initialed by the patient and by all persons responsible for evidence collection. After the kit is sealed it must be secured in a staff-only access area until collected by the evidence technician. All staff must sign in and out of that area any time the kit-containing cabinet is opened.

Most states require police notification for all reported sexual assaults. Because police involvement is often difficult for the victim, the practitioner must protect everyone's best interest through proper documentation.

If the patient refuses notification of law enforcement agencies, make a clear notation of this refusal in the chart. A complete refusal documentation must state:

1. that the patient is coherent and thought competent to make the choice,
2. that the patient specifically stated "Do not notify police,"
3. documentation that you counseled the patient to notify law enforcement agencies,
4. documentation that notification was not made by you to protect the patient confidentiality, and
5. documentation that the patient's safety has been considered and that patient is in no imminent danger.

When law enforcement *is* notified, document:

1. the date and time of call,
2. the name of the person making the call,
3. the name of the person receiving the call, and
4. information provided to police.

Even if the patient refuses police notification, *still attempt* to gather the evidence kit. It may be kept refrigerated for up to 2 weeks and still be valid for evidence. At the end of 2 weeks, document the patient's continued refusal to submit evidence and destroy the kit. If patient refuses kit collection, document this in the same way that police refusal is documented.

Patients cannot be protected from the tragedy of sexual assault. But once the assault has occurred, physicians can serve an essential function and can help to speed recovery. The survivor of sexual assault has many interwoven medical, psychological, and legal needs. With just a few basic skills, the clinician can start the healing, prevent further damage, and help the victim resume a healthy, productive life.

References

1. Hicks J. Sexual assault. In Nichols DH, Evrard JR, eds. *Obstetrics and Gynecology*. Philadelphia, Pa: Harper and Row; 1985:473–487.
2. Beebe, DK. Emergency management of the adult female rape victim. *Am Fam Phys.* 1991:2041–2046.
3. United States Department of Justice, Federal Bureau of Investigation. Uniform Crime Reports for the United States. Washington, DC: U.S. government Printing Office; 1988:46–48.
4. Hicks DJ. Sexual battery: management of the rape victim. In: Sciaria JJ, ed. *Gynecology and Obstetrics*. Philadelphia, Pa: Harper and Row; 1990:1–11.
5. Ledray LE. *Recovering from Rape*. New York: Henry Holt & Co; 1986.
6. Tucker S, Clair E, Ledray L, et al. Sexual assault evidence collection. *Wisconsin Med J.* 1990;407–411.
7. Heinrich L. Care of the female rape victim. *Nurse Practitioner*. 1987:9–27.
8. Kilpatrick D. Rape victims: detection, assessment and treatment. *Clinical Psychologist*. 1983;36:92–95.
9. Burgess AW, Holmstrom LL. Rape trauma syndrome. *Am J Psychiatry*. 1974;131:981–986.
10. American Psychological Association Committee on Nomenclature and Statistics: Diagnostic and Statistical Manual of Mental Disorders. 3rd ed. Washington, DC: American Psychological Association, 1980.
11. Block AP. Rape trauma syndrome as expert testimony. *Arch Sex Behav*. 1990;19:309–323.
12. Martins CA, Warfield MC, Braen GR. Physicians' management of the psychological aspects of rape. *JAMA*. 1983;249:501–503.
13. Martin PY, Dinitto DM. The rape exam: beyond the hospital emergency room. *Women & Health*. 1987;12:5–28.
14. Geist RF. Sexually related trauma. *Emerg Med Clin North AM*. 1988;6:439–466.
15. Cartwright Peter S. Factors that correlate with injury sustained by survivors of sexual assault. *Obstet Gynecol*. 1987;70:44–46.
16. Solola A, Scott C, Severs H, et al. Rape: management in a noninstitutional setting. *Obstet Gynecol*. 1983;61:373–380.
17. Everett RB, Jimerson GI. The rape victim: a review of 117 consecutive cases. *Obstet Gynecol*. 1977;50:88–90.
18. Schwarcz S, Whittington W. Sexual assault and STDs: detection and management in adults and children. *Reviews of Infectious Disease*. 1990;12:S682–S690.
19. Lacey HB. Sexually transmitted disease and rape: the experience of a sexual assault centre. *Int J STD & AIDS*. 1990;1:405–409.
20. Murphy SM. Rape, sexually transmitted diseases and human immunodeficiency virus infection. *Int J STD & AIDS*. 1990;1:79–82.
21. Groth AN, Burgess AW. Sexual dysfunction during rape. *New Engl J Med*. 1977;297:764–766.
22. Hayman CR, Lanza C. Sexual assault on women and girls. *Am J Obstet Gynecol*. 1971;109:480–486.
23. Kaufman A, Vandermeer J, Divasto P, et al. Follow-up of rape victims in a family practice setting. *South Med J*. 1976;69:1569–1571.
24. Forster GE, Pritchard J, Munday PE, et al. Incidence of sexually transmitted diseases in rape victims during 1984. *Genitourin Med*. 1986;62:267–269.
25. Glaser JB, Hammerschlag MR, McCormack WM. Sexually transmitted diseases in victims of sexual assault. *New Engl J Med*. 1986;315:625–627.

26. Jenny C, Hooton TM, Bowers A, et al. Sexually transmitted diseases in victims of rape. New Engl J Med. 1990;322:713–716.
27. Beckman CR, Groetzinger LL. Treating sexual assault victims: a protocol for health professionals. *Female Patient.* 1989;14:78–83.
28. Wertheimer A. Examination of the rape victim. *Postgrad Med.* 1982;71:173–180.
29. Murphy, S, Kitchen V, Harris JRW, et al. Rape and subsequent seroconversion to HIV. *BMJ.* 1989;299:718.
30. Gellert GA, Mascola L. Rape and AIDS. *Pediatrics.* 1989;83:S644–S645.
31. Horsburgh CR Jr, Ou CY, Jason J, et al. Duration of HIV infection before detection of antibody. *Lancet.* 1989;2:637–640.
32. Foster IM, Bartlett S. Anti-HIV substances for rape victims. *JAMA.* 1989;262:2090–2091.
33. Centers for Disease Control. 1989 Sexually transmitted disease treatment guidelines. *MMWR.* 1989;38(suppl)5–8.
34. Glaser J, Hammerschlag M, McCormack W. Epidemiology of sexually transmitted diseases in rape victims. *Review of Infectious Diseases.* 1989;11:246–254.
35. *Brownfield* v *Daniel Freeman Memorial Hospital*, 256 Cal. Rptr. (1989).
36. Brushwood DB. *Am J Hosp Pharm.* 1990;47:395–396.
37. Seusabaugh GF, Bosliewski J, Blake ET. The laboratory's role in investigating rape. *Diang Med.* 1985;46–53.
38. Soules MR, Pollard AA, Brumn J, et al. The laboratory's role in investigating rape. *Am J Obstet Gynecol.* 1978;130:142–147.
39. Short J, DeLuca M, Divosto P, et al. Detection of sperm in victims of rape. *New Engl J Med.* 1978;229:424.
40. Greydanus DE, Shaw RD, Kennedy EL. Examination of sexually abused adolescents. *Sem Adol Med.* 1987;3:59–65.
41. Graves H, Seusabaugh G, Rhule E. Postcoital detection of a male-specific semen protein. *New Engl J Med.* 1985;312:338–343.
42. McCloskey K, Muscillo G, Noardurer B. Prostatic acid phosphatase activity in the postcoital vagina. *J Forensic Sci.* 1975;20:630–636.
43. McCauley J, Gyzinski G, Welch R, et al. Toluidine blue in the corroboration of rape in the adult victim. *Am J Emerg Med.* 1987;5:105–108.
44. Slaughter L, Brown C. Colposcopy to establish physical findings in rape victims. *Am J Obstet Gynecol.* 1991;83–86.

26
Surveillance of Gynecologic Malignancies

Suzanne Bergen

Introduction

Unfortunately, a significant number of women treated for gynecologic malignancies will develop persistent or recurrent cancer. Surveillance of these women after primary therapy is necessary for the early detection of disease. This is especially important when early detection can lead to palliation of symptoms and even more so when long-term survival can be achieved with additional therapy. If long-term survival cannot be achieved with additional therapy, then the clinician must weigh the potential benefits of early disease detection against the rigors of invasive follow-up.

Rationale for Cancer Surveillance

In addition to the detection of recurrent or persistent disease, other reasons exist for closely following patients treated for gynecologic cancers (Table 26.1). Close follow-up allows the clinician to detect and manage complications of the primary therapy. Some patients who are disease-free are forced to live with complications of treatment. Women treated with chemotherapeutic agents can develop secondary malignancies that manifest themselves years later.[1] The long-term effects of radiation—including radiation cystitis, enteritis, and proctitis; small bowel obstructions; rectovaginal and vesicovaginal fistulas; ureteral obstruction due to fibrosis; loss of ovarian function and distortion of the vagina; and the development of secondary uterine malignancies—are well documented.[2-14] Surgical therapy can be associated with long-term sequelae, including bladder and bowel dysfunction following a radical hysterectomy, small bowel obstruction and chronic pelvic pain attributable to adhesions, and vasomotor symptoms associated with castration.[15-17] Seeing the patient at regular intervals allows for the early recognition of these complications and for treatment or palliation of symptoms. If treatment or palliation are not possible, then simple reassurance that these problems do not represent recurrent cancer can be very meaningful to the patient.

The clinician following a woman for gynecologic cancer may be the only physician the patient is seeing; thus, regular follow-up visits allow the clinician to monitor the patient's overall well-being and health. This includes identification of other unassociated medical problems and the institution of therapy for these diseases. If the clinician is not trained to

TABLE 26.1. Rationale for surveillance of gynecologic malignancies.

1. Detection of persistent or recurrent disease
2. Identification and management of complications of primary therapy
3. Identification and management of unassociated medical problems
4. Screening for secondary malignancies
5. Emotional support and reassurance

TABLE 26.2. American Cancer Society recommendations for the early detection of cancer.

Procedures	Age 20–40	Age 40 & Over
Physical: an examination of the mouth, thyroid, breasts, rectum, ovaries, skin, lymph nodes and a pelvic exam and health counseling to teach proper diet, exercise, health habits, breast self-examination and how to stop smoking.	At least every 3 years.	Every year.
Mammography	Age 35–40, an initial x-ray to profile later changes.	Age 40–49, every year or two. Over age 50, every year.
Pap test for cervix	Annually for all women sexually active or 18 years and older. After three or more consecutive normal annual examinations, as often as the physician advises, regardless of patient's age.	Same as ages 20–40.
Tissue sample of endometrium	For higher risk women at menopause. Higher risk includes obesity, abnormal uterine bleeding, estrogen therapy, history of infertility.	Same as ages 20–40.
Stool blood test for colon & rectum	As often as physician advises for higher risk patients.*	Age 40–50, as often as physician advises, if higher risk patient. After 50 every year.*
Procto exam	As often as physician advises for higher risk patients.*	Age 50+, annually for 2 years. Then every 3–5 years as physician advises.

*Higher risk includes personal or family history of colon or rectal cancer, personal or family history of polyps in colon or rectum, ulcerative colitis.

manage these problems, then referral to the appropriate physician for management can be accomplished. The physician should also implement cancer screening for secondary primaries, using the guidelines put forth by the American Cancer Society (Table 26.2).

Cancer surveillance and follow-up also provide a time of contact between the physician and the patient during which the patient can express fears and concerns. Once the patient is diagnosed with cancer, each ache, pain, or unusual feeling can be perceived as recurrent cancer. Many patients have concerns related to altered self-esteem, self-perception, or sexual dysfunction. Follow-up visits provide an opportunity for the physician to offer emotional support as well as reassurance that the patient's feelings are normal. The physician can also suggest ways for the patient to alleviate or cope with these problems. Referrals can be made to appropriate counselors or support groups. Many of these issues are difficult for the patient to talk about, and the clinician must give the patient adequate time to express her concerns and fears. At times, the physician may be able to do no more than simply listen to the patient discuss her problems, giving her comfort by letting her know that she is not alone and that the physician is there when needed. This is just as important, but often overlooked, a component of follow-up as attending to the physical issues. Time must be set apart to address these issues.

Surveillance of Gynecologic Malignancies

There are no absolute guidelines offered by the American Cancer Society, the Society of Gynecologic Oncology, or the American College of Obstetrics and Gynecology concerning how frequently a patient should be seen and followed after she has been treated for a gynecologic malignancy. Nor are there specific recommendations regarding the appropriate surveillance procedures and laboratory tests. In this chapter, the author recommends some general guidelines for follow-up. The clinician must consider each patient's disease process and modify these guidelines where appropriate, shortening the interval between surveillance for high-risk patients and prescribing additional testing where appropriate.

Intervals for Follow-Up

From 70% to 85% of recurrences are seen within the first 2 years after primary therapy.[18–20] Patients should be closely scrutinized during the first 2 years following therapy for evidence of recurrent disease. The standard recommendation for follow-up is an exam at the completion of therapy and then every 3 months for the next 2 years. Since most recurrences are seen within 5 years, the patient should be seen every 6 months for the ensuing 3 years and then yearly thereafter. Of course, patients with specific problems or at high risk of recurrence should be seen more frequently. Surveillance throughout the patient's life should be continued, as a risk of late recurrence still exists and the risk of developing a second primary malignancy is present. Women should be encouraged to see their physician immediately if a new symptom or problem develops.

History and Physical Examination

The first step for any surveillance is taking an appropriate history and asking a detailed review of systems, focusing on symptoms that may suggest recurrence or complications of therapy. Questions should be tailored to the type of cancer the patient has had and the initial therapy she received (Table 26.3).

The second step in cancer surveillance is a thorough physical examination. This should include a measurement of the patient's weight and comparison with her previous weight. Any

TABLE 26.3. Pertinent symptoms of patients with persistent or recurrent disease.

General:
  Weight loss or weight gain
  Fatigue
  Weakness
Pulmonary:
  Cough
  Hemoptysis
  Dyspnea
  Chest pain
Gastrointestinal:
  Abdominal pain
  Nausea and vomiting
  Change in bowel movements
  Constipation
  Diarrhea
Genitourinary:
  Dysuria
  Frequency of urination
  Urinary incontinence
  Blood in urine
  Difficulty emptying bladder
  Vaginal bleeding or discharge
  Back or thigh pain
  Pelvic pain or pressure
  Blood in stool
Musculoskeletal:
  Extremity or back pain
  Swelling in legs or arms

loss or gain should be documented. Auscultation of the lungs and heart should be performed, listening especially for basilar rales or decreased breath sounds indicating pleural effusions. A breast examination should be performed. A thorough abdominal examination, taking note of any hepatic enlargement, ascites, or abdominal masses, is required. Careful attention should be given to the lymph node survey, especially the supraclavicular and inguinal nodal chains. A thorough evaluation of the extremities, particularly noting any leg edema, should be performed.

The pelvic examination should include careful inspection and palpation. The vulva, vagina, and cervix should be visually scrutinized for any abnormal growths or lesions. The examiner should always have a high degree of suspicion and should perform a biopsy of any unusual or abnormal area. If extensive vaginal vault necrosis is present, caution should be exercised in performing biopsies. Vault necrosis precedes the development of a vesico-vaginal fistula in more than 79% of patients, and careless biopsies will contribute to fistula formation.[21] A careful bimanual examination is then performed, noting areas of nodularity, irregularity, or asymmetry. A bimanual pelvic examination after extensive radiation and surgery can be difficult to interpret because varying degrees of radiation fibrosis and scarring may develop. An inexperienced examiner may mistake radiation fibrosis for recurrent tumor. Conversely, recurrent tumor can be hidden in a pelvis that has radiation fibrosis extending from sidewall to sidewall. When any doubt exists or before any treatment is initiated, histologic documentation of recurrent or persistent disease is mandatory. This can be done by either fine-needle aspiration or Tru-Cut needle biopsies if no gross mucosal lesion is seen. To be complete, a pelvic examination must always include a careful rectal-vaginal exam. Often, tumor will be located in the cul-de-sac or rectal-vaginal septum; without a rectal exam, this would go undetected. Rectal strictures due to tumor or radiation will be discovered only by this simple exam. At the conclusion, the stool on the examiner's glove can be tested for occult blood with a simple stool blood test.

Surveillance of Cervical Cancer

For surveillance of a patient with cervical cancer, an aggressive Pap smear should be performed at each visit. An abnormal Pap smear suggestive of carcinoma is the first sign of recurrence in 3% to 50% of patients and may precede clinical evidence of disease by 3 to 24 months.[22,23] The differentiation between dysplastic cells, invasive cancer, and repair cells is very difficult to make after radiation therapy to the cervix and vagina. Thus, the clinician must be cautious in making the diagnosis of persistent disease too early following therapy. If dysplastic cytology is noted, there is a 56% chance of identifying recurrent disease. If dysplastic cells are found within 3 years of therapy, there is a 71% chance of recurrence.[24] If an abnormal Pap smear is obtained and a gross lesion is not seen, colposcopic evaluation of the cervix and vagina is indicated. Radiation therapy can markedly alter colposcopic patterns due to radiation-induced endarteritis obscuring normal areas. It has been the author's experience that a course of intravaginal estrogen cream can improve the atrophic changes, making the colposcopy more accurate. A biopsy should be performed of any colposcopically abnormal area. If the colposcopy is not diagnostic and the area of dysplastic cells is not identified, multiple random biopsies or a partial vaginectomy may be indicated.

In addition to a Pap smear, various diagnostic tests have been suggested to be routine in the surveillance of cervical cancer. Photopulos et al performed an IVP, chest X ray, barium enema, cystoscopy, proctoscopy, and bone scan at regular intervals in 77 patients with cervical cancer.[25] Of the 77 patients, 9 (12%) had abnormal studies, and 8 of these 9 patients died of recurrent cancer. Studies were abnormal in 6 of 10 patients suspected of recurrences but in only 3 of 63 not suspected of recurrence. Photopulos et al concluded that, with the exception of the IVP and chest X ray, these tests should not be routinely performed because most recurrences were diagnosed by physical examination.[25]

Unilateral ureteral obstruction on an IVP can indicate a pelvic sidewall recurrence, but it is rarely encountered in patients with a normal bimanual pelvic examination. Because central or lateral pelvic recurrences following primary surgical treatment may be cured with radiation, it has been recommended that an IVP be performed at 6 months and again at 1 year in postoperative patients considered at high risk for failure.[26]

Following primary radiation therapy, an IVP should be obtained if symptoms suggest ureteral obstruction or if pelvic findings suggest recurrence. Because a small percentage of patients may develop radiation-induced ureteral strictures,[27] the diagnosis of recurrence must always be documented either histologically or cytologically.

In the author's practice, annual chest X rays have not been routinely performed in all patients with cervical cancer. Their use has been reserved for symptomatic patients or when pelvic recurrence is documented. Because there is no curative chemotherapy, the patient benefits little from detection of pulmonary metastases early after primary therapy.

Surveillance of Endometrial Cancer

The minimum surveillance for endometrial carcinoma should include a careful abdominal and pelvic exam each visit accompanied by a Pap smear, because the vaginal vault is the most frequent site of recurrent disease. Pulmonary metastases are common. Thus, a chest X ray should be performed at least annually. Patients who develop complaints related to the gastrointestinal, urologic, or skeletal systems should be evaluated with the appropriate radiographic and laboratory tests.

Recent data indicate that CA-125 levels can be elevated in patients with endometrial cancer and are in fact elevated in 78% of patients with Stage IV and recurrent disease.[28] Some authors have suggested that CA-125 levels are useful in following patients with endometrial carcinoma. Patsner et al[29] measured CA-125 levels at 4-month intervals in 125 patients with Stages I and II endometrial carcinoma. Thirteen patients developed tumor recurrence. None of six patients with isolated vaginal recurrences had elevated CA-125 levels, but all seven patients with distant metastases had elevated serum levels at the time of diagnosis. False elevations were observed in patients who received radiation therapy and patients who developed small-bowel obstructions.[29] The role of CA-125 levels in the surveillance of patients with endometrial cancer still must be defined. Further study is required before routine serial CA-125 determinations are made on all patients with endometrial cancer.

Surveillance of Ovarian Cancer

As with cervical and endometrial cancer, the first step in surveillance of patients treated for epithelial ovarian cancer is a thorough review of systems and physical examination. The physician should pay particular attention to the supraclavicular and inguinal nodal areas and auscultation of the lungs. A thorough abdominal and pelvic examination, assessing for the presence of ascites or masses, is also necessary. A Pap smear should be performed annually for routine surveillance.

A chest X ray should be obtained if the patient develops pulmonary symptoms or physical examination findings suggest pulmonary disease. Patients with Stage IV disease and a history of pulmonary metastases have an increased risk of relapse in the chest, and chest X rays should be performed in these patients at regular intervals. CT scans of the abdomen and pelvis should be reserved for evaluation of the symptomatic patient, for confirmation of physical examination findings, or to search for disease in a patient with an elevated CA-125 level.

The CA-125 assay has become the hallmark of surveillance for epithelial ovarian cancer. If the CA-125 level was elevated prior to initial therapy, it is an excellent tool for cancer surveillance. Elevated CA-125 levels have been observed in more than 80% of patients with epithelial ovarian cancer.[30] These elevated levels have been found to predict tumor recurrence in patients who are clinically free of disease. Elevated CA-125 levels have been observed before clinical recurrence in more than 90% of patients, with a median lead time of 3 months.[31] More than one-third of these patients had lead times of more than 6 months. Other authors have observed elevated CA-125 levels in 93% of patients who relapsed. Again, the median lead time was 3 months.[32] CA-125 levels appear to be a strong and reliable predictor of preclinical tumor recurrence. They should be drawn at each follow-up visit. If a rising trend or a frankly elevated CA-125 level is documented, a thorough evaluation for recurrent disease, including chest X ray and CT scan of the abdomen and pelvis, is warranted. Histologic confirmation of recurrent disease by fine-needle aspiration, or laparoscopy, or laparotomy where indicated, prior to initiation of second-line therapy, is warranted.

Germ cell tumors of the ovary represent a unique group of neoplasms that can produce various tumor markers based on the cell types comprising the tumor (Table 26.4). Elevated

TABLE 26.4. Biologic markers in germ cell tumors of the ovary.

Type	AFP	hCG
Dysgerminoma	−	+/−
Immature teratoma	+/−	−
Endodermal sinus tumor	+	−
Mixed germ cell tumor	+/−	+/−
Choriocarcinoma	−	+
Embryonal carcinoma	+/−	+
Polyembryoma	+/−	+

CA-125 levels have also been observed in more than 80% of germ cell tumors,[33] and elevations of the glycolytic enzyme lactic dehydrogenase (LDH) have been documented with dysgerminoma.[34] Patients with these tumors are generally much younger than those with epithelial ovarian cancers. These tumors can behave very aggressively; therefore, close surveillance is necessary to document early relapse. If elevated tumor markers are found prior to initial therapy, they should be performed at routine intervals. In addition, standard screening tools as outlined for epithelial ovarian cancer to detect preclinical recurrence of disease should be performed.

Surveillance of Vulvar Cancer

As with other gynecologic cancers, more than 80% of recurrences will be within the first 2 years following primary therapy, requiring the closest scrutiny during those years. More than half of the recurrences are local and near the site of the primary lesions.[35]

A careful review of systems and physical examination are the first steps in surveillance. The examination should include a careful palpation of the inguinal area and the vulva. Suspicious masses or nodules in the groin can be evaluated in the office setting with fine-needle aspiration. If results are negative or equivocal, histologic evaluation is advised. Any suspicious lesions on the vulva *must* be histologically evaluated. This is easily done by infiltrating the abnormal area with 1% lidocaine and using a Keyes or Baker punch biopsy instrument to excise a small plug of tissue. Hemostasis can usually be achieved by the application of silver nitrate to the biopsy bed. Usually, no suture is required. A thorough speculum evaluation of the vagina and cervix and a bimanual and rectovaginal exam should also be performed. A Pap smear should be performed at each visit because patients with squamous lesions of the vulva are at some increased risk of developing a squamous neoplasm on the cervix or vagina, especially those patients who have tumors associated with human papilloma virus.

Chest X ray, CT scans, and bone scans should be performed as the patient's symptoms or physical findings indicate.

References

1. Kaldor JM, Day, NE, Pettersson F, et al. Leukemia following chemotherapy for ovarian cancer. *New Engl J Med.* 1990;322:1–6.
2. Villasanta U. Complications of radiotherapy for carcinoma of the uterine cervix. *Am J Obstet Gynecol.* 1972;114:717–726.
3. Strockbine MF, Hancock JE, Fletcher GH. Complications in 831 patients with squamous cell carcinoma of the intact uterine cervix treated with 3,000 rads or more whole pelvis radiation. *American Journal of Radiology.* 1970;108:293–304.
4. Martinbeau PW, Kjorstad KE, Iversen T. Stage IB carcinoma of the cervix: Norwegian Radium Hospital, results of treatment in major complications. II. Results when pelvic nodes are involved. *Obstet Gynecol.* 1982;60:215–218.
5. Kottmeier HL. Complications following radiation therapy in carcinoma of the cervix and their treatment. *Am J Obstet Gynecol.* 1964;88:854–866.
6. Palmer JA, Bush RS. Radiation injuries to the bowel associated with the treatment of carcinoma of the cervix. *Surgery.* 1976;80:458–464.
7. Maruyama Y, van Nagell RJ Jr, Utley J, et al. Radiation and small bowel complications in cervical carcinoma therapy. *Radiology.* 1974;112:699–703.
8. Coutsoftides T, Favio VW. Small intestine cutaneous fistulas. *Surg Gynecol Obstet.* 1979;149:333–336.
9. Alert J, Jimenez J, Beldarrian L, et al. Complications from irradiation of carcinoma of the uterine cervix. *Acta Radiologica Oncologica.* 1980;19:13–15.

10. Boronow RC, Rutledge F. Vesicovaginal fistula, radiation and gynecologic cancer. *Am J Obstet Gynecol.* 1971;111:85–90.
11. Burns BC, Upton RT. The management of urinary tract complications of treatment for carcinoma of the uterine cervix. *Cancer of the Uterus and Ovary.* Chicago, Ill: Yearbook Medical Publishers; 1966:157–268.
12. Lang EK, Wood M, Brown R, et al. Complications in the urinary tract related to treatment of carcinoma of the cervix. *Southern Medical Journal.* 1973;66:228–236.
13. Abitbol MM, Davenport JH. Sexual dysfunction after therapy for cervical carcinoma. *Am J Obstet Gynecol.* 1974;119:181–189.
14. Abitbol MM, Davenport JH. The irradiated vagina. *Obstet Gynecol.* 1974;44:249–256.
15. Christ F, Wagner U, Debus G. Early bladder function disorders following Wertheim surgery: causes and therapeutic consequences. *Geburtshilfe und Frauenkeilkunde.* 1983;43:380–383.
16. Kristensen GB, Frimodt-Moller TC, Poulsen HK, et al. Persistent bladder dysfunction after surgery and combination therapy of cancer of the cervix uteri Stages IB and 2A. *Gynecol Oncol.* 1984;18:38–42.
17. Seski JC, Diokno AC. Bladder dysfunction after radical abdominal hysterectomy. *Am J Obstet Gynecol.* 1977;128:643–647.
18. Krebs HB, Helmkamp BF, Sevin BU, et al. Recurrent cancer of the cervix following radical hysterectomy and pelvic node dissection. *Obstet Gynecol.* 1982;59:422–427.
19. Munnell EW, Bonney JA Jr. Critical points of failure in the therapy of cancer of the cervix. *Am J Obstet Gynecol.* 1960;81:521–534.
20. van Nagell JR Jr, Rayburn W, Donaldson ES, et al. Therapeutic implications of patterns of failure in cancer of the uterine cervix. *Cancer.* 1979;44:2354–2361.
21. Burns BC, Upton RT. *Cancer of the Uterus and Ovary.* Chicago, Ill: Yearbook Medical Publishers; 1966.
22. Campos J. Persistent tumor cells in the vaginal smears and prognosis of cancer of the radiated cervix. *Acta Cytologica.* 1970;14:519–522.
23. Rayburn WF, van Nagle JR Jr. Cervicovaginal cytology in the diagnosis of recurrent carcinoma of the cervix uteri. *Surg Gynecol Obstet.* 1980;151:15–16.
24. Wentz WB, Reagan JW. Clinical significance of postirradiation dysplasia of the uterine cervix. *Am J Obstet Gynecol.* 1970;106:812–817.
25. Photopulos GJ, Shirley REL Jr, Ansbacher R. Evaluation of convential diagnostic tests for detection of recurrent carcinoma of the cervix. *Am J Obstet Gynecol.* 1977;129:533–535.
26. Shingleton HN, Orr JW. Post-treatment surveillance. In: *Cancer of the Cervix Diagnosis and Treatment.* Edinburgh, Scotland: Churchill-Livingston; 1987:213.
27. Hatch KD, Parham G, Shingleton HN, et al. Ureteral strictures and fistulas following radical hysterectomy. *Gynecol Oncol.* 1984;19:17–23.
28. Niloff JM, Klug TL, Schaetzel E, et al. Elevation of serum 125 in carcinomas of the fallopian tube, endometrium and endocervix. *Am J Obstet Gynecol.* 1984;148:1057–1058.
29. Patsner B, Orr JW Jr, Mann JW Jr. The use of serum CA-125 measurements in post-treatment surveillance of early-stage endometrial carcinoma. *Am J Obstet Gynecol.* 1990;162:427–429.
30. Bast RC Jr, Klug TL, St John E, et al. A radioimmunoassay using a monoclonal antibody to monitor the course of epithelial ovarian cancer. *N Engl J Med.* 1983;309:883–887.
31. Niloff JM, Knapp RC, Lavin PT, et al. The CA-125 assay as a predictor of clinical recurrence in epithelial ovarian cancer. *Am J Obstet Gynecol.* 1986;155:56–60.
32. Vergote IB, Bormer OP, Abeler VM. Evaluation of serum CA-125 level in the monitoring of ovarian cancer. *Gynecol Oncol.* 1987;157:88–92.
33. Altaris MM, Goldberg GL, Levin W, et al. The value of cancer antigen 125 as a tumor marker in malignant germ cell tumors of the ovary. *Gynecol Oncol.* 1986;25:150–159.
34. Awais GM. Dysgerminoma and serum lactic dehydrogenase levels. *Obstet Gynecol.* 1983;62:99–101.
35. Disaia PJ, Creasman WT. *Clinical Gynecologic Oncology.* 3rd ed. St. Louis, Mo: CV Mosby Co; 1989:262.

Index

Abortion
   pelvic inflammatory disease and, 78
ACP, 301
Adenocarcinoma of the endometrium
   estrogen therapy and, 170
Adenomyosis, 138–139
Adhesive pelvic pain, 138
Adrenal
   hyperplasia and hirsutism, 290, 292
   steroidogenesis and hirsutism, 290
   suppression and hirsutism, 292
Adrenocorticotropic hormone (ACTH)
   hirsutism and, 289–290, 293
AIDS
   candidiasis and, 101
   cervical neoplasia and, 103
   contraception and, 104
   genital ulcer disease and, 102
   herpes and, 103
   HIV and, 97, 100–107
   menstrual abnormalities and, 104
   pelvic inflammatory disease (PID) and, 102
   pregnancy and, 104–106
   sexual dysfunction and, 214
   syphilis and, 102–103
Alcohol(ism)
   lesbian women and, 284
   osteoporosis and, 169
   sexual dysfunction and, 214
Amenorrhea, see also Secondary amenorrhea
   ectopic pregnancy and, 115
   exercise induced, 161
   hyperestrogenic, 63–64
   hypergonadotropic, 63–64, 71
   hyperprolactinemic, 72–73
   hypothalamic, 71–72
   nipple discharge and, 185, 186
   osteoporosis and, 168
   pituitary, 72
   weight related, 287
   with androgen or estrogen excess, 73
Androgenicity
   of oral contraception, 153
Anemia
   iron deficiency, 146
Anorexic nervosa, 70, 287
Anovulation
   chronic and hirsutism, 289
Antiandrogens
   hirsutism and, 290–291
Asherman syndrome, 70
Atrophy of the pelvic floor, 168
   menopause and, 167

Bacterial vaginosis, see also Chronic vaginitis, 39
   clinical manifestations and diagnosis, 41–42
   epidemiology, 40–41
   infectious sequellae of, 42
   laboratory diagnosis, 42
   pathophysiology, 39–40
   treatment, 42–43
   treatment failures and recurrence, 43
Baldness, 287
Bartholin duct cysts and abscesses, 266–274
   anatomy of, 266
   anesthetic choices for, 269–270

Bartholin duct cysts and abscesses (cont.)
  aspiration of, 270–271
  carbon dioxide laser and, 273–274
  carcinoma of the, 274
  dyspareunia and, 268
  etiology of abscess formation in, 267
  evolution of various treatments for, 267–269
  gonorrhea and, 272
  marsupialization of, 268, 269, 271–274
  typical patient presentation of, 266–267
  word catheter and, 273
Bethesda classification system, see Classification systems
Bimanual exam, 126–128
Bladder dysfunction
  menopause and, 167
Bleeding
  abnormal uterine, 49–54
  breakthrough bleeding and oral contraceptives, 161
  postmenopausal, 130–131
Breast biopsies, 178–182
  aspiration for cytology with a fine needle, 179–180, 187
  aspiration for suspected breast cysts, 178–179
  core biopsy for histology, 189
  excisional, 181–182
  incisional, 180–181
Breast carcinoma, 10–12
  breastfeeding and, 205
  cigarette smoking and, 171
  estrogen therapy and, 168, 169, 173
  nipple discharge and, 186
  oral contraceptives and, 158–159
  risk factors, 11–12
  screening guidelines, 11–12
  screening methods, 11
  screening protocol, 12
  the American diet and, 171
Bone demineralization, see also Osteoporosis, 292
Breast disease, benign, 146
Breastfeeding, see also Lactation
  advantages of, 191
  an adopted child, 205–206
  breast abscess, 201
  breast carcinoma, 205
  breast rejection, 201
  complications, 190–207
  contraindications to, 204–205
  diabetic mothers and, 197–202
  drug therapy and, 205
  engorgement, 200
  epileptic mothers and, 204
  follow-up, 196–197
  galactocele, 201
  galactosemia and, 205
  gastrointestinal diseases and, 203
  hepatitis and, 205
  herpes and, 204
  HIV-infection and, 105–106, 204–205
  hypertension and, 203–204
  latch on breast, 194–195
  let-down failure and, 198
  mastitis, 200–201
  milk supply problems, 202
  nipple discharge and, 192
  nipple problems and, 198–199
  PBB contamination and, 205
  PCB contamination and, 205
  plugged ducts, 200
  prenatal vitamins and, 193
  resource guide, 206
  technique, 194–195
  tuberculosis and, 205
  working mother and, 204

CA-125 serum marker, 8–9
Calcium
  kidney stones and, 172
  osteoporosis and, 168, 171–172
Candidiasis
  AIDS and, 101
Carbon dioxide laser
  bartholin duct cysts and abscesses and, 273–274
Carcinoma in situ, 16
Cardiovascular disease
  estrogen and, 169
  menopause and, 167, 169, 173
  oral contraception and, 156
Cervical carcinoma, 3–4, 159, 310–311
Cervical cytology
  management of atypical, 5
Cervical neoplasia, 159
  AIDS and, 103
  human papilloma virus (HPV) and, 16
  pap smears and, 159
  oral contraceptives and, 159

Cervical intraepithelial neoplasia (CIN), 16
  homosexual women with, 281
  human papilloma virus (HPV) and, 16
  laser vaporization in treating, 22-23
  management of, 20-22
  methods of treatment, 22-24
  percentages that are aneuploid, 16
  post treatment follow-up, 24
  prereqisites for the treatment of, 20-21
  risk factors for, 16
  treatment methods for, 20
Cervix
  anatomy and pathology of the, 15-17
Chemical irritants
  resulting in vulvovaginitis, 44
Chlamydia trachomatis, 31
  rape and, 299, 300
Cholesterol
  estrogen and, 169
  oral contraceptives and, 154-156
Chronic pelvic pain (see also Adhesive pelvic pain, Dysmenorrhea, Endometriosis, and Leiomyomata), 135-144
  conization and, 22
  diagnostic evaluation of, 139-143
  differential diagnosis of, 137-139
  endometriosis, 137-138
  hypermenorrhea and, 143
  imaging studies, 141-142
  laboratory studies, 141
  nongynecologic, 137
  nonsteroidal antiinflammatory drugs and, 143
  perception, 136
  physical examination, 140-141
  psychogenic, 137
  surgical treatment of, 143
  syndrome, 92
  treatment of, 142
Chronic salpingitis, 141
Chronic vaginitis, see also Bacterial vaginitis, Cytolytic vaginitis, Trichomonas vaginitis, Vaginitis and, Vulvovaginal candidiasis, 30-45
  bacterial vaginosis, 39-43
  chemical irritants resulting in, 44
  diagnosis of, 31
  indigenous flora and, 30
  miscellaneous and less common causes of, 43-45
  physiologic discharge and, 30-31
  trichomonas vaginitis, 36-37
  vulvar conditions unrelated to, 44
  vulvovaginal candidiasis, 32-43
Cigarette smoking
  breast cancer and, 171
  lesbian health issues and, 280
  oral contraception and, 156, 158, 159
  osteoporosis and, 168-169
  PID risk and, 81
Classification systems
  Bethesda classification system, 5-6
  for reporting pap smear results 5-6, 17
  Papanicolaou's, 5
  Richard's descriptive cytologic classification, 6
  revised Bethesda classification system, 18
Cimetidine
  hirsutism and, 291
Clitoromegaly, 288
  hirsutism and, 288
Clomiphene citrate (CC) therapy, 72, 74
Coital
  climax, 212
  graph, 212
Coloscopy, 19-21
  technique of, 19-20
  terminology, 20
Condyloma acuminata, 16
Congenital abnormalities
  sexual dysfunction and, 214
Conization
  cold-knife, in treating CIN, 22
  indication for diagnostic, 21
  laser, in treating CIN, 22
Contraception
  AIDS and, 104
  PID and, 80-81
Coronary atherosclerosis, 156
Critical weight theory, 72
Cryosurgery
  in treating CIN, 23-24
Cushing's syndrome
  hirsutism and, 288
  masculine habitus and, 68
Cyclosporine, 287
Cyproterone
  hirsutism and, 291
Cystitis
  lesbian women and, 282
Cystometry, 239-242

Cytolytic vaginitis, 44–45
Cytoscope, 238

DES-exposure
  cervical carcinoma and, 3
  lesbianism and, 278
Detrusor instability, see Urinary incontinence
Detrusor hyperflexia, see Urinary incontinence
Diabetes
  breastfeeding mother with, 197–202
  oral contraception and, 155, 156
  vulvovaginal infection and, 33
Diet
  breast carcinoma and the American, 171
  breastfeeding and, 193
  lactation and, 193
Doderlein's cytolysis, see Cytolytic vaginosis,
Dominant breast mass, see also Breast cancer 175–183
  aspiration for cytology with a fine needle, 179–180
  aspiration for suspected breast cysts, 178–179
  benign breast diseases and, 176
  core biopsy for histology, 180
  excisional biopsy, 181–182
  incisional biopsy, 180–181
  medical history and, 176
  physical examination and, 177
  steps in diagnosing, 175–176
Drug therapy
  breastfeeding and, 205
Duct ectasia, 186
Dysplastic cells, 15–16
Dysmenorrhea, see also Chronic pelvic pain, 135, 142–144
  dyspareunia and, 142
Dyspareunia
  bartholin duct cysts and, 268
  dysmenorrhea and, 142
  dyspareunia, 168
  menopause and, 167
  sexual disorders and, 214, 219
  vestibulitis and, 140

Ectopic pregnancy, 111–118
  ancillary diagnostic aids, 116–118
  evaluation and diagnosis of, 114–118
  location, 114

  oral contraception and, 146
  pathogenesis of, 112–113
  pathology, 113–114
  PID and, 76, 79, 81, 82, 85, 87, 92, 112–113
  ultrasonography and, 129
Electrocoagulation, 23
Endocervicitis, 83
Endocrine
  abnormalities and sexual dysfunction, 214
  causes of secondary amenorrhea, 66
Endometrial carcinoma, 6–8
  estrogen therapy and, 168, 169
  oral contraception and, 146
  risk factors, 8
  screening criteria, 7–8
  screening methods, 7
  surveillance of, 311
Endometriosis
  lesbian women and, 282
  pelvic pain and, 137–138, 141, 140, 144
  sexual dysfunction and, 214
Entamoeba histolytica gastroenteritis, 299
Epilepsy
  breastfeeding and, 204
Estradiol
  hirsutism and, 288–289
Estrogen
  cardiovascular disease and, 169
  cholesterol and, 169
  hirsutism and, 290–291
  level of oral contraceptives, 153
Estrogen therapy, 169–173
  abnormal uterine bleeding and, 169
  add-back, 292
  adenocarcinoma of the endometrium and, 170
  breast carcinoma and, 168, 169, 173
  contraindications for, 169, 171
  endometrial carcinoma and, 168, 169
  gallbladder disease and, 169, 171
  hirsutism and, 291
  hypertension and, 169, 170
  migraine headaches and, 169
  vascular thrombosis and, 169

Fallopian tubes
  anatomy of the, 111–112
Fertility, see also Infertility
  return of after oral contraceptives, 158

F

Fibrocystic
  changes, 186
  disease, 171, 175
Fitz-Hugh-Curtis syndrome, 79, 84, 140
Flat condyloma, 16
Fluorides
  in the management of menopause, 173
Flutamide, 291–292

G

Galactorrhea
  causes of, 185
  hirsutism and, 291
  nipple discharge and, 185–187, 189
Galactosemia
  infant and breastfeeding, 205
Gallbladder disease, 169, 171
  estrogen therapy and, 169
Gastrointestinal diseases
  breastfeeding and, 203
Genital ulcer disease
  AIDS and, 102
Genital mycoplasmas
  in the etiology of PID, 77
Genuine stress incontinence, see Urinary incontinence
Germ cell tumors
  of the ovary, 311–312
Giardia lamblia, 299
Glucocorticoids
  hirsutism and, 292
Gonorrhea, 31
  bartholin duct cysts and, 272
  nipple discharge and, 185
  rape and, 299, 300

H

Heart disease
  menopause and, 168
  oral contraception and, 160
Hepatitis, 97–100
  breastfeeding and, 205
  delta hepatitis, 99
  effects on pregnancy and the fetus, 98
  hepatitis A, 97–98
  hepatitis B, 98–99
  hepatitis C, 99
  hepatitis E, 99–100
  rape and, 297, 299
Herpes, 31, 257–259
  AIDS and, 103
  asymptomatic viral shedding, 258
  breastfeeding and, 204
  clinical findings, 257–258
  diagnosis, 258
  etiology, 257
  incidence, 257
  lesbian women and, 282
  management, 258–259
  management of pregnant women with, 259
  rape and, 300
Hirsutism, 287–293
  adrenal hyperplasia and, 290, 292
  adrenal steroidogenesis and, 290
  adrenal suppression and, 292
  antiandrogens and, 290–291
  biochemical evaluation of, 289
  cautions and limitations, 292–293
  chronic anovulation and, 289
  cimetide and, 291
  clinical assessment of, 289
  Cushing's syndrome and, 288
  cyproterone and, 291
  clitoromegaly and, 288
  distinguished from virilism, 288
  estrogens and, 290–291
  galactorrhea, 291
  glucocorticoids and, 292
  infertility and, 290
  ketoconazole, 292
  mechanical therapy and, 290
  medication that may cause, 287
  menstrual dysfunction and, 146, 288
  not caused by hyperandrogenism, 287
  pathophysiology of, 288
  pharmacotherapy, 290
  ovarian suppression and, 292
  reductase inhibitors and, 292
  treatment for, 287
  virilism and, 288
  weight and, 290, 291
HIV infection
  breastfeeding and, 105–106, 204–205
  cervical intraepithelial neoplasia (CIN) and, 16
  lesbianism and, 282
  pap smears and, 16, 104
  rape and, 297, 299
  vulvovaginal candidiasis and, 33, 35

Hormonal replacement therapy, see Estrogen therapy
Human papilloma virus (HPV), 16, 19
  cervical neoplasia and, 16
  pap smears and, 16
  three categories of, 16
Human immunodeficiency virus (HIV) and AIDS, see also AIDS, 97, 100–107
  care of infected women, 100–101
  course of infection, 100
  diagnosis of neonatal infection, 106
  management of specific problems in infected women, 101–107
  occupational exposure to, 106–107
  perinatal transmission of, 105
  pregnancy and, 104–105
  primary care of disease, 101
  vulvar disorders and, 263
Hyperandrogenicity, 287, 289, 290
Hyperestrogenetic amenorrhea, see Amenorrhea
Hypergonadotropic amenorrhea, see Amenorrhea
Hyperlipidemia, 156
Hypermenorrhea
  chronic pelvic pain and, 143
Hyperplastic dystrophy, 3
Hyperprolactinemia, 157–158
Hyperprolactinemic amenorrhea, see Amenorrhea
Hypertension, 169
  breastfeeding and, 203–204
  estrogen therapy and, 169, 171
  oral contraception and, 156–157
Hypertrichosis, 287
Hypogonadotropic anovulation, 71–72
Hypothalamic
  amenorrhea, 71–72
  chronic anovulation, 68
  dysfunction, 68–70
Hypothalamus
  secondary amenorrhea and, 62–63
Hypothyroidism
  sexual disorders and, 216
Hyperlactinemia
  sexual disorders and, 216
Hypothyroidism, 287
Hysterectomy, 143
  versus myomectomy, 138
Hysterosalpingogram, 68, 142

Hysteroscopy
  as a screening test for endometrial carcinoma, 8

Infertility
  hirsutism and, 290
  nipple discharge and, 185, 186
  PID and, 87, 93
  tubal, 80
  ultrasound and, 132
Insulin resistance, 289
Intraductal papilloma, 185
Intrauterine synechiae, 70

Ketoconazole
  hirsutism and, 292

Lactation, see also Breastfeeding
  after reduction mammoplasty, 191
  antepartum management, 190–192
  complications, 190–206
  nipple preparation, 193
  nutrition for, 193
  possibility for full, 205
  postpartum management, 193–194
Laparoscopy
  chronic pelvic pain, 142
Laser vaporization
  in treating CIN, 22–23
Latent disease, 16
Leiomyomata, 138–139
  posterior uterine, 141
Lesbianism, 276–284
  adolescent factors and, 278–279
  alcoholism and drug abuse, 284
  behavior patterns, 284
  cervical intraepithelial neoplasia (CIN) and, 16
  cystitis and, 282
  definition of terms, 276–277
  DES exposure and, 278
  doctor patient dialogues and, 280–281
  endometriosis, 282
  etiology of, 277–278
  gender identity and, 279
  gynecologic care and, 279–280
  herpes and, 282

Index

HIV infection and, 282
incidence of, 277
medical provider approach toward, 280
motherhood, 283
neuroendocrine factors and, 278
pap smear and, 281
parental factors and, 278
physician and patient attitudes, 280
pregnancy and, 279, 282
psychologic and psychiatric problems, 283-284
psychologic considerations and, 279
sexual dysfunction and, 284
vaginal infections and, 281-282
Let-down failure, see also Lactation, 198
Lichen sclerosis, 3, 260-262
Lipid metabolism
 contraceptive steroids and, 154-155
Lipid hypothesis, 155-156
Lobular neoplasia, 177
Loop excision procedure, 22
 in treating CIN, 22
Luteinizing hormone, 288-289

Mammary duct ectasia, 177
Mammogram, 11-12, 176
Masturbation, 214
 exercises, 219
Menopause
 atrophy of the pelvic floor and, 167
 bladder dysfunction and, 167
 dyspareunia and, 167
 long term complications of, 168-169
 management of, 167-173
 osteoporosis and, 167, 168
Menstruation (see also Amenorrhea, Bleeding, Menstrual cycling, Menstrual dysfunction)
 abnormalities of and AIDS, 104
 cessation of and hirsutism, 289
 early onset of, 176
 physiology of, 62-64
 retrograde, 78, 85
Menstrual cycling
 neoplasms, 50
 normal and abnormal, 49-50
 ovarian cysts and, 50
 PID and, 81
 prostaglandins and, 50

Menstrual dysfunction, 146, 288
 hirsutism and, 288, 290
 oral contraceptives and, 158
Metastases
 pulmonary, 311
Migraine headaches, 169
 estrogen therapy and, 169
Mondor's disease, 177
Myocardial infarction
 oral contraception and, 156

Neoplasm
 hirsutism and, 288
 menstrual cycling and, 50
 vulvar intraepithelial, 260, 264
Nipple discharge, 184-189
 characteristics of, 184
 clinical evaluation and management of, 186-189
 galactorrhea and, 185-187, 189
 gonorrhea and, 185
 infertility and, 185, 186
 nonphysiologic (pathologic), 185-186
 physiologic nipple discharge, 184
 preparation for breastfeeding, 192
 problems, 198
 oral contraceptives and, 184
Nonsteroidal antiinflammatory drugs (NSAIDs)
 chronic pelvic pain and, 143

Office ultrasonography of the pelvis, 121
 bimanual exam and, 126-128
 equipment for, 121-123
 orientation and, 125-126
 patient preparation, 123-124
 probe preparation, 123
 rationale for, 121
 routine office uses, 128
 training and cost, 132-133
Oral contraception, 146-162
 cardiovascular disease and, 156, 160
 cervical neoplasm and, 159
 changing trends and, 159-160
 cholesterol and, 154-156
 cigarette smoking and, 156, 158, 159
 contraindications to use, 160-161
 diabetes and, 155, 156
 ectopic pregnancy and, 146

Oral contraception (*cont.*)
   endometrial carcinoma and, 146
   female reproductive tract cancers and, 158–159
   health advantages, 146
   historical perspective, 146, 147
   hypertension and, 156–157
   menstrual dysfunction and, 158
   metabolic effects, 153–155
   myocardial infarction and, 156
   nipple discharge and, 184
   pharmacology of, 150–153
   prescribing, 160–161
   scientific foundation of, 147–149
Osteoporosis, 146, 292
   cigarette smoking and, 168–169
   medical therapy of, 172
   menopause and, 167, 168
   risk factors for, 168–169
   therapy of postmenopausal, 171–173
Ovarian, see also Ovarian carcinoma, Ovaries
   cyst formation, 146
   polycystic disease (PCOD) and, 67–68, 73–74
   suppression and hirsutism, 292
   wedge resection, 74
Ovarian carcinoma, 9–10, 132, 146
   risk factors of, 10
   screening criteria in, 10
   surveillance of, 311–312
   ultrasound and, 132
Ovaries, see also Ovarian, Ovarian carcinoma, and Ovaries
   secondary amenorrhea and, 64, 66, 71
   polycystic, 73–74, 67–68
Overflow incontinence, see Urinary incontinence

Pad test
   urinary leakage and, 242
Paget's disease, 177
Painful intercourse, 83
Pap smear(s), see also classification systems 1, 3, 4
   as a screening test for endometrial cancer, 7–8
   cervical neoplasia and, 159
   cytologic screening and, 17
   HIV infections and, 16, 104
   HPV and, 16
   interpretation of results, 5–6
   lesbian health and, 281
   management, 22
   management of the abnormal, 15–24
   rape and, 301, 303
   screening criteria, 4–5
   screening interval, 17
   screening recommendations, 18
   sexual disfunction and, 217
   specimen collection, 4
   technique, 18–19
   terminology, 17
Papanicolaou's classification system, 5
Papillomatosis, 186
PBB contamination
   breastfeeding and, 205
PCB contamination
   breastfeeding and, 205
Pelvic inflammatory disease (PID), 76–93
   AIDS and, 102
   cigarette smoking and, 5
   contraception and, 80–81
   criteria for clinical diagnosis of, 86
   criteria for hospitalization of outpatients with, 90
   cryosurgery in treating, 23–24
   diagnostic algorithm, 87
   diagnosis, 81–86
   dysmenorrhea and chronic pelvic pain and, 135, 141–144
   ectopic pregnancy and, 76, 79, 81, 82, 85, 87, 92, 112–113
   estimated indirect costs of, 76
   etiology, 77–78
   follow-up treatment, 92
   genital mycoplasmas in the etiology of, 77
   infertility and, 87, 93
   menstrual cycling and, 81
   oral contraception and, 146
   pathophysiology, 78–79
   recommendations for prevention of, 82
   risk factors for, 80
   sequelae of, 92–93
   silent, 83
   treatment, 86–92
   ultrasonography and, 85, 91
Perihepatitis, 140
Perinatal transmission of, 105
Peritonitis, 141

Pituitary
  adenoma and nipple discharge, 186
  amenorrhea, 72
  secondary amenorrhea, 63–64
  tumor and nipple discharge, 186
Pneumocystogram, 179
Polycystic ovarian disease (PCOD), 67–68, 73–74
Postmenopause
  evaluation of pelvic mass and, 9
  screening and, 131–132
  therapy for, 171–173
  ultrasound and, 131–132
Postoperative adhesive disease, 144
Pregnancy, see also Ectopic pregnancy
  AIDS and, 104–106
  chronic vaginitis and, 30
  hepatitis and, 98
  HIV and, 104, 105
  lesbian women and, 279, 282, 283
  prevention of after rape, 300–301
  prophylaxis, 300
  rape and, 295
  recurrent urinary tract infections and, 225–227
  ultrasound and, 128–130
Preventive medicine, 1, 2
  pelvic inflammatory disease and, 82
  pregnancy and, 300–301
  screening and, 1–2
  STD and, 82
  three categories of, 1–2
Progesterone challenge test, 68
  secondary amenorrhea and, 68
Progestogens, 150–152
Prophylaxis, 300
Purulent discharges, 186

Radiation
  long-term effects of, 307
Rape, 295–305
  chlamydia trachomatis and, 299–300
  evidence collection, 302, 304
  evidence exam, 301–302
  gonorrhea and, 299, 300
  hepatitis and, 297, 299
  herpes and, 300
  HIV and, 297, 299
  incidence, 295
  induced problems, 295
  office preparation for, 295
  pap smear and, 301, 303
  physical exam, 298–303
  pregnancy and, 295
  prevention of pregnancy after, 300–301
  proof of force, 302
  required tests, 299
  sexually transmitted diseases and, 298–299
  trauma syndrome, 297–298
  victim's history, 296
Recommendations
  for the early detections of cancer, 308
Recurrent disease, 307–312
  pertinent symptoms of, 309
Reductase inhibitors
  hirsutism and, 292
Relactation, 205
Rheumatoid arthritis, 146
Richard's descriptive cytologic classification system, 6

Screening
  criteria for endometrial carcinoma, 7–8
  criteria for pap smear, 4–5
  cytologic, 17–19
  endometrial cancer, 6–8
  guidelines in breast cancer, 11–12
  gynologic cancer, 2–6
  gynecological health care and, 1–12
  history of gynecological, 1
  methods for breast cancer, 11
  methods for endometrial carcinoma, 7
  methods for ovarian carcinoma, 9–10
  ovarian carcinoma, 10
  postmenopause, 131–132
  preventive medicine and, 1–2
  principles of, 2
Secondary amenorrhea, 62–74
  endocrine causes of, 65–66
  evaluation of the patient and, 66
  history of patient and, 66–67
  hypothalamus and, 71–72, 62–63
  laboratory tests and, 68
  ovaries and, 64, 66, 71
  physical examination and, 67–68
  pituitary, 63–64
  progesterone challenge test and, 68,

Secondary amenorrhea (*cont.*)
    treatment of, 70–74
    ultrasonography and, 68
    uterus and, 64–65, 70
Semen-specific markers, 301
Sensory urge incontinence, see Urinary incontinence
Sex therapy, 217–219
Sexual disorders, 211–219
    dyspareunia, 214
    etiology of, 214–215
    female, 211–217
    hypoactive sexual desire disorder, 214–216
    inhibited female orgasm, 214
    lesbian women and, 284
    morbidity and, 211–214
    other disorders, 215–217
    primary anorgasmia, 215
    sexual aversion disorder, 214
    vaginismus, 215
Sexual history
    sexual problems and, 218
Sexual responses, 212–214
Sexually transmitted diseases, see also AIDS, Gonorrhea, Herpes, and Syphilis
    rape and, 298–299
Sonography
    as a screening tool for endometrial carcinoma, 8
Sprionolactone, 291
Squamous intraepithelial lesion (SIL), 16
    low and high grade, 17
    three courses of, 16
STD
    recommendations for individuals to prevent, 82
    risk factors for, 80–81
Steroid
    effects on lipids, 155
Stress test
    urinary leakage and, 242
Subclinical disease, 16
Surveillance of gynecological malignancies, 307–312
    cervical cancer and, 310–311
    endometrial cancer and, 311
    history and physical exam, 309–310
    intervals for follow up, 309
    ovarian cancer and, 311–312
    rational for, 307–308
    symptoms of recurrent diseases, 309
    vulvar cancer and, 312
Supplemental nutrition system, 205
Syphilis
    AIDS and, 102–103
    rape and, 299, 300

Thyroid disorders, 287
Tietz's syndrome, 177–178
Trichomonas vaginitis, see also Chronic vaginitis, 36–37, 40
    adverse reactions, 39
    diagnosis of, 37–38
    routine treatment for, 38
    treatment failures, 38–39
Tuberculosis
    breastfeeding and, 205

Ultrasonography, see also Office ultrasonography of the pelvis
    assessing pelvic structures, 9
    disadvantages as a screening tool, 9
    dominant breast mass and, 176
    ectopic pregnancy and, 129
    for assessment of endometrial thickness, 8
    infertility and, 132
    needle, 176
    ovarian carcinoma and, 132
    PID and, 85, 91
    pregnancy and, 128–130
    secondary amenorrhea and, 68
    to screen for early cancer of the ovary, 9
Unconsummated marriage, 217
Urethroscope, 238
Urinary incontinence, 230–251
    assessment of urethral mobility and, 236–237
    behavior modification and, 249
    cystometry, 239–242
    detrusor hyperflexia, 231, 247–248, 250
    detrusor instability, 231, 247–249
    endoscopy, urethroscopy, and cystoscopy and, 238–239
    epidemiology of, 230
    etiology of, 230–232
    genuine stress incontinence, 230–231, 245, 249–250

leak point pressure, 246
neurologic examination and, 235–236
office evaluation and treatment of, 230–251
office treatment of, 247–251
overflow incontinence, 232
pad test, 242
pathophysiology of female, 232–247
pelvic floor stimulation and, 251
pharmacologic treatment of, 247
physical examination and, 234–235
sensory urge incontinence, 231
stress test, 242
urethral closure pressure profiles, 244–246
urethrocystometry, 242–244
urodynamic evaluation, 242
uroflowmetry, 237–238
urogynecologic history and, 233–234
videocystourethrography, 247
voiding pressure studies, 246–247

Urinary tract infections
antibiotic regimens for treatment of, 225
diagnosis of, 224
incidence of, 222–223
microbiology of, 223
pathophysiology of, 223–224
recurrent, 221–226
recurrent in pregnancy, 225–227
terminology, 221–222
treatment of, 224–225

Uterus
estrogen therapy and abnormal bleeding of the, 169
secondary amenorrhea and, 64–65, 70

Vaginal carcinoma, 3
Vaginismus
sexual disorders and, 216, 219
Vaginitis, see also Bacterial vaginitis, Chronic vaginitis, Cytolytic vaginitis, Trichomonas vaginitis, and Vulvovaginal candidiasis, 216–217
cytolytic, 44–45
sexual dysfunction and, 214

Vascular thrombosis, 169
estrogen therapy and, 169
Vestibulitis, 140
Videocystourethrography, 247
Virilism, 287–293
distinguished from hirsutism, 288
Virilizing tumors, 68
masculine habitus and, 68
Vulvar cancer, see also Vulvar disorders, 2–3
surveillance of, 312
Vulvar disorders, see also Vulvar cancer, and Vulvovaginal candidiasis, 260–265
chemical irritants resulting in, 4
dystrophy, 3
human papilloma virus, 263
lichen sclerosis, 260–262
limited vulvar condyloma, 263
neoplastic epithelial disorders, 260
refractory and extensive vulvar condyloma, 263–264
squamous cell hyperplasia, 262–263
technique of vulvar biopsy, 265
vulvar intraepithelial neoplasia, 260, 264
vulvar vestibulitis, 264
Vulvovaginal candidiasis, see also Bacterial vaginitis, Chronic vaginitis, Trichomonas vaginitis, and Vaginitis, 31–43
clinical presentation, 33–34
diabetes and, 33
epidemiology of, 32
etiologic factors, 32–33
etiology of, 34–36
HIV infection and, 33, 35
mechanism of disease production, 32
therapy of, 34
therapy of recurrent, 35–36

Window operation
for treating bartholin duct cysts and abscesses, 268–269
Word catheter
bartholin duct cysts and abscesses, 273